The Role of Fungi in Biodeterioration of Cultural Heritage: New Insights for Their Control

The Role of Fungi in Biodeterioration of Cultural Heritage: New Insights for Their Control

Editors

Filomena De Leo
Daniela Isola

MDPI • Basel • Beijing • Wuhan • Barcelona • Belgrade • Manchester • Tokyo • Cluj • Tianjin

Editors
Filomena De Leo
University of Messina
Italy

Daniela Isola
University of Tuscia
Italy

Editorial Office
MDPI
St. Alban-Anlage 66
4052 Basel, Switzerland

This is a reprint of articles from the Special Issue published online in the open access journal *Applied Sciences* (ISSN 2076-3417) (available at: https://www.mdpi.com/journal/applsci/special_issues/Fungi_Biodeterioration_Control).

For citation purposes, cite each article independently as indicated on the article page online and as indicated below:

LastName, A.A.; LastName, B.B.; LastName, C.C. Article Title. *Journal Name* **Year**, *Volume Number*, Page Range.

ISBN 978-3-0365-6495-1 (Hbk)
ISBN 978-3-0365-6496-8 (PDF)

Cover image courtesy of Daniela Isola

Contents

About the Editors

Filomena De Leo

Prof. Filomena De Leo graduated in Biological Sciences, specialized in "Applied Microbiology" and obtained a Postgraduate Degree (PhD) in Microbial Biotechnologies. She is currently an Associate Professor of Microbiology at the Department of Chemical, Biological, Pharmaceutical and Environmental Sciences of the University of Messina, Italy; professor of the "General Microbiology", degree Course CdL in "Biotechnology", and professor of the "Nutrimicrobiomic" degree Course CdLM in "Biology of Health, Applied Technologies and Nutrition". She is a member of the Council of PhD in "Applied Biology and Experimental Medicine" of the University of Messina, Italy. She is a member of the Reference Scientific Community of the "Stazione Zoologica Anton Dohrn", Naples, Italy (http://www.szn.it/index.php/it/chi-siamo/comunita-scientifica-di-riferimento, accessed on 7 October 2022). She is a member of the University Research Center for the study of extreme environments and extremophiles (CUR AEE) and of the National Consortium for Marine Sciences (CoNISMa). Her main interests are the study of the biodiversity, ecology and genetic diversity of bacterial and fungal populations involved in the biodeterioration of cultural heritage (especially stone monuments), and more recently the diversity of halophilic micro-organisms in deep-sea environments. One of her particular focuses is the taxonomy of black fungi, which are among the most harmful micro-organisms causing the biodeterioration of stone monuments. She has identified two new fungal species and four new bacterial species isolated from monuments. She has participated in numerous national and international research projects both as the scientist responsible and as a member of the Research Unit. Prof Filomena De Leo has published 51 articles in journals, 16 monographs (or book chapters), 19 Proceedings of International Congress and 54 posters and oral communications. Her publications are visible on "Researchgate" at https://www.researchgate.net/profile/Filomena_De_Leo, accessed on 7 October 2022.

Daniela Isola

Prof. Daniela Isola is a contract Professor of Botany Applied to Cultural Heritage at the University of Tuscia. She graduated in Biological Sciences at the University of Cagliari and received postgraduate specializations in Neuropharmacology and Environmental Hygiene (University of Cagliari), and Bioinformatics (La Sapienza University of Rome). She obtained a PhD in Biological and Biochemical Evolution at the University of Tuscia. Since then, she has focused on fungi from extreme environments and "black fungi". Her main interests are the biodeterioration of cultural heritage and biodegradation of aromatic hydrocarbons. Particular attention has been paid to the identification, diversity, phylogeny, and metabolic skills of the involved organisms, and low-impact methods to control biodeterioration. She is the author of 4 genera and 11 fungal species, more than a thousand sequences deposited in GenBank (covering bacteria, fungi, viruses and higher plants) and about 40 scientific papers, with an H-index of 22.

applied sciences

MDPI

Editorial

The Role of Fungi in Biodeterioration of Cultural Heritage: New Insights for Their Control

Filomena De Leo [1] and Daniela Isola [2,*]

[1] Department of Chemical, Biological, Pharmaceutical and Environmental Sciences, University of Messina, 98122 Messina, Italy
[2] Department of Economics, Engineering, Society and Business Organization (DEIM), University of Tuscia, Largo dell'Università Snc, 01100 Viterbo, Italy
* Correspondence: isola@unitus.it

Citation: De Leo, F.; Isola, D. The Role of Fungi in Biodeterioration of Cultural Heritage: New Insights for Their Control. *Appl. Sci.* **2022**, *12*, 10490. https://doi.org/10.3390/app122010490

Received: 8 October 2022
Accepted: 13 October 2022
Published: 18 October 2022

Publisher's Note: MDPI stays neutral with regard to jurisdictional claims in published maps and institutional affiliations.

Introduction

Fungi are nature's major decomposers, and they play an essential role in biogeochemical cycles. Thanks to their wide enzyme repository, they can break down organic matter and hard-to-degrade biopolymers such as lignin, cellulose, and chitin. They also produce several inorganic and organic metabolites, organic acids, and chelating agents of fundamental importance in geological processes, leading to chemical transformations at local and global scales [1]. Though frequently overlooked, fungi are ubiquitous and can be found even in environments that are not so suitable for life due to their ability to metabolize xenobiotics [1]. Therefore, is not surprising that fungi are considered the most detrimental threat to both indoor and outdoor artefacts of historical and/or artistic value; their presence can lead to physical, chemical, and aesthetical damages [2,3].

Cultural heritage conservation refers to the measures, protocols, and methods taken to extend the life of artefacts, monuments, and sites. Biological deterioration of historical and artistic artefacts has gained significant attention in recent decades, evidencing how limited our knowledge of biodeteriogens is when we consider the wide range of heritage materials used and the environments in which they are kept. In conservation practices, precise and reliable identification of biodeteriogens is useful; however, further research is needed to assess their ecological requirements and metabolic profiles. Indeed, the environmental conditions favoring various taxonomic groups, and their limited lives, are essential in assessing the risk for artefacts and designing indirect (preventive) and direct (biocide treatments) control methods.

In this Special Issue, we present seven study cases and two reviews addressing the deterioration of different materials, such as stone, textiles, wood, and scientific instruments, exposed to outdoor, confined, or semi-confined environments.

From the UNESCO heritage sites of Cuma, Ercolano, Nola, Oplonti, and Pompei, Petraretti and colleagues [4] performed a deep sampling on frescoes, marble, mortars, plaster, and tuff to assess the biodiversity of the culturable fungal fraction. In addition to isolating fungal strains with detrimental potential, the authors suggest the importance of collecting detrimental fungi for future research on cultural heritage.

The study performed by Isola and colleagues focused on a selection of black meristematic fungal strains belonging to the Culture Collection of Fungi from Extreme Environments (CCFEE, Viterbo, Italy). All strains were isolated from marble monuments of the Bonaria Cemetery (Cagliari, Italy) and, using plate assays, the main ecological and metabolic traits and tolerance to traditional biocides were assessed [5]. The preservation of traditional Romanian clothes was the focus of the Ilies study group [6]. In detail, different essential oils were applied to different materials comprising coats (e.g., wool, cotton, leather) to control the detrimental fungal species. Other artefacts from museums include optical lenses studied by Ngo and co-workers [7]. This study evidenced the surface corrosion produced by fungi, here favored by the high humidity which is typical of North Vietnam's climate.

The study performed by Jurado and colleagues is of interest, instead, for the conservation of show caves [8]. In this study, the microclimatological and aerobiological monitoring gave insight into the seasonal dynamics of airborne fungi in the Nerja cave (Spain).

Barboux and colleagues [9] proposed an alternative instrumental method (FTIR) for detecting *Serpula lacrymans*; this method could be especially useful for protecting wooden structures from the dry rot fungi which are prevalent in indoor environments, especially in the holarctic regions. The study performed by Chlebicki and colleagues was devoted to testing the in vitro inhibitory effect exerted by bioactive ions exposed to galvanic systems with changing electrode distance [10].

This Special Issue also presents two reviews. The study performed by Zucconi and colleagues addressing wall paintings accounts for 60 years of fungal biodeterioration reports [11]. An exhaustive list of identified fungi was produced. However, the technical and methodological improvements that occurred within that period evidenced some limitations in the fungal identification (mainly morphological) and culturingprotocols, posing the basis for future improvements.

De Leo and colleague reviewed 109 papers published within the last 30 years to critically summarize the current knowledge on black fungi associated with the biodeterioration of stone monuments with a look at control methods and future perspectives [12].

This Special Issue offers an insight into the vast world of fungal deteriogens. More research is necessary in this field; however, with the incoming improvements in instrumental technologies, molecular diagnoses, and sustainable control methods, the research prospects in this field are exciting.

Author Contributions: Conceptualization, D.I. and F.D.L.; writing—original draft preparation, D.I.; writing—review and editing, D.I. and F.D.L. All authors have read and agreed to the published version of the manuscript.

Funding: This research received no external funding.

Acknowledgments: We wish to thank the editorial team of Applied Sciences for the support given during this editorial project.

Conflicts of Interest: The authors declare no conflict of interest.

References

1. Gadd, G.M.; Rhee, Y.J.; Stephenson, K.; Wei, Z. Geomycology: Metals, actinides and biominerals. *Environ. Microbiol. Rep.* **2012**, *4*, 270–296. [CrossRef] [PubMed]
2. Sterflinger, K.; Pinzari, F. The revenge of time: Fungal deterioration of cultural heritage with particular reference to books, paper and parchment. *Environ. Microbiol.* **2012**, *14*, 559–566. [CrossRef] [PubMed]
3. Isola, D.; Selbmann, L.; Meloni, P.; Maracci, E.; Onofri, S.; Zucconi, L. Detrimental rock black fungi and biocides: A study on the Monumental Cemetery of Cagliari. In *Science and Technology for the Conservation of Cultural Heritage*; Rogerio-Candelera, M.A., Lazzari, M., Cano, E., Eds.; CRC Press: London, UK, 2013; pp. 83–86.
4. Petraretti, M.; Duffy, K.; Del Mondo, A.; Pollio, A.; De Natale, A. Community Composition and Ex Situ Cultivation of Fungi Associated with UNESCO Heritage Monuments in the Bay of Naples. *Appl. Sci.* **2021**, *11*, 4327. [CrossRef]
5. Isola, D.; Bartoli, F.; Meloni, P.; Caneva, G.; Zucconi, L. Black Fungi and Stone Heritage Conservation: Ecological and Metabolic Assays for Evaluating Colonization Potential and Responses to Traditional Biocides. *Appl. Sci.* **2022**, *12*, 2038. [CrossRef]
6. Ilieș, D.; Hodor, N.; Indrie, L.; Dejeu, P.; Ilieș, A.; Albu, A.; Caciora, T.; Ilieș, M.; Barbu-Tudoran, L.; Grama, V. Investigations of the Surface of Heritage Objects and Green Bioremediation: Case Study of Artefacts from Maramureş, Romania. *Appl. Sci.* **2021**, *11*, 6643. [CrossRef]
7. Ngo, C.; Nguyen, Q.; Nguyen, T.; Quach, N.; Dudhagara, P.; Vu, T.; Le, T.; Le, T.; Do, T.; Nguyen, V.; et al. Identification of Fungal Community Associated with Deterioration of Optical Observation Instruments of Museums in Northern Vietnam. *Appl. Sci.* **2021**, *11*, 5351. [CrossRef]
8. Jurado, V.; Del Rosal, Y.; Liñan, C.; Martin-Pozas, T.; Gonzalez-Pimentel, J.; Saiz-Jimenez, C. Diversity and Seasonal Dynamics of Airborne Fungi in Nerja Cave, Spain. *Appl. Sci.* **2021**, *11*, 6236. [CrossRef]
9. Barboux, R.; Bousta, F.; Di Martino, P. FTIR Spectroscopy for Identification and Intra-Species Characterization of Serpula lacrymans. *Appl. Sci.* **2021**, *11*, 8463. [CrossRef]

10. Chlebicki, A.; Spisak, W.; Lorenc, M.; Śliwa, L.; Wołowski, K. Electromagnetic Field as Agent Moving Bioactive Cations. A New Antimicrobial System in Architecture Technology. *Appl. Sci.* **2021**, *11*, 8320. [CrossRef]
11. Zucconi, L.; Canini, F.; Isola, D.; Caneva, G. Fungi Affecting Wall Paintings of Historical Value: A Worldwide Meta-Analysis of Their Detected Diversity. *Appl. Sci.* **2022**, *12*, 2988. [CrossRef]
12. De Leo, F.; Marchetta, A.; Urzì, C. Black Fungi on Stone-Built Heritage: Current Knowledge and Future Outlook. *Appl. Sci.* **2022**, *12*, 3969. [CrossRef]

Short Biography of Authors

Prof. Filomena De Leo graduated in Biological Sciences, specialized in "Applied Microbiology" and Post graduated Degree (PhD) in Microbial Biotechnologies. Currently is Associate Professor of Microbiology at the Department of Chemical, Biological, Pharmaceutical and Environmental Sciences of the University of Messina, Italy: professor of "General Microbiology", degree Course CdL in "Biotechnology", and professor of "Nutrimicrobiomic" degree Course CdLM in "Biology of Health, Applied Technologies and Nutrition". She is a member of the Council of PhD in "Applied Biology and Experimental Medicine" of the University of Messina, Italy. She is member of the Reference Scientific Community of the "Stazione Zoologica Anton Dohrn" Naples, Italy (http://www.szn.it/index.php/it/chi-siamo/comunita-scientifica-di-riferimento, accessed on 7 October 2022). She is member of the University Research Center for the study of extreme environments and extremophiles (CUR AEE) and she is member of the National Consortium for Marine Sciences (CoNISMa). Her main interest is the study of biodiversity, ecology and genetic diversity of bacterial and fungal population involved in the biodeterioration of Cultural Heritage (especially stone monuments) and more recently in the diversity of halophilic microorganisms in deep sea environments. One special task is the taxonomy of black fungi that are among the most harmful microorganisms causing biodeterioration of stone monuments. She described two new fungal species and four new bacterial species isolated from monuments. She has participated in numerous National and International research projects both as Scientific Responsible and as a member of the Research Unit. Prof Filomena De Leo pubblished 51 articles in Journals, 16 monographs (or book chapter), 19 Proceedings of International Congress and 54 Poster and oral communication. The pubblications are visible on "Researchgate" link https://www.researchgate.net/profile/Filomena_De_Leo, accessed on 7 October 2022.

Prof. Daniela Isola is a contract Professor of Botany Applied to Cultural Heritage at the University of Tuscia. She graduated in Biological Sciences at the University of Cagliari and received postgraduate specializations in Neuropharmacology and Environmental Hygiene (University of Cagliari), and Bioinformatics (La Sapienza University of Rome). She obtained PhD in Biological and Biochemical Evolution at the University of Tuscia. Since then, she has focused on fungi from extreme environments and "black fungi". Her main interests are the biodeterioration of cultural heritage and biodegradation of aromatic hydrocarbons. Particular attention has been paid to the identification, diversity, phylogeny, metabolic skills of the involved organisms, and low-impact methods to control biodeterioration. She is the author of 4 genera and 11 fungal species, more than a thousand of sequences deposited in GenBank (covering bacteria, fungi, viruses and higher plants) and about 40 scientific papers with an H-index of 22.

applied sciences

MDPI

Article

Community Composition and Ex Situ Cultivation of Fungi Associated with UNESCO Heritage Monuments in the Bay of Naples

Mariagioia Petraretti [1,*], Karl J. Duffy [1], Angelo Del Mondo [1,2], Antonino Pollio [1] and Antonino De Natale [1]

1 Dipartimento di Biologia, Università di Napoli Federico II, Complesso Universitario Monte S. Angelo, Via Cintia 4, 80126 Napoli, Italy; karljoseph.duffy@unina.it (K.J.D.); angelo.delmondo@szn.it or angelo.delmondo@unina.it (A.D.M.); antonino.pollio@unina.it (A.P.); denatale@unina.it (A.D.N.)

2 Stazione Zoologica Anton Dohrn, Istituto Nazionale Di Biologia, Ecologia e Biotecnologie Marine, Villa Comunale, 80121 Napoli, Italy

* Correspondence: mariagioia.petraretti@unina.it

Abstract: The Bay of Naples, Italy, is renowned for its archaeological heritage. However, this heritage is threatened by the combination of weathering and the biological activity of microorganisms. Fungi are among the major agents of microbial deterioration of cultural heritage since they can cause cracks and lesions in monuments due to the penetrating force of their hyphae. Such biodeterioration may weaken the stone structures and threaten the longevity of these culturally important monuments. To address this, we collected, identified, and maintained in culture filamentous fungi that colonize the external surface of monuments at five important archaeological sites near Naples, namely Cuma, Ercolano, Nola, Oplonti, and Pompei. We isolated a total of 27 fungal taxa, all of which can be cultivated in the laboratory, and form a part of our reference collection. Many of the described fungal taxa we found belong to groups that are involved in stone biodeterioration and can thus be considered as model organisms for in vitro studies. These results emphasize the importance of identifying and cultivating fungal stock cultures for non-invasive studies on biodeterioration. Our newly developed reference collection represents a useful resource that is available to other researchers to rapidly identify potentially hazardous fungi on other monuments.

Keywords: fungi; ex situ collection; biodeterioration; biodegradation; cultural heritage

Citation: Petraretti, M.; Duffy, K.J.; Del Mondo, A.; Pollio, A.; De Natale, A. Community Composition and Ex Situ Cultivation of Fungi Associated with UNESCO Heritage Monuments in the Bay of Naples. *Appl. Sci.* **2021**, *11*, 4327. https://doi.org/10.3390/app11104327

Academic Editors: Filomena De Leo and Daniela Isola

Received: 12 March 2021
Accepted: 6 May 2021
Published: 11 May 2021

Publisher's Note: MDPI stays neutral with regard to jurisdictional claims in published maps and institutional affiliations.

1. Introduction

Fungi play an important role in the deterioration of buildings. Deterioration caused by fungal colonization involves both physical and chemical damage of stone surfaces, and in most cases, they take place simultaneously [1]. Physical damage is related to the ability of fungal hyphae to penetrate into the substratum, where pores and fissures provide a useful microhabitat for fungal growth. The pressure exerted by fungal growth leads to further damage due to cell turgor pressure and exopolysaccharide formation that, in addition to fungal adhesion on stone surfaces, increases mechanical pressure [2]. Moreover, chemical damage occurs due to byproducts of fungal metabolism that leads to corrosion and discoloration of stone surfaces. Fungi are able to excrete a large variety of organic acids that act as metal-chelators [3] and mediate the precipitation of secondary minerals produced through the reaction of anions from excreted acids with cations from the stone. The formation of secondary minerals, such as carbonates, oxalates, and phosphates, can cause blistering, scaling, granular disintegration, and flaking or "spalling" of outer layers, leading to stone decay [4]. There is a close relationship between material and colonizing organisms [5]; indeed, the degree of fungal colonization of a stone surface also depends on the structure, wetness, and chemical and mineralogical compositions of the substrata as well as environmental conditions [6,7]. Different lithotypes, e.g., brick, limestone, marble, tuff, and porphyry, provide a diverse range of substrates that fungi may use to acquire nutrients and grow.

Despite the large number of studies in the literature in which damage to cultural heritage is directly associated with fungi colonization, the occurrence of fungi on cultural heritage monuments does not necessarily mean that these fungi cause the loss of chemical and physical properties of the substrate; indeed, filamentous fungi as well as lichens could protect colonized materials, especially against environmental parameters [8], or they could bear no influence on the material properties. The multifaceted role of fungi in biodeterioration can be effectively assessed on the basis of preliminary in vitro tests, particularly recommended in the issues of monument protection, that require ex situ conservation strategies for fungal strains isolated from monuments. Ex situ collections may significantly improve our knowledge of the role of fungi in stone cultural heritage biodeterioration, providing the basis for an appropriate and effective maintenance and restoration strategy. The importance of maintaining a broad range of taxa in collections for ex situ conservation accessible to researchers prompted us to perform a survey campaign along the archaeological remains of Campania, Italy. Campania hosts a large number of works of art and monuments made of different stone materials, spanning the last three thousand years. Despite this unique cultural heritage, a deep sampling aimed to assess the biodiversity of cultivable fungi in these historical areas has never been conducted.

Using a combination of microscopical, genetic, and culture techniques, here we describe the taxonomic diversity of fungi that occur in the UNESCO heritage sites of Cuma, Ercolano, Nola, Oplonti, and Pompei, with the overall aim of developing an ex situ collection of fungal strains from these archeological sites.

2. Materials and Methods

2.1. Sampling

The sampling campaign in this study was carried out in March 2018 at some of the most important cultural heritage sites in Campania, namely the Sibyl Caves in Cuma, the Suburban Baths in Ercolano, the Roman Amphitheater in Nola, the House of Poppea in Oplonti, and the House of Fauno and the House of Castricio in Pompei (Figure 1).

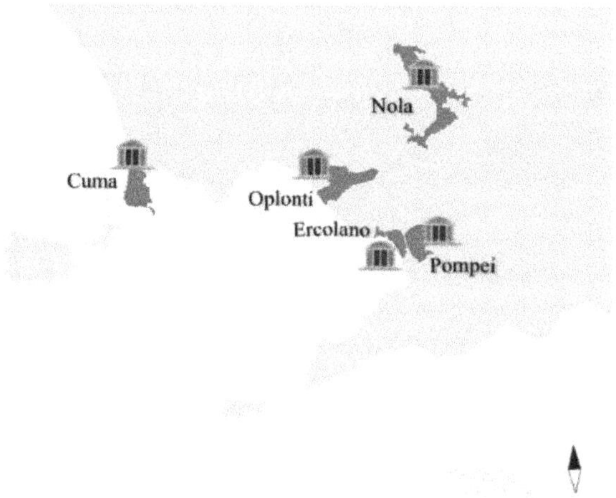

Figure 1. Location of the sampling site of UNESCO heritage monuments in the bay of Naples, Campania, Italy.

At every site, we measured ecological parameters, e.g., temperature and relative humidity, using a thermo-hygrometer (model HI 9564, Hanna® Instruments, Smithfield, RI, USA) and light intensity using a Climalux N light meter (Laboratori di Strumentazione

Industriale S.p.a., Milan, Italy). We measured pH on substrates at sampling points using a pH test paper strip. All the environmental parameters are shown in Table 1. The sampling points were chosen on the basis of the visibility of the fungal presence on the surface. Biofilm samples were taken by gently scraping the walls of the sampling sites with a sterile scalpel and adhesive tape strips were also used as a non-destructive sampling method [9]. The materials were deposited into sterile vials, until arrival at the laboratory.

Table 1. Values of environmental parameters (light, pH, relative humidity, and temperature) at each sampling site.

Location	Light (μmol m^{-2} s^{-1})	pH	Relative Humidity	Temperature (°C)
Cuma, Sibyl Caves	0.8± 0.01	7/8	50 ± 1.2%	19.2 ± 0.3
Ercolano, Suburban Baths	30 ± 0.6	7/8	94 ± 1.7%	15.3 ± 0.9
Nola Roman Amphitheater	130.84	7/8	90 ± 1.2%	16.7 ± 1.2
House of Poppea, Oplonti	129.95	7/8	95 ± 0.9%	13.2 ± 0.9
House of Fauno, Pompei	46 ± 0.6	7/8	58.8 ± 0.9%	18.2 ± 0.9
House of Castricio, Pompei	8.97	7/8	90 ± 1.2%	18.3 ± 0.9

2.2. Confocal Laser Scanning Microscope Analysis

The recorded adhesive tape samples were cut into small sections (approximately 1 × 1 cm, [9]), placed on a glass slide, and observed on a Confocal Laser Scanning Microscope (CLSM), Zeiss LSM 700 (Carl Zeiss AG, Munich, Germany, using the software Zen 2011), by capturing images with a 63× water immersion objective. Images were acquired in three channels simultaneously: the red channel was used to discriminate phototrophs containing autofluorescence pigments (chlorophyll a and phycobilins), with excitation beams at 488 and 639 nm and emissions at 590–800 nm; the green channel was used to detect extrapolymeric matrix (EPS) using concanavalin-A with Alexa 488, with the excitation beams at 488 nm and emissions at 553–636 nm; and calcofluor-white was used to evidence the bacteria and hyphae with the excitation beams at 405 and 488 nm and emissions at 415–506 nm (blue channel) [10].

2.3. Isolation of Fungal Strains

After the sampling campaign, samples were inoculated on agar medium, such as Potato Dextrose Agar (PDA) prepared according to Samson et al. [11], Bold's Basal Medium (BBM) [12] added to sucrose (12 g/L) according to Jeger et al. [13], and Malt-Yeast Extract-Sucrose Agar (MEA, Difco™) prepared according to Skaar and Stenwig [14]. Incubation was carried out at 22 ± 2 °C for 30 days. At the end of the incubation period, enumeration of microorganisms as cfu/g of sample was carried out and the several mycelia obtained were isolated with the aid of a stereomicroscope. Afterwards, fungi were separately cultivated on PDA and finally observed with a stereomicroscope.

2.4. Identification of Fungal Isolates

Fungal strains were identified through a polyphasic approach that is an integrated approach of identification based on morphological and molecular features of microorganisms [15]. According to Barnett and Hunter [16] and Fassatiovà and Ellis [17], the morphological identification of fungi was based on the macroscopic features of colonies growing on agar plates and the micromorphology of the reproductive structure. The morphological analysis was then confirmed by molecular analysis. For each fungal isolate, the following procedure was applied: DNA was extracted with a modified DNA extraction protocol [18] and used for a Polymerase Chain Reaction with primers targeting the internal transcribed spacer region (ITS) (primer forward, 5′-TCCGTAGGTGAACCTGCGG-3′; primer reverse, 5′-TTCAAAGATTCGATGATTCAC-3′). The ITS is the region spanning ITS1, 5.8S rRNA, and ITS2 was recently elected to be the universal barcode marker for fungi [19]. This DNA region has enough gaps between the intraspecific and interspecific variation across the kingdom Fungi and has been shown to have a high amplification success rate in various fungal taxa, e.g., it can discriminate the majority of species in Mucorales [20]. The barcode region together with a well-curated database of DNA sequences may constitute a reliable and fast tool for culture collection in the task of providing certification of fungal cultures. The amplification reaction was carried out in a reaction volume of 25 μL containing 2.5 μL of 10× reaction buffer, 1.5 μL of $MgCl_2$, 2 μL of dNTP, 1.5 μL of each of the primers, and 0.2 μL of Taq polymerase (EconoTaq, Lucigen, Middleton, WI, USA). An amount of DNA, approximately 100 ng, was added to each reaction mixture in a PCR tube. The profile used was the same described by Del Mondo 2017 [21]. Amplification was run in an Applied Biosystem 2720 thermal cycler. The amplification product was then evaluated on 1.2% (*w/v*) agarose gel in an electrophoretic purified with a QIAquick® PCR Purification kit (Qiagen Inc, Valencia, CA, USA). The sequence reaction was obtained with the BigDye Terminator Cycle Sequencing technology (Applied Biosystems, Foster City, CA, USA), purified automatically using the Agencourt CleanSEQ Dye terminator removal Kit (Agencourt Bioscience Corporation, 500 Cummins Center, Suite 2450, Beverly, MA, USA) and a robotic station Biomek FX (Beckman Coulter, Fullerton, CA, USA). The product was analyzed on an Automated Capillary Electrophoresis Sequencer 3730 DNA Analyzer (Applied Biosystems). The amplification primers were used as the sequencing primers. The obtained sequence was searched for in BLAST version 2.0 (National Center for Biotechnology Information databases) and identified. The ITS sequences obtained in this study have been deposited in GenBank (the accession numbers are listed in Table 2).

Table 2. Identification of the fungal species complex level based on ITS sequences with the description of their sampling site and lithic substrate. ACUF Collection Codes and Gene Bank Accession numbers are given for each strain.

Identified Species Complex Level	Sites	Source	ACUF Collection Code	Gene Bank Accession Number
Alternaria section *Alternata*	Pompei	Mortar	033f	MW881067
	Pompei	Mortar	032f	MW881066
Alternaria section *Alternata*	Nola	Marble	053f	MW881087
Alternaria section *Alternata*	Ercolano	Plaster	017f	MW881054
Alternaria sp.	Ercolano	Mortar	020f	MW881053
Aspergillus section *Aeni*	Pompei	Mortar	039f	MW881073
Aspergillus section *Usti*	Ercolano	Mortar	029f	MW881060
Aspergillus section *Usti*	Cuma	Tuff	012f	MW881047
	Cuma	Tuff	022f	MW881049
Aspergillus section *Nigri*	Ercolano	Plaster	007f	MW881065
	Ercolano	Plaster	019f	MW881062

Table 2. *Cont.*

Identified Species Complex Level	Sites	Source	ACUF Collection Code	Gene Bank Accession Number
Aspergillus section *Circumdati*	Pompei	Frescos	015f	MW881099
	Pompei	Frescos	026f	MW881100
Aspergillus sp.	Cuma	Tuff	008f	MW881051
Cladosporium sp.	Pompei	Mortar	041f	MW881075
Clonostachys sp.	Oplonti	Mortar	005f	MW881093
	Oplonti	Mortar	056f	MW881095
	Oplonti	Mortar	010f	MW881098
	Oplonti	Mortar	021f	MW881097
Clonostachys sp.	Nola	Marble	042f	MW881076
	Nola	Marble	043f	MW881077
	Nola	Marble	044f	MW881078
Curvularia geniculata species complex	Ercolano	Plaster	023f	MW881052
Fusarium oxysporum species complex	Ercolano	Plaster	031f	MW881064
Fusarium section *Discolor*	Cuma	Tuff	009f	MW881048
Fusarium oxysporum species complex	Pompei	Frescos	014f	MW881102
Fusarium oxysporum species complex	Ercolano	Plaster	018f	MW881055
	Ercolano	Plaster	024f	MW881056
	Ercolano	Plaster	025f	MW881057
	Ercolano	Plaster	028f	MW881059
Fusarium oxysporum species complex	Oplonti	Mortar	054f	MW881090
	Oplonti	Mortar	055f	MW881091
	Oplonti	Mortar	001f	MW881089
Fusarium solani species complex	Ercolano	Plaster	016f	MW881063
Fusarium sp.	Oplonti	Mortar	006f	MW881088
Fusarium tricinctum species complex	Oplonti	Mortar	002f	MW881094
Lecanicillium sp.	Pompei	Frescos	013f	MW881101
Lecanicillium sp.	Ercolano	Mortar	027f	MW881058
Lecanicillium sp.	Ercolano	Plaster	030f	MW881061
Neofusicoccum parvum species complex	Cuma	Tuff	011f	MW881050
Penicillium section *Fasciculata*	Pompei	Mortar	036f	MW881070
	Pompei	Mortar	038f	MW881072
	Pompei	Mortar	035f	MW881069

Table 2. *Cont.*

Identified Species Complex Level	Sites	Source	ACUF Collection Code	Gene Bank Accession Number
	Pompei	Mortar	040f	MW881074
Penicillium sp.	Pompei	Mortar	034f	MW881068
	Pompei	Mortar	037f	MW881071
Penicillium section *Aspergilloides*	Nola	Marble	046f	MW881080
	Nola	Marble	047f	MW881081
	Nola	Marble	048f	MW881082
	Nola	Marble	049f	MW881083
	Nola	Marble	051f	MW881085
Purpureocillium sp.	Oplonti	Frescos	004f	MW881096
Talaromyces section *Talaromyces*	Oplonti	Mortar	003f	MW881092
Talaromyces section *Talaromyces*	Nola	Marble	045f	MW881079
Trichoderma sp.	Nola	Marble	050f	MW881084
	Nola	Marble	052f	MW881086

2.5. Fungal Preservation for Ex Situ Conservation

In according to the World Federation for Culture Collection Guidelines, more than one method was applied for each fungal strain for successful preservation. Our fungal strains are stored by different methods: (a) PDA on Petri dishes at a temperature range of between 22 °C and 25 °C in darkness; (b) in a corked glass tube with sterilized water at room temperature [22]; and (c) at −80 °C in glycerol (selected strains only). This method, namely cryopreservation together with freeze drying, is considered to be a long-term preservation method [23]. To maintain fungi in a viable state, to evaluate purity, and to avoid devitalization, monthly checking and refreshment of cultures were performed. All the fungal strains are maintained in the Algal Culture Collection (ACUF) at the Department of Biology, University of Naples Federico II, Italy. This collection, traditionally devoted to the maintenance of aero-terrestrial microalgae and cyanobacteria [24], has been enriched with a special section devoted to the maintenance of fungal strains isolated directly from archeological sites in Campania. Each strain is included in a private database with all the information regarding sampling sites, origin substrate, data on collection, ecological notes, cultivation and maintenance methods, phenotypic characteristics, and genomic analysis. In order to maintain the safety of the data associated with each of the strains preserved in the collection, all computer files are duplicated and kept in a separate area. Furthermore, we deposited our isolates at the Mycotheca Universitatis Taurinensis, Turin, Italy (MUT), a renowned collection specialized in fungal preservation.

3. Results

3.1. Description of Damage and Substrate Change

Biological growth on stone can result in changes in surface color and structure depending on the identity of the organism and their growth and behavior. At our sampling sites, the biological colonization on stone surfaces assumed the forms of epilithic formations with a patina aspect. In particular, these organisms formed a subaerial biofilm, which is a type of biofilm that occurs at the atmosphere–rock interface. This type of biofilm has been frequently reported in the literature on hypogean monuments, such as catacombs [25], and on walls, statues, and wetlands. These formations may have a colored patinas aspect, depending on the type of biocenosis and of the growth phase of the prevailing species. As shown in Figure 2b,d,e, the stone surface appears with a green and greenish stain, probably due to the presence of organic pigments (e.g., chlorophylls, carotenoids, melanins) [26]. In Figure 2h,k,n, the stone surface appears with a black stain and this is related to the mixed association of different fungal groups.

Figure 2. Archeological sites: Sibyl Caves in Cuma (**a**), Suburban Baths in Ercolano (**d**), the Roman Amphitheater in Nola (**g**), the House of Poppea in Oplonti (**j**), and the House of Fauno (**m**) and the House of Castricio (**p**) in Pompei; the visible alteration at the same sites (**b**,**e**,**h**,**k**,**n**,**q**); the recorded adhesive tape samples observed on the CLSM (**c**,**f**,**i**,**l**,**o**,**r**; scale bar, 50 μm).

3.2. *Confocal Laser Microscopy*

All the samples analyzed by CLSM revealed that many cells contained chlorophyll and phycobilin (red auto-fluorescence), which were ascribed to algae and cyanobacteria and polysaccharide polymers (e.g., cellulose and chitin) in their cell walls (blue color), which were ascribed to fungi (Figure 2c,f,i,l,o,r).

3.3. *Molecular Identifications*

Table 2 shows the identification of the isolated fungal species retrieved from the sampled UNESCO monuments, together with sites and source sampled as well as the collection code linked to the fungal strains and the GenBank accession numbers of the obtained sequences.

A total of 18 fungal taxa, belonging to 3 different Classes, 5 different Orders, and 10 different Families, were obtained (Figure 3) and kept in culture.

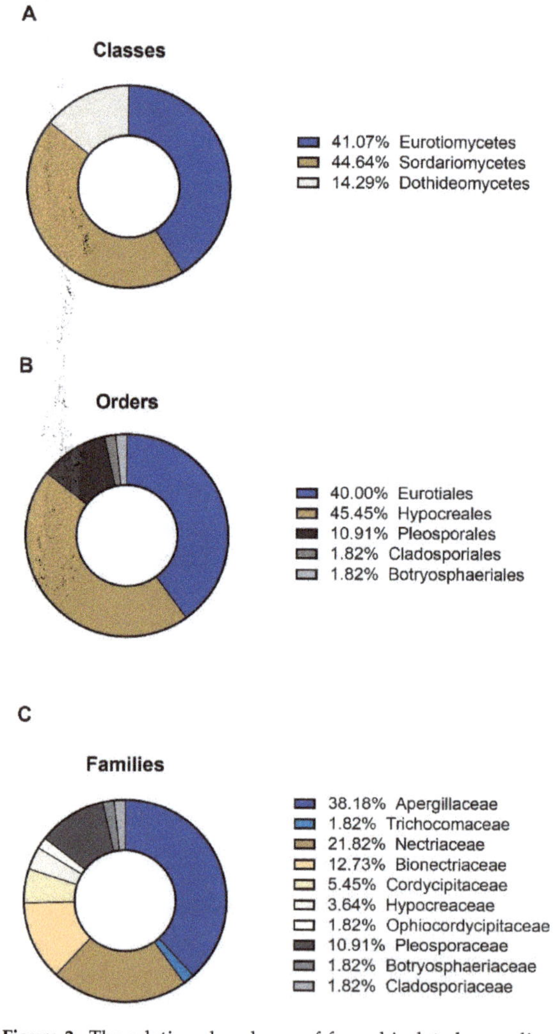

Figure 3. The relative abundance of fungal isolated according to (**A**) classes, (**B**) orders, and (**C**) families.

Overall, the most common genera were *Aspergillus* in Cuma (60%), *Fusarium* in Ercolano (42.86%), *Penicillium* in Nola (41.67%), *Fusarium* in Oplonti (45.45%), *Penicillium* (60%) in Pompei, Fauno, and *Aspergillus* (50%) in Pompei, Castricio (Figure 4).

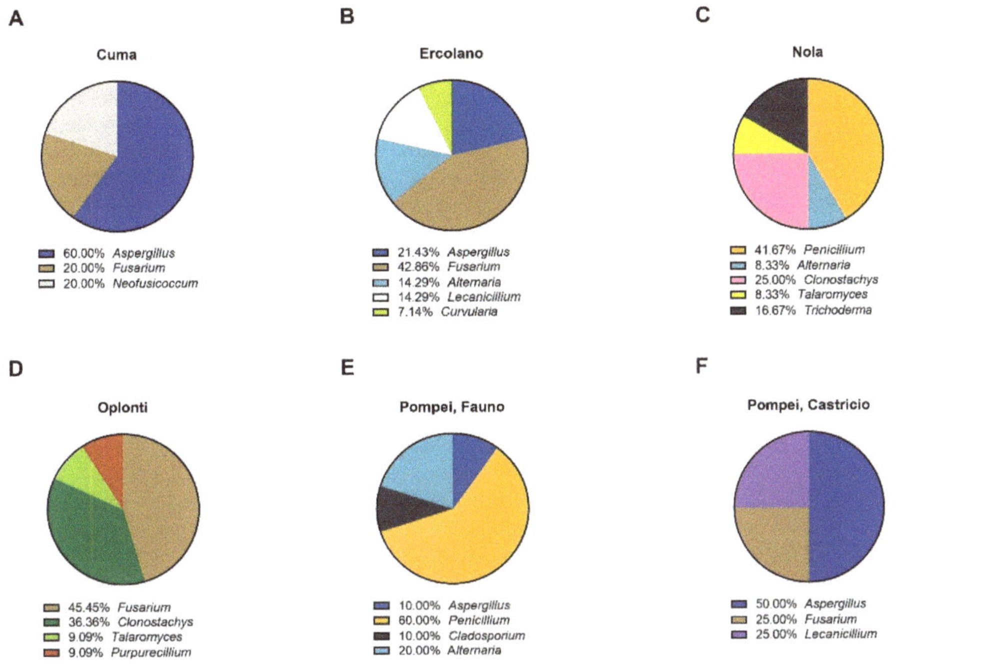

Figure 4. The relative prevalence of fungal genera found in: Sibyl Caves in Cuma (**A**), Suburban Baths in Ercolano (**B**), the Roman Amphitheater in Nola (**C**), the House of Poppea in Oplonti (**D**), and the House of Fauno (**E**) and the House of Castricio (**F**) in Pompei.

4. Discussion

The protection of cultural heritage often involves the study of the bio-receptivity of building materials and the biodegradation potential of microorganisms involved in deterioration, which can be primarily assessed through laboratory studies. In this work, a non-destructive method was used to sample microorganisms at UNESCO cultural heritage sites in Campania, Italy. In particular, we sampled using adhesive tape sampling coupled with microscopical analysis to identify the constituent microorganisms in biofilms of these monuments, which allows us to examine the existing relationships between the surface and the colonizing microorganisms [9]. The observations made using confocal light microscopy demonstrated fungal colonization in all the adhesive tape samples analyzed as a network of filamentous structures. Furthermore, we observed the presence of cyanobacteria and algae closely connected with filamentous structures, demonstrating that fungi actively colonize the rock as essential compounds of the biofilm sampled and not as contaminants. The isolation of fungi through culture-dependent methods, followed by maintenance of strains in culture, was aimed at obtaining quality-controlled isolates for further studies on biodeterioration processes and to develop innovative strategies for their control. For each strain, we collected data related to the substrate and environmental conditions at the sampling location and this information can be used to develop future laboratory experiments simulating specific environmental conditions under which these fungi can grow. Indeed, future perspectives are directed towards using our fungi as models to perform in vitro experiments for understanding the patterns of microbial colonization of stone materials [21]. The fungal isolates in this study are widespread, frequently associated with soil particles and plant material, which is probably due to their broad tolerance to different environmental conditions and allows them to colonize a large array of terrestrial habitats [27]. Molecular identification of sampled strains was performed using the internal transcribed spacer (ITS) rDNA area, which is the most widely used marker for fungi [19]. Unfortunately, for many Ascomycota genera, such as *Penicillium* and *Aspergillus*, the ITS is not variable enough to allow for species-level identification [28]. Because of the limitations associated with the chosen molecular marker, herein we considered the species-complex level for an overview of fungal diversity on stone monuments. We recognize the necessity of further identification using taxon-specific markers (e.g., SSRs) for identifying isolates to species level as the collection grows. The genera isolated in our sampling include *Alternaria, Aspergillus, Cladosporium, Clonostachys, Curvularia, Fusarium, Lecanicillium, Neofusicoccum, Penicillium, Purpureocillium, Talaromyces*, and *Trichoderma*, which are known to be ubiquitous filamentous fungi of soil and are often airborne. Some of them were already described as colonizer, or occasionally pioneer, taxa of deteriorated monuments [29]. For most of these species, there is no representative strain of the wide range of morphology and physiology expressed within that species and therefore it is necessary to maintain a number of representative strains. Some collections, such as the CABI Bioscience Genetic Resource Collection, retain on average five strains for each species, but in some cases this number is not sufficient. This is the case for host specificity in the plant pathogen, such as *Fusarium oxysporum*, which has a large number of genetic variants [30]. Thus, ex situ conservation of microorganisms collected from monuments can ensure that all strains with their unique properties are preserved and maintained over time. In accordance with other studies, we observed that the genera most frequently identified as colonizers of several stone substrates are *Aspergillus, Fusarium*, and *Penicillium* [31]. The action of these fungi, included in the orders Capnodiales and Pleosporales, could lead to aesthetic alteration and biopitting of stone materials [32]. Moreover, some of the fungal isolates belonging to *Alternaria* section *Alternata, Cladosporium sp., Fusarium solani* species complex, and *Penicillium* section *Aspergilloides* are known to contribute to acidification and the dissolution of stone by excreting organic acids. In particular, oxalic acid secreted by fungi can dissolve limestone calcium carbonate, producing calcium oxalates, one of the most severe biodeterioration processes affecting limestone monuments [33]. In addition, recently it has been shown that the genus *Purpureocillium* may have halotolerant characteristics [34], which may further broaden the impact these fungi have on stone structures.

5. Conclusions

The collection and molecular identification of fungal strains and their associated ecological data, describing their site of sampling, type of substrate, and morphological diagnostic characteristics, represent a key resource for the development of biotechnological approaches devoted to the conservation of cultural heritage. The ex situ conservation of fungi sampled from bio-deteriorated environments can ensure that isolates are preserved to maintain their integrity and long-term survival. This is essential for future research on the preservation of historical monuments, including the ecological differentiation of fungal communities according to sampling sites and the production of desirable end products applicable for bioremediation.

Author Contributions: Conceptualization, M.P. and A.D.M.; methodology, M.P.; software, M.P. and A.D.M.; validation, A.D.M., A.P. and A.D.N.; data curation, M.P.; writing—original draft preparation, M.P.; writing—review and editing, K.J.D., A.D.N. and A.P.; supervision, A.P. All authors have read and agreed to the published version of the manuscript.

Funding: This research received no external funding.

Institutional Review Board Statement: Not applicable.

Informed Consent Statement: Not applicable.

Data Availability Statement: Data are contained within the article.

Acknowledgments: The authors thank the Herculaneum Archaeological Park, the Pompei Archaeological Park, the Phlegrean Fields Archaeological Park, Soprintendenza Archeologia Belle Arti e Paesaggio della città metropolitana di Napoli, and the Società dei Naturalisti in Napoli (Naples, Italy) for the assistance with research. The authors gratefully thank Serena Di Lecce and Alessandra Di Leva for their valuable technical support.

Conflicts of Interest: The authors declare no conflict of interest.

References

1. Salvadori, O.; Municchia, A.C. The Role of Fungi and Lichens in the Biodeterioration of Stone Monuments. *Open Conf. Proc. J.* **2016**, *7*, 39–54. [CrossRef]
2. Burford, E.P.; Fomina, M.; Gadd, G.M. Fungal involvement in bioweathering and biotransformation of rocks and minerals. *Miner. Mag.* **2003**, *67*, 1127–1155. [CrossRef]
3. Sterflinger, K. Fungi as Geologic Agents. *Geomicrobiol. J.* **2000**, *17*, 97–124. [CrossRef]
4. Wright, J.S. Geomorphology and stone conservation: Sandstone decay in Stoke-on-Trent. *Struct. Surv.* **2002**, *20*, 50–61. [CrossRef]
5. Turick, C.E.; Berry, C.J. Review of concrete biodeterioration in relation to nuclear waste. *J. Environ. Radioact.* **2016**, *151*, 12–21. [CrossRef] [PubMed]
6. Caneva, G.; Nugari, M.P.; Nugari, M.P.; Salvadori, O. *Plant Biology for Cultural Heritage: Biodeterioration and Conservation*; Getty Publications: Los Angeles, CA, USA, 2008.
7. Gaylarde, C.; Silva, M.R.; Warscheid, T. Microbial impact on building materials: An overview. *Mater. Struct.* **2003**, *36*, 342–352. [CrossRef]
8. Pinna, D. Biofilms and lichens on stone monuments: Do they damage or protect? *Front. Microbiol.* **2014**, *5*, 133. [CrossRef]
9. Urzì, C.; De Leo, F. Sampling with adhesive tape strips: An easy and rapid method to monitor microbial colonization on monument surfaces. *J. Microbiol. Methods* **2001**, *44*, 1–11. [CrossRef]
10. Larson, C.; Passy, S.I. Spectral fingerprinting of algal communities: A novel approach to biofilm analysis and biomonitoring. *J. Phycol.* **2005**, *41*, 439–446. [CrossRef]
11. Samson, R.A.; Hoekstra, E.S.; Frisvad, J.C.; Filtenborg, O. *Introduction to Food-Borne Fungi*; Centraalbureau voor Schimmelcultures: Delft, The Netherlands, 1996.
12. Nichols, H.W.; Bold, H.C. Trichosarcina polymorpha Gen. et Sp. Nov. *J. Phycol.* **1965**, *1*, 34–38. [CrossRef]
13. Jeger, M.J.; Lamour, A.; Gilligan, C.A.; Otten, W. A fungal growth model fitted to carbon-limited dynamics of Rhizoctonia solani. *New Phytol.* **2008**, *178*, 625–633. [CrossRef] [PubMed]
14. Skaar, I.; Stenwig, H. Malt-yeast extract-sucrose agar, a suitable medium for enumeration and isolation of fungi from silage. *Appl. Environ. Microbiol.* **1996**, *62*, 3614–3619. [CrossRef]
15. Passarini, M.R.Z.; Santos, C.; Lima, N.; Berlinck, R.G.S.; Sette, L.D. Filamentous fungi from the Atlantic marine sponge Dragmacidon reticulatum. *Arch. Microbiol.* **2012**, *195*, 99–111. [CrossRef] [PubMed]
16. Barnett, H.L.; Hunter, B.B. *Illustrated Genera of Imperfect Fungi, Mycol*, 3rd ed.; Burgess Publishing Company: Minneapolis, MN, USA, 1972.
17. Bushell, M.E. (Ed.) Fassatiovà O Moulds and filamentous fungi in technical microbiology. In *Progress in Industrial Microbiology 22*; Elsevier: Amsterdam, The Netherlands, 1986.
18. Doyle, J. DNA Protocols for Plants. In *Molecular Techniques in Taxonomy*; Springer: Berlin/Heidelberg, Germany, 1991; pp. 283–293. [CrossRef]
19. Schoch, C.L.; Seifert, K.A.; Huhndorf, S.; Robert, V.; Spouge, J.L.; Levesque, C.A.; Chen, W.; Fungal Barcoding Consortium. Nuclear ribosomal internal transcribed spacer (ITS) region as a universal DNA barcode marker for Fungi. *Proc. Natl. Acad. Sci. USA* **2012**, *109*, 6241–6246. [CrossRef] [PubMed]
20. Walther, G.; Pawłowska, J.; Alastruey-Izquierdo, A.; Wrzosek, M.; Rodriguez-Tudela, J.; Dolatabadi, S.; Chakrabarti, A.; De Hoog, G. DNA barcoding in Mucorales: An inventory of biodiversity. *Pers. Mol. Phylogeny Evol. Fungi* **2013**, *30*, 11–47. [CrossRef]
21. Del Mondo, A.; Pinto, G.; De Natale, A.; Pollio, A. In vitro colonization experiments for the assessment of mycelial growth on a tuff substratum by a fusarium solani strain isolated from the Oplonti (Naples, italy) archaeological site. *Int. J. Cons. Sci.* **2017**, *8*, 651–662.
22. Smith, D.; Ryan, M.J.; Day, J.G. *The UKNCC Biological Resource: Properties, Maintenance and Management*; UKNCC Secretariat: Egham, UK, 2001; 382p.
23. OECD. *Biological Resource Centres: Underpinning the Future of Life Sciences and Biotechnology*; OECD Publications: Paris, France, 2001; p. 66.
24. D'Elia, L.; Del Mondo, A.; Santoro, M.; De Natale, A.; Pinto, G.; Pollio, A. Microorganisms from harsh and extreme environments: A collection of living strains at ACUF (Naples, Italy). *Ecol. Quest.* **2018**, *29*, 1–16. [CrossRef]
25. Gorbushina, A.A. Life on the rocks. *Environ. Microbiol.* **2007**, *9*, 1613–1631. [CrossRef]
26. Warscheid, T.; Braams, J. Biodeterioration of stone: A review. *Int. Biodeterior. Biodegrad.* **2000**, *46*, 343–368. [CrossRef]

27. Isola, D.; Zucconi, L.; Onofri, S.; Caneva, G.; De Hoog, G.S.; Selbmann, L. Extremotolerant rock inhabiting black fungi from Italian monumental sites. *Fungal Divers.* **2016**, *76*, 75–96. [CrossRef]
28. Seifert, K.A.; Samson, R.A.; Dewaard, J.R.; Houbraken, J.; Lévesque, C.A.; Moncalvo, J.-M.; Louis-Seize, G.; Hebert, P.D.N. Prospects for fungus identification using CO1 DNA barcodes, with Penicillium as a test case. *Proc. Natl. Acad. Sci. USA* **2007**, *104*, 3901–3906. [CrossRef] [PubMed]
29. Del Mondo, A.; De Natale, A.; Pinto, G.; Pollio, A. Correction to: Novel qPCR probe systems for the characterization of subaerial biofilms on stone monuments. *Ann. Microbiol.* **2019**, *69*, 1097–1106. [CrossRef]
30. Kirk, P.M.; Cannon, P.F.; David, J.C.; Stalpers, J.A. *Dictionary of the Fungi*, 9th ed.; CABI Publishing: Wallingford, UK, 2001; p. 55.
31. Saarela, M.; Alakomi, H.-L.; Suihko, M.-L.; Maunuksela, L.; Raaska, L.; Mattila-Sandholm, T. Heterotrophic microorganisms in air and biofilm samples from Roman catacombs, with special emphasis on actinobacteria and fungi. *Int. Biodeterior. Biodegrad.* **2004**, *54*, 27–37. [CrossRef]
32. Sterflinger, K. Fungi: Their role in deterioration of cultural heritage. *Fungal Biol. Rev.* **2010**, *24*, 47–55. [CrossRef]
33. Gadd, G.M. Geomycology: Biogeochemical transformations of rocks, minerals, metals and radionuclides by fungi, bioweathering and bioremediation. *Mycol. Res.* **2007**, *111*, 3–49. [CrossRef] [PubMed]
34. Arpini, C.M.; Nóbrega, Y.C.; Castheloge, V.D.; Neves, D.S.; Tadokoro, C.E.; Da Costa, G.L.; Oliveira, M.M.E.; Santos, M.R.D.D. Purpuriocillium lilacinum infection in captive loggerhead sea turtle hatchlings. *Med Mycol. Case Rep.* **2019**, *23*, 8–11. [CrossRef]

applied sciences

MDPI

Article

Black Fungi and Stone Heritage Conservation: Ecological and Metabolic Assays for Evaluating Colonization Potential and Responses to Traditional Biocides

Daniela Isola [1,2,*], Flavia Bartoli [2], Paola Meloni [3,4], Giulia Caneva [2] and Laura Zucconi [1]

1 Department of Ecological and Biological Sciences (DEB), University of Tuscia, 01100 Viterbo, Italy; zucconi@unitus.it
2 Department of Sciences, Roma Tre University, 00146 Rome, Italy; flavia.bartoli@uniroma3.it (F.B.); giulia.caneva@uniroma3.it (G.C.)
3 Department of Mechanical, Chemical and Materials Engineering (DIMCM), University of Cagliari, 09123 Cagliari, Italy; paola.meloni@unica.it
4 Laboratorio Colle di Bonaria, University of Cagliari, 09125 Cagliari, Italy
* Correspondence: daniela.isola@uniroma3.it

Abstract: Identifying species involved in biodeterioration processes is helpful, however further effort is needed to assess their ecological requirements and actual activity. Black fungi (BF) represent one of the most underestimated threats to stone cultural heritage in the Mediterranean basin; they are difficult to kill or remove due to their ability to grow inside the rock and cope with several stresses. Despite this, little is known about BF and factors favoring their growth on stone surfaces. Eighteen BF species were here investigated for temperature and salt tolerance, and metabolic traits by plate assays. The relation between some highly damaged monuments and their BF settlers was assessed using X-ray diffraction analysis, mercury intrusion porosimetry, and SEM. The sensitiveness to four commonly used traditional biocides was also tested. All strains were able to grow within the range of 5–25 °C and in the presence of 3.5% NaCl. Instrumental analyses were fundamental in discovering the relation between halophilic strains and weathered marble sculptures. The acid, cellulase, esterase, and protease production recorded proved BF's potential to produce a chemical action on carbonate stones and likely affect other materials/historical artefacts. Besides, the use of carboxymethylcellulose and Tween 20 should be evaluated in restoration practice to prevent tertiary bioreceptivity. Agar diffusion tests helped identify the most resistant species to biocides, opening the perspective of its use as reference organisms in material testing procedures.

Keywords: benzalkonium chloride resistance; biodeterioration control; *Exophiala*; halophilic fungi; fungal acid production; *Knufia*; monuments salt weathering; QAC biocides; *Salinomyces thailandicus*; marble XRD; MIP investigations

Citation: Isola, D.; Bartoli, F.; Meloni, P.; Caneva, G.; Zucconi, L. Black Fungi and Stone Heritage Conservation: Ecological and Metabolic Assays for Evaluating Colonization Potential and Responses to Traditional Biocides. *Appl. Sci.* **2022**, *12*, 2038. https://doi.org/10.3390/app12042038

Academic Editor: Gyungsoon Park

Received: 27 January 2022
Accepted: 11 February 2022
Published: 16 February 2022

1. Introduction

All materials exposed to outdoor conditions are inevitably subject to deterioration processes leading to biotic and abiotic alterations. Surface colonization and the related biodeterioration phenomena greatly vary depending on the substrate's bioreceptivity, macro- and microclimatic conditions (e.g., water availability, sun exposure, shading, and orientation), and nutrients supply [1–4].

Among the stone biodeteriogens, rock black fungi are considered one of the most remarkable issues for cultural heritage conservation, particularly in arid and semi-arid environments, where they live at their ecological optimum [5–7]. Black fungi (BF), also known as rock-inhabiting fungi (RIF), black yeasts, or microcolonial fungi (MCF), are a polyphyletic poikilotolerant morpho-ecological group with remarkable adaptations allowing them to cope with changes in extreme temperatures, drought, starvation, osmotic stress, and high solar and UV-radiation as it occurs on rock surfaces [8–12].

When colonizing rocks, BF can induce a chemical deterioration by secreting siderophore-like compounds, however the most relevant damages are believed to be due to a mechanical action. Hyphal morphology and the strong mechanical turgor pressure (up to 12.39 bar in penetrating silicate and carbonate rocks) allow them to dig cavities on the heritage stone surfaces at depths ranging from a few hundred microns to several millimeters [7,13,14]. Furthermore melanin, the most important factor in BF stress resistance, is also responsible of the aesthetic alterations imparting a dark, blackish-brown appearance to the lithic surface [15–17].

Knowledge of the deteriogens' diversity is undoubtedly useful, however more efforts should be given to outline the ecological needs of the occurring species and their involvement in deterioration processes. The metabolic and ecological profiles, such as the environmental conditions, which favor the various taxonomic groups, are crucial in defining the risk for the artefacts and in designing both preventive/indirect and corrective/direct (e.g., biocide treatments) control measures [1,18]. In fact, the awareness on the tolerance range of the involved species for each environmental factor has a great practical relevance in preventing undesirable growth [1,19]. Such information is also relevant when considering biocide treatments, due to the need to customize the treatment selecting the most efficient product against specific microorganisms [20].

Corrective measures against fungal spreading are often required and such interventions are mainly carried out using biocidal products both to remove the biomass in the early stages of artwork treatment and later to prevent re-colonization [21–23]. In stone treatments, BF represent a challenge for restorers being able to survive and re-colonize artworks after restoration treatments with biocides or synthetic polymeric coatings [6,11,24–27]. To date, very little information is available on their sensitiveness to biocides and on their metabolic and eco-physiological profile, mainly limited to a few model fungi [28,29]. In this light, traditional culture dependent assays can be useful to deepen in detail their ecological traits and biodeteriorative potential [30].

Based on the multidisciplinary investigation surveys performed on the deteriogenic phenomena affecting the monumental Cemetery of Bonaria in Cagliari since 2010, we focused on a selection of 18 BF strains isolated from some of its marble monuments.

The goals of the present study were: (i) the selection and identification of the BF strains of interest; (ii) the assessment of the main fungal traits with respect to thermal and salt growth preferences; (iii) the possible relation between the recorded fungal traits and the substrate in selected monuments; (iv) the detrimental potential of the investigated species; and finally (v) their sensitiveness to a few traditional biocides. The achieved results will also give the opportunity of testing the best techniques of evaluation of metabolic evaluation and control. Data will also provide new insights to design more efficient preventive and control protocols, as well as on the leading forces driving the colonization processes.

2. Materials and Methods

2.1. Black Fungal Strains Selection and Identification

The ancient monumental cemetery of Bonaria (39°12′36.96″ N 9°07′26.13″ E) in Cagliari, Italy, has been object of multi-disciplinary studies aimed at favoring its conservation, rediscovery as urban historical heritage, and valorization [5,17,31–34]. The climatic conditions of the site are characterized by hot dry summers, very mild winters, and humidity may be high due the sea proximity. Over the period 2011–2014, we performed a few samplings from some Carrara marble funerary monuments (Figure 1), thanks to the collaboration of the Superintendence of Heritage Landscape, Historical, Artistic and Ethno-Anthropological Heritage for the metropolitan city of Cagliari and for the provinces of Oristano, Medio Campidano, and Municipality Cemeteries Direction of Cagliari.

Forty BF strains were isolated and most of them were identified through phylogenetic analysis, and new species described [31]. Three strains not considered previously, namely CCFEE 5778, 5945, and 6327, were identified by internal transcribed spacer (ITS) and LSU sequencing (in case of poor ITS identities) followed by BLASTn comparison. The primer set used for amplifications were ITS4-ITS5 and LR0R-LR7, respectively. In short, the reactions were performed in a total volume of 25 μL using 5 pmol of each primer, Bioline BioMix (Bioline Reagents, London, UK), and about 30 ng of genomic DNA [35]. The PCR protocols consist of an initial denaturation and final extension and 35 cycles of amplification with annealing at 55 °C (ITS) and 52 °C (LSU). After identification, representative strains of the whole BF diversity found in the site have been subject to selection in order to give the wider representativeness possible with precedence to type strains (when available).

Figure 1. The sampled Carrara marble funerary monuments of the ancient monumental cemetery of Bonaria—Cagliari entitled to (**A**) Avv. Giuseppe Todde; (**B**) Antonio Viganigo; (**C**) A. M. Frau Carta; (**D**) Francesca Warzee; (**E**) Rossino Bolla spouse; (**F**) Giuseppina Ara dei conti Ciarella; (**G**) Zelina Ferrà Gastaldi Millelire; (**H**) Ignazio Ruda Roych conte di San Lorenzo; (**I**) Unknown Burial.

All strains are part of the Culture Collection of Fungi from Extreme Environments (CCFEE, Viterbo, Italy) and new sequences were deposited in GenBank.

2.2. Thermal Preferences, Salt Tolerance, and Metabolic Assays

Thermal preferences and salt tolerance were assessed by inoculating small fragments of fungal mycelia onto malt agar plates (MA: 30 g/L malt extract, 15 g/L bacteriological agar; VWR) and incubating them for a month within the range 0–40 °C with 5 °C intervals. The growth at 37 °C was also assessed.

Similarly, salt tolerance was determined inoculating small fragments of mycelia on MEA 2% plates (malt extract 20 g/L, 15 g/L bacteriological agar) with increasing concentration of NaCl (0, 3.5, 7, 10, 12.5, 15, 20% *w*/*w*), incubated at 20 °C for a month.

The detrimental potentials of isolates were assayed through plate trials for acid ($CaCO_3$ Agar, ACID), amylase (AMY), lipase (LIP), and protease (namely caseinase; Skim Milk Agar, SM) production as previously described [36]. The cellulase and pectinase activities were assessed using CMC agar (CMC) and pectinase screening agar medium (PSAM), respectively. CMC was prepared using $(NH_4)H_2PO_4$ 1 g; KCl 0.2 g; $MgSO_4 \cdot 7H_2O$ 1 g; yeast extract 1 g; carboxymethylcellulose low-density 26 g, and bacteriological agar 3 g per liter of solution. PSAM was instead prepared using pectin 1 g; $(NH_4)_2$ HPO_4 0.3 g; KH_2PO_4 0.2 g; K_2HPO_4 0.3 g; $MgSO_4$ 0.01 g; and bacteriological agar 2.5 g per 100 mL of solution, pH 5.5. Plates were incubated at 20 °C for a month. To readability, for AMY, CMC, ACID, and PSAM plates were flooded with Lugols' iodine solution and read after 15 min incubation and washed with saline (NaCl 0.9%). The colonies' diameters were taken weekly and the halos, expressed as measure from the colony border, were taken at the end of the experiment. All measures were expressed as average of three replicas and standard deviation (SD).

2.3. Marble Substrate Investigations

To deepen the possible relation between the fungal traits recorded and the marble substrate, from which they were isolated, a selection of highly weathered monuments (e.g., Ruda Rojch and Viganigo) were analyzed by X-ray diffraction analysis (XRD), mercury intrusion porosimetry (MIP), light refraction optical microscopy and scanning electron microscopy (SEM).

In detail, detached and not replaceable scales were micronized in an agate jar, to obtain powders passing to 0.63 µm sieve and pressed in a glass holder. XRD spectra were acquired on powders using a diffractometer Rigaku Miniflex II, equipped with a graphite monochromator, $Cu_{k\alpha}$ wavelength in the following instrumental conditions: 15 kV, 30 mA, Ni filter, scan from 4 to 70°2θ, step sampling 0.01°θ. The identification of minerals was carried out using the search Crystal Impact MATCH! software v. 3.10, which uses, for comparison, both the JCPDS Database (Joint Committee on Powder Diffraction), and the COD Database (Crystallography Open Database) [37]. These results were compared to a quite unaffected old marble from a slab (Zelina Ferrà Gastaldi Millelire).

Total open porosity was measured by mercury intrusion porosimetry (MIP) on 1 cm^3 (c.a.) degraded and control marble fragments oven-dried at 50 ± 5 °C until reaching constant mass. An AutoPore IV 9500 (Micromeritics Instrument Corporation, Norcross, GA, USA) operating at 2200 bar, and 10 s of equilibration time was used. Results were compared to unaffected-new Carrara white marble fragments (Statuarietto lithotype) used as control.

Specimens of about 0.5 × 0.5 cm, were preliminarily observed in light reflection mode using a Motic™ binocular microscope (Kowloon, Hong Kong) to evaluate the conservation conditions of samples and to select the portions of major interest for SEM analysis. These fragments were then glued in epoxy resin (Struers Ltd., Catcliffe, UK), cut in thin sections (30 µm thick) and finally polished. Other fragments were coated with a graphite layer (10 to 20 nm thick) and observed with a Zeiss EVO LS 15 environmental scanning electron microscope (Carl Zeiss SMT AG, Oberkochen, Germany), LaB6 cathode, EHD 5 kV, WD 13.5 mm.

2.4. Selected Chemicals

Among the available traditional biocidal formulations [20,21], we selected four products. Two of them contain quaternary ammonium compounds (QAC): Benzalkonium Chloride (CTS srl, Altavilla Vicentina, Italy; benzalkonium chloride 90%; BZC), and Preventol RI50 (Bresciani srl, Milano, Italy; benzalkonium chloride 50%; PREV). Otherwise, Lichenicida 264 (Bresciani srl, Milano, Italy; LICH) has *N,N*-dimethyl-*N′*-phenyl-*N′*-(fluorodichloromethyl)-sulphamide (also known as dichlofluanid) as active molecule, and Biotin R (CTS srl, Altavilla Vicentina, Italy; BioR) contains 3-iodo-2-propynyl butyl carbamate (IPBC), and 2-*n*-octyl-4-isothiazolin-3-one (OIT) dissolved in 2-(2-butoxyethoxy) ethanol. BZC and PREV were diluted in distillated water, LICH in alcohol [38], and BioR in white spirit.

2.5. Sensitiveness to Biocides

The sensitiveness of fungal isolates to biocides was measured using the agar diffusion tests (ADT) method. Due to their dimorphic clumped growth habit, the strong but variable melanization, and the absence of spores, the BF suspension was standardized as dry weight as follows: a MA slant culture was suspended scaping its surface in 3 mL of sterile saline solution (NaCl 0.9%), transferred into a 5 mL conical tube, homogenized by sterile pestle, and used to inoculate a 250-mL Erlenmeyer flask with malt extract broth (3% *w*/*v*; 30 mL). The incubation was performed at room temperature and 180 rpm on a rotary shaker for 1–3 days. ADTs were performed, in triplicate using 90 mm-Petri dishes. For each plate 200 μL of fungal suspension (8 ± 0.2 mg/mL dry weight) were added to 15 mL of MA by inclusion; a cellulose sterile disk (6 mm diam) was placed on the culture medium and soaked with 6 μL of biocide solution. Biocides were tested at the following concentrations: 0.5, 1, 1.5, 2, and 3% for BZC and PREV; 0.25, 0.5, and 1% for LICH, and 0.5, 1, 2, 3, 4, and 5% for BioR.

2.6. Statistical Analyses

PAST software (PAleontological STatistics, ver. 4.06b, [39]) was used to perform a Principal Component Analysis (PCA) assessing similarities/differences among the strains in study with respect of their salt tolerance, metabolic traits, and resistance to biocides.

Two-way ANOVA with Tukey post hoc test comparison ($p < 0.05$) was performed using GraphPad Prism version 9.3.1 (GraphPad Software, San Diego, CA, USA) to assess significance between biocides increasing doses.

3. Results

3.1. Identification and Selection of BF Species

Based on GenBank sequence match, the three strains were identified as follows. Strain CCFEE 5778 has been reported as *Saccotheciaceae* sp. due to the ITS (OM390240) best match with *Saccothecium rubi* MFLUCC 14-1171 (98.16%), and LSU (OM346745) match with *Selenophoma linicola* CBS 468.88 (98.90%) and *Aurobasidium pullulans* AFTOL-ID 912 (98.70%). Due to his very poor ITS match (OM568835) with *Diaporthe longispora* CBS 194.36 (coverage below 40%) and 97.21% LSU (OM346746) identity with *Phaeotrichum benjaminii* CBS 541.72, CCFEE 5945 was referred as "Unknown dothideomycete". Differently, CCFEE 6327 (OM390241) was identified as *Exophiala oligosperma* due to its 99.83% identity with the species type strain CBS 725.88.

The selected strains were in total 18, one for each species found, and among them were 4 *Knufia*, 3 *Neodevriesia*, and 2 *Exophiala* species (Table 1).

Table 1. The 18 strains in study, collection number (CCFEE), and monuments of isolation. (T) indicates type strain. In bold the new identified strains. All monuments were built in white Carrarese marble.

Species	CCFEE	Funerary Monument
Aureobasidium pullulans	5736	Giuseppina Ara dei conti Ciarella
Coniosporium uncinatum	5737	Francesca Warzee
Exophiala bonariae (T)	5792	Zelina Ferrà Gastaldi Millelire
Exophiala oligosperma	6327	Anna Maria Frau Carta
Knufia karalitana	6001	Unknown tombstone
Knufia marmoricola	6204	Anna Maria Frau Carta
Knufia mediterranea	5710	Avv. Giuseppe Todde
Knufia petricola	5776	Giuseppina Ara dei conti Ciarella
Neodevriesia bulbillosa	5704	Ignazio Ruda Roych conte di San Lorenzo
Neodevriesia capensis	6200	Anna Maria Frau Carta
Neodevriesia sardiniae (T)	6202	Anna Maria Frau Carta
Neophaeotheca triangularis	5703	Ignazio Ruda Roych conte di San Lorenzo
***Saccotheciaceae* sp.**	5778	Zelina Ferrà Gastaldi Millelire
Salinomyces thailandicus *	5723	Antonio Viganigo
Saxophyla tyrrhenica (T)	5935	Unknown burial
Unknown dothideomycete	5945	Rossino Bolla's spouses
Vermiconidia calcicola (T)	5770	Giuseppina Ara dei conti Ciarella
Verrucocladosporium dirinae	5707	Ignazio Ruda Roych conte di San Lorenzo

* this new combination has been proposed by Czachura and colleagues as synonym of *Hortaea thailandica* [40].

3.2. Main Ecological Traits Assessment: Thermal and Salt Tolerance

All the tested strains were able to grow within the range of 5–25 °C (Figure 2, Table S1); the majority can grow at 0 °C (13/18), less than a half (7/18) at 35 °C; only *E. oligosperma* CCFEE 6327 at 37 °C, and none at 40 °C. Most of the strains grew best at 20 °C, except *K. petricola* CCFEEE 5776, *N. capensis* CCFEE 6200, and *E. oligosperma* CCFEE 6327 having an optimum at 25 °C.

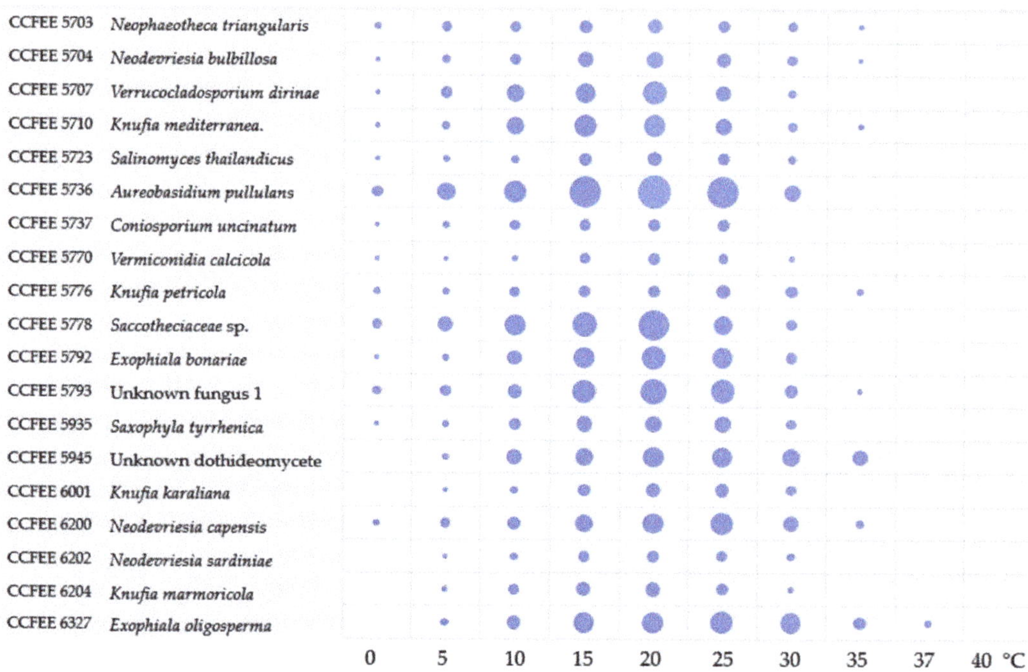

Figure 2. Thermal preference chart. The bubble size is proportional to the average growth recorded at the end of the experiment (1 month). Table S1 reports the full average values and SD recorded.

Growth rates were extremely variable among the studied strains; for instance, after one month of incubation strains *A. pullulans* CCFEE 5736, *Saccotheciaceae* sp. 5778, *E. bonariae* CCFEE, 5792, and *E. oligosperma* CCFEE 6327 grew from about four (3.6) to seven times more than *V. calcicola* CCFEE 5770 and *C. uncinatum* CCFEE 5737.

Different degrees of salt tolerance were recorded (Figure 3, Table S2). *Verr. dirinae* CCFEE 5707, *N. bulbillosa* CCFEE 5704, *Sal. thailandicus* CCFEE 5723, and *Neoph. triangularis* CCFEE 5703 grow better in the presence of 3.5%. Five strains (namely the previous and *A. pullulans*) were able to grow at 20% NaCl, the highest concentration tested. *A. pullulans* CCFEE 5736, although able to grow in the presence of salt, drastically reduced its colony size up to 46% at 3.5% and 95.6% at 20% of NaCl. A high sensitivity to salt was instead found for the unknown dothideomycete CCFEE 5945 with a dramatic decrease in colony diameter of up to 74% in the presence of NaCl 3.5%, and no growth at 7%. Similarly, the slow-growing *C. uncinatum* CCFEE 5737 was unable to grow at 7%. The remaining strains showed instead an intermediate tolerance having *K. petricola* CCFEE 5776 as the most tolerant of the strains able to grow up to 10% NaCl.

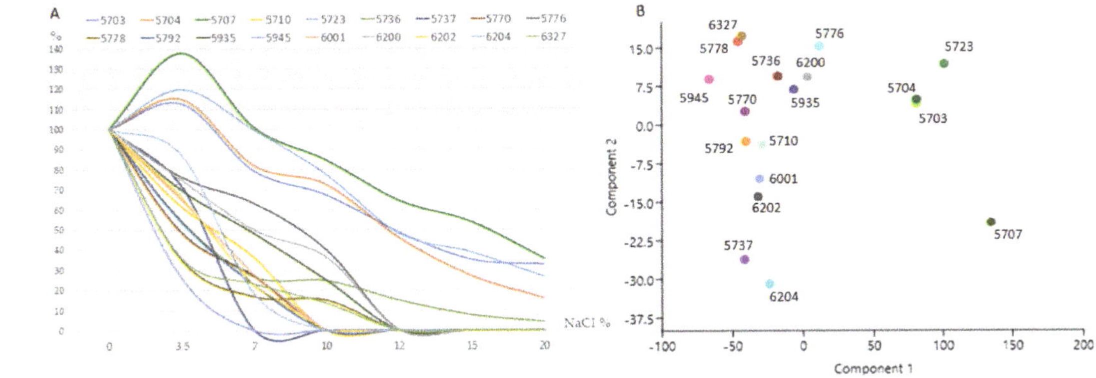

Figure 3. Salt tolerance tests results. (**A**) The measures taken after one-month incubation at 20 °C are calculated as a percentage compared to the control plate (0% NaCl). (**B**) The PCA plot evidenced in the right side of the separated group of halophilic strains, and others (from right to left side) ordered according to their salt tolerance.

3.3. Mineralogical and Physical Investigations

Similar XRD diffraction patterns were recorded in the two deteriorated marbles affected by granular disintegration and crumbling (namely Antonio Viganigo, AV; and Ruda Roych, RR-data not shown). Significative differences were instead evidenced when the diffraction pattern of the AV deteriorated and soluble salt affected marble (Figure 4B) was compared to a gypsum film from an old compact marble monument (Ferrà Gastaldi Millelire; FGM, Figure 4A). Soluble salts and other decay phases were found in different amounts. In the grey film of FGM monument gypsum ($CaSO_4 \cdot 2H_2O$) and calcite ($CaCO_3$) prevailing on whewellite ($CaC_2O_4 \cdot H_2O$), quartz (SiO_2), halite (NaCl), sylvine (KCl) and titanite ($CaTiSiO_5$). On a more weathered monument (AV), affected by granular disintegration and crumbling, the whitish crust shows mineralogical phases related to natural pollution (marine and soil particles) as halite, quartz, albitic plagioclase, as well gypsum and calcite.

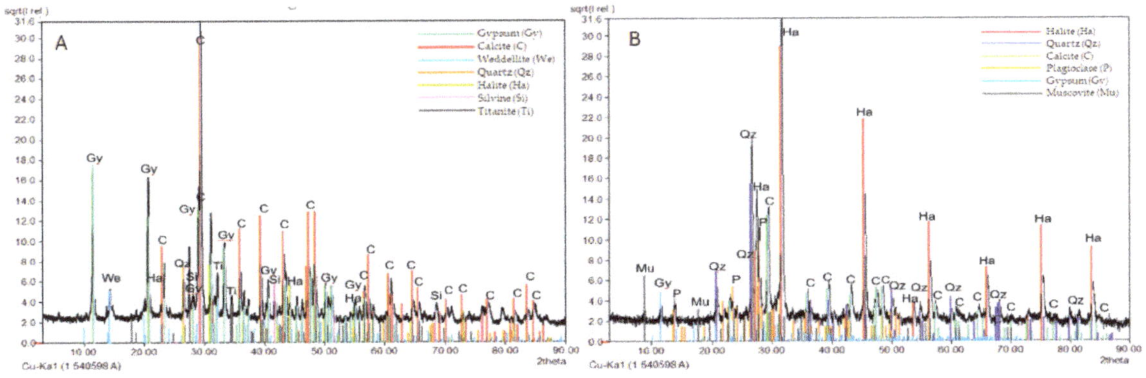

Figure 4. XRD analysis results. XRD patterns of gypsum grey film on old compact marble (FGM) (**A**) and whitish crust on crumbling-granular disintegration marble (AV) (**B**).

MIP porograms (Figure 5A,B) evidenced high differences when comparing a new "Statuarietto" Carrara marble having a porosity of 0.7% ± 0.2 (black), and historical marble affected by granular decoesion 5% ± 0.2 (AV, gray) and high granular decoesion 8% ± 0.2 (RR, red). The porograms show macro- and micro-pores; the former are represented by microcracks or confluent calcitic crystal dislocation. The secondary porosity, acquired by physical and chemical decay processes, represents a key indicator of biodeterioration/bioreceptivity.

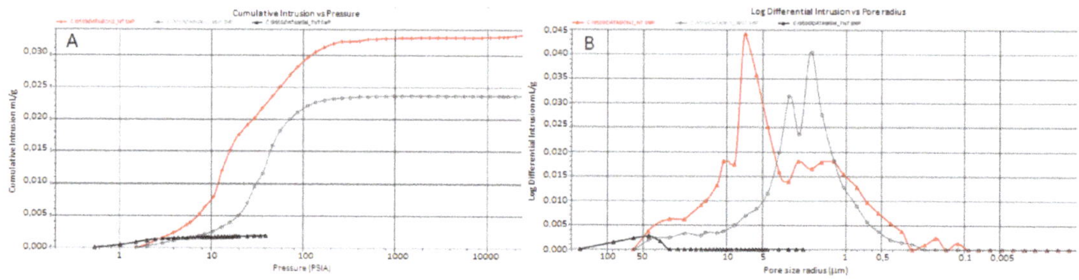

Figure 5. MIP porosity results. MIP porosity of an unaffected-new "Statuarietto" marble (black), a decohesed (gray, AV), and highly decohesed marble (red, RR). (**A**) Cumulative Hg intrusion vs. pressure and (**B**) Log. differential intrusion vs. pore size distribution.

Light microscopy evidenced the presence of salts and other pollution materials in dark crusts (Figure 6B,D). SEM investigation confirmed the deep decohesion of calcitic crystals clasts, the presence of gypsum (desert rose twinning, Figure 6E), and halite crystals already evidenced in the previous analyses; these latter trials give an impacting measure of the phoenomena being marble crystals completely covered by salt (Figure 6H).

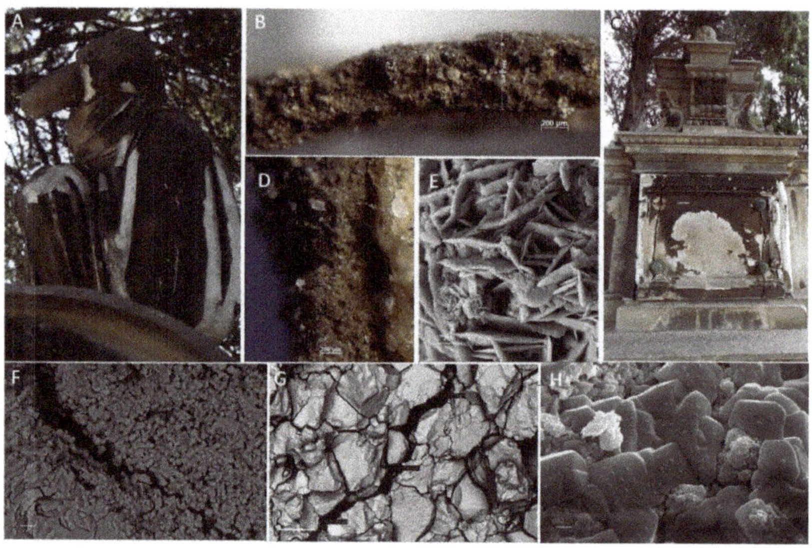

Figure 6. Granular disintegration marble: a salty niche. Particular to the monuments affected by granular disintegration into sand and powder devoted to (**A**) Ruda Roych (RR) and (**C**) Antonio Viganigo (AV). (**B**,**D**) Dark sulfation crust of AV monument by light reflection optical microscopy: (**C**) cross section, (**D**) surface and fracture on dark crust and salt crystals (NaCl) deposition. (**E**–**H**) RR sample SEM observation. (**E**) gypsum crystals; (**F**,**G**) granular disintegration, (**H**) salt crystals (NaCl) on marble surface. The bar corresponds to 10 µm in (**F**), 200 µm in (**G**); and 2 µm in (**H**).

3.4. *Detrimental Potential of the Investigated Species: The Metabolic Screenings*

As reported in Table 2, acid production was observed after iodine pouring in nine out eighteen investigated strains. No production was recorded for the two *Exophiala* species, while positive records were found for all the *Knufia* species. All tested strains were positive for the esterase activity (LIP) being able to use Tween 20 as a sole carbon source and precipitate calcium salt crystals. The majority of the tested strains showed amylase (14/18), and cellulase (15/18) activity, while caseinase and pectinase activities were recorded in half of them (9/18).

Table 2. Results of the metabolic assays performed on selected BF. The activities tested were for acid (ACID), lipase (LIP), amylase (AMY), protease (SM), pectinase (PSAM), and cellulase (CMC) production.

Strain No.	Species	ACID	LIP	AMY	SM	PSAM	CMC
5703	*Neophaeotheca triangularis*	−	++	+	++++	+	++
5704	*Neodevriesia bulbillosa*	−	+++	+++	−	++++	+++
5707	*Verrucocladosporium dirinae*	−	++	+	++++	+	++
5710	*Knufia mediterranea.*	+++	+++	++++	+++	++	+++
5723	*Salinomyces thailandicus*	+	+	++	−	−	++++
5736	*Aureobasidium pullulans*	+	+	++	++	−	++++
5737	*Coniosporium uncinatum*	−	+++	+	+	+	−
5770	*Vermiconidia calcicola* (T)	−	+++	+	−	−	−
5776	*Knufia petricola*	+++	+++	+	++	++	+
5778	*Saccotheciaceae* sp.	+	++	++++	+	++	++++
5792	*Exophiala bonariae* (T)	−	++++	−	−	+++	−
5935	*Saxophyla tyrrhenica* (T)	−	+	−	−	−	+
5945	Unknown dothideomycete	−	++	+	−	−	+
6001	*Knufia karalitana*	++	++	++	−	−	++
6200	*Neodevriesia capensis*	+++	++++	++	+	+	++
6202	*Neodevriesia sardiniae* (T)	++	++	+	+	−	+
6204	*Knufia marmoricola*	+	++	−	−	−	++
6327	*Exophiala oligosperma*	−	++	−	−	−	−

(−) negative; increasing positive responses expressed in mm measured from the colony border: (+) 0.1–5 mm, (++) 5.1–9.9 mm, (+++) 10–14.99 mm, (++++) 15–20 mm.

PCA analysis performed on halos occurred after metabolic assays evidenced in the right side the most responsive strains (Figure 7). All the halophilic strains take place within the most responsive strains. Interestingly *K. mediterranea* CCFEE 5710, due to the intense halos recorded, occupied quite a far position in respect to the other *Knufia* species in the study (namely *K. petricola* CCFEE 5776, *K. karalitana* CCFEE 6001, and *K. marmoricola* CCFEE 6204).

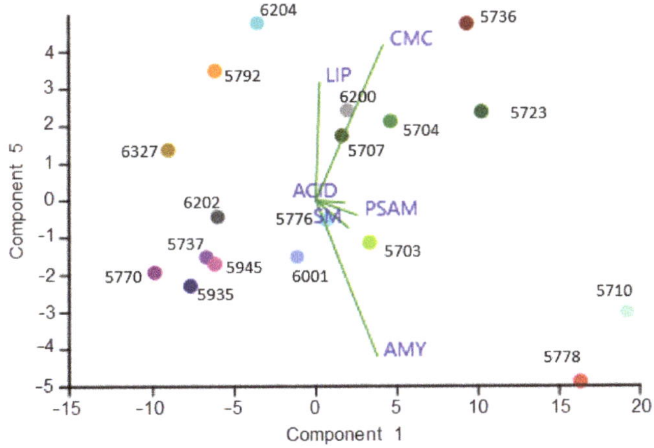

Figure 7. PCA analysis performed on halos recorded for each plate test.

Moreover, during metabolic screenings it was possible to collect additional information such as the loss of melanization by the dothidealean fungi (namely *A. pullulans* CCFEE 5736 and *Saccotheciaceae* sp. CCFEE 5778) when grown in SM (Figure 8L), the volcano-like growth showed by *K. mediterranea* CCFEE 5710 in presence of high concentration of glucose (Figure 8B), and the inhibitory action of *K. marmoricola* and *K. petricola* against a cladosporiacean plate contaminant (Figure 8M,N, respectively).

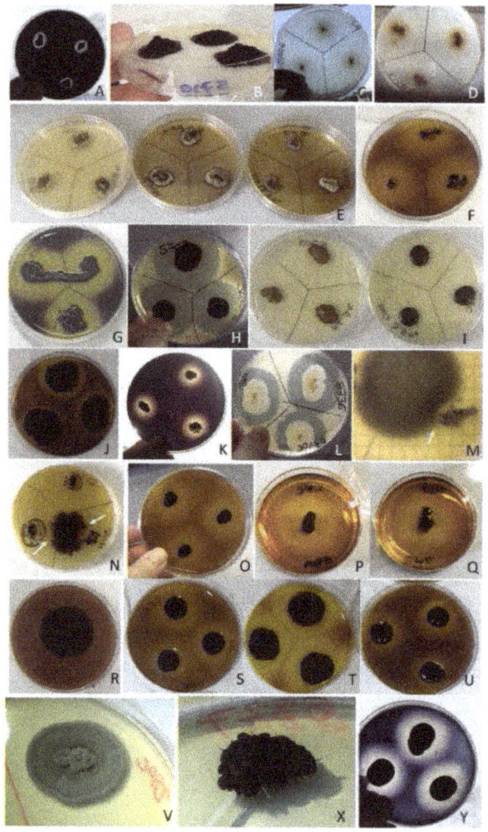

Figure 8. Plate assays outcomes. (**A**) *C. uncinatum* CCFEE 5737 AMY test; (**B**) *K. mediterranea* CCFEE 5710 on $CaCO_3$ agar: the colonies raise in the center and detach from plate; (**C**) *K. petricola* CCFEE 5776 esterase activity test (LIP); (**D**) *E. oligosperma* CCFEE 6327 esterase activity test (LIP); (**E**) *N. bulbillosa* CCFEE 5704 culture media darkening in presence of salt: from left to right side NaCl 0%, 3.5% and NaCl 7%; (**F**) *K. mediterranea* CCFEE 5710 PSAM test; (**G**) *Neoph. triangularis* CCFEE 5703 amylase test (AMY); (**H**) *Neoph. triangularis* CCFEE 5703 SM test; (**I**) *Neoph. triangularis* CCFEE 5703 colonies color shift in presence of salt from left to the right side NaCl 0% and 3.5%; (**J**) *Verr. dirinae* CCFEE 5707 PSAM test; (**K**) *V. calcicola* CCFEE 5770 AMY test; (**L**) *A. pullulans* CCFEE 5736 SM test; (**M**) Contaminated PSAM plate with *K. marmoricola* CCFEE 6204 creating inhibition halos (white arrow). (**N**) Contaminated PSAM plate with *K. petricola* CCFEE 5776 creating inhibition halos (white arrows); (**O**) *K. petricola* CCFEE 5776 acid production plate; (**P**) *N. bulbillosa* CCFEE 5704 cellulase test plate (CMC); (**Q**) *K. mediterranea* CCFEE 5710 cellulase test plate; (**R**) *Verr. dirinae* CCFEE 5707 negative acid production test; (**S–U**) *K. karalitana* CCFEE 6001, *K. mediterranea* CCFEE 5710, *K. marmoricola* CCFEEE 6204 positive acid production plate test; (**V,X**) *E. oligosperma* CCFEE 6327 colony morphology shifts from flat (**V**) grown at 25 °C to meristematic (**X**) grown at 35 °C; (**Y**) *N. capensis* CCFEE 6200 AMY test.

3.5. Sensitiveness to Traditional Biocides

The tested black fungal strains showed different degrees of sensitiveness to biocides (Table 3) and no linear response to incremental doses (Table S3). Two successive increasing biocide doses frequently did not significantly increase inhibitory halos (Table S4).

Table 3. Inhibition halos recorded after the application of the highest dose suggested by suppliers. The different colors represent the different sensitiveness to biocides, as indicated in the legend below, where h is the halo recorded in millimeters. The complete results are shown in Table S3.

CCFEE Strain No.	Species	Biocides			
		PREV 3%	BZC 3%	LICH 1%	BioR.5%
5703	*Neoph. triangularis*	22.8 ± 0.86	29.5 ± 0.52	45.3 ± 0.62	90 ± 0
5704	*N. bulbillosa*	57.2 ±0.4	59.3 ± 0.61	76.12 ± 0.52	90 ± 0
5707	*V. dirinae*	21.7 ± 1.1	20.6 ± 0.7	39.5 ± 1.01	74.5 ± 0.5
5710	*K. mediterranea*	54.3 ±0.79	41.5 ± 0.33	45.3 ± 0.75	90 ± 0
5723	*Sal. thailandicus*	76.7 ±0.6	49.3 ± 0.5	63.86 ± 0.81	90 ± 0
5736	*A. pullulans*	26.3 ±1.1	13.6 ± 0.91	38.82 ± 0.63	46.52 ± 0.6
5737	*C. uncinatum*	56.7 ±0.8	30.1 ± 0.55	61.875 ± 0.9	76.85 ± 1.02
5770	*V. calcicola*	58.8 ±1.03	63.5 ± 0.76	59.61 ± 0.62	90 ± 0
5776	*K. petricola*	57.8 ±0.92	35.2 ± 1.03	40.32 ± 0.3	79.23 ± 0.5
5778	*Saccotheciaceae* sp.	49 ± 0.7	30.4 ± 0.98	39.82 ± 0.71	62.03 ± 1.03
5792	*E. bonariae*	26.5 ±1.1	26.5 ± 0.98	17.53 ± 1.03	16.06 ± 0.2
5935	*S. tyrrhenica*	56.5 ± 0.7	39.28 ± 0.25	68.89 ± 1.05	90 ± 0
5945	Unknown dothideomycete	32.3 ±0.84	28.8 ± 0.8	46.53 ± 0.63	90 ± 0
6001	*K. karalitana*	64.13 ± 0.71	29.68 ± 0.33	90 ± 0	90 ± 0
6200	*N. capensis*	47.6 ± 0.2	68.7 ± 1.1	67.58 ± 0.62	90 ± 0
6202	*N. sardiniae*	56.26 ± 0.08	40.3 ± 0.74	76.74 ± 0.72	84.1 ± 0.1
6204	*K. marmoricola*	50.6 ± 0.3	36.8 ± 0.42	53.7 ± 0.94	90 ± 0
6327	*E. oligosperma*	29.49 ± 0.68	14.97 ± 0.43	29.49 ± 0.68	66.78 ± 0.8

0>h ≤ 11.2	11.2>h ≤ 22.5	22.5>h ≤ 33.7	33.7>h ≤ 45	45>h ≤ 56.2	56.2>h ≤ 67.5	67.5>h ≤ 78.7	78.7>h ≤ 90

The majority of strains evidenced a low sensitiveness to both BZC and PREV, and others, such *E. bonariae* CCFEE 5792, proved quite tolerant to all the biocides tested.

PCA analysis performed considering all responses to the different doses of biocides applied (Table S3), evidenced a sort of gradient where *N. bulbillosa* CCFEE 5704, *V. calcicola* CCFEE 5770, and *S. tyrrhenica* CCFEE 5935 resulted the most sensitive species to the tested biocides (Figure 9A).

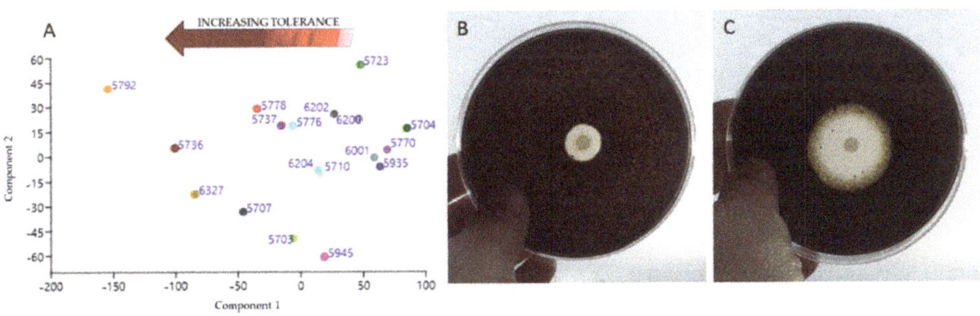

Figure 9. Results of sensitiveness to traditional biocides. (**A**) PCA plot based on the ADT halos recorded using four different biocides at all the tested doses; from the left to the right there is an increasing sensitiveness, in the opposite sense an increasing tolerance to biocides. (**B**,**C**) ADT performed on *E. oligosperma* CCFEE 6327: (**B**) sharp inhibition halo obtained using BZC 3%, (**C**) double halo obtained when BioR 4% was used. There are notable zones of total (next to the disc), and partial inhibition (external) where growth is slowed.

Interestingly BZC always produced sharp halos (Figure 9B) while the other biocides (PREV, LICH, and BioR), especially at low doses, produced thin total inhibition halos surrounded by halos of partial inhibition (Figure 9C) as the growth was mainly slowed down.

4. Discussion

Biodeterioration processes are rarely caused by only one group of microorganisms, and more frequently are due to polymicrobic associations coexisting at the same time, in the same place, on a given artifact. To study them, two main approaches are available: the high throughput sequencing and the traditional culture-based approach. The biomolecular approach has become the dominant method of studying the microbial diversity in the last decade [41]. In fact, the main part of environmental microorganisms are unculturable and the behavior of a single species can frequently be explained in a community context [41]. However, beyond the exponentially increasing information, there is not an equally rapid increase in knowledge regarding the mechanisms and the main actors of the deterioration phenomena [42]. Moreover, the identification and characterization of monument settlers from a physiological and ecological point of view is of the utmost importance in order to design preventive control methods and biocide treatments [21,36,43–45]. From this perspective, an increasing number of papers propose traditional metabolic plate assays as a useful, easy and cost-effective tool to unveil the potential risk tied to some cultural heritage settlers [30,36,46–50]. Such a proposal is confirmed in this work.

4.1. Selection and Identification of the BF Strains of Interest

Even if they were all isolated from a single site, the eighteen strains considered in this study could represent other BF lithic communities even at different latitudes. *Knufia*, *Neodevriesia*, *Exophiala*, *Coniosporium*, and *Vermiconidia* are indeed well known genera due to their frequent finding and broad distribution on stone monuments in urbanized areas and natural crops.Their presence has been recorded in the Mediterranean countries as well as in Russia, Ukraine, Austria, Germany, Poland, China and, not least, Antarctica [51–53]. Information on their environmental preferences and metabolic traits could be helpful in the future to better understand their geographical distribution.

4.2. Temperature and Osmotic Tolerance as Indicators of Environmental Preferences

The thermal trials yielded important information on the optimum temperature of the studied strains, their growth range, and also on their growth rate. The majority of the strains grew within the range 5–30 °C, 39% of them instead within 0–35 °C, and only one, *E. oligosperma* CCFEE 6327, at 37 °C. With respect to their growth rate, the studied strains can be clustered into three groups: fast-; mid-fast-; and slow-growing black fungi. The majority of the slow-growing fungi (e.g., *V. calcicola* CCFEE 5770, *C. uncinatum* CCFEE 5737, and *Knufia* species) have been recorded from stone surfaces only; the others, namely the fast-growing (*A. pullulans* CCFEE 5736, and *Saccotheciaceae* sp. CCFEE 5778) and the mid-fast- growing species (e.g., *E. bonariae* and *E. oligosperma*, Unknown CCFEE 5945, *Verr. dirinae* CCFEE 5707, and *N. capensis* CCFEE 6200), can be considered at least in part to be occasional. *A. pullulans* is considered, indeed, a generalist that can inhabit different habitats without substantial specialization at the genomic level [54], and *E. oligosperma* is a well-known opportunist isolated from a number of household environments such as tap water and diesel car tanks [41,55,56]. This confirms the two survival strategies in fungi from extreme environments: they can be either widespread generalists, thriving in both extreme and moderate conditions, or specialists more or less confined to extremes [57–59].

Growth rate can even be drastically affected when salt is addicted to culture medium. All the tested strains were able to grow at 3.5% of NaCl. This trait increases their fitness for the site, being close to the sea with saltiness periodically deposited and washed away from surfaces with rains. Anyway, different degrees of tolerance were recorded, and strains tested grouped with respect to their response to increasing salt concentration and to their limits. The lowest tolerance has been recorded for *C. uncinatum* CCFEE 5737 and the unknown CCFEE 5945. The majority of the strains have their limits between 7 (6/18) and 10% (5/18), respectively. Five, instead, can grow up to 20%. While *A. pullulans* showed a broad-range tolerance to salt, *Verr. dirinae* CCFEE 5707, *N. bulbillosa* CCFEE 5704, *Sal. thailandicus* CCFEE 5723, and *Neoph. triangularis* CCFEE 5703 proved to be halophilic. The halophilic trait for a part of them was already reported [60,61], however, it was not for *N. bulbillosa* CCFEE 5704 and *Verr. dirinae* CCFEE 5707, which ability was suggested by indirect evidences only. For example, *N. bulbillosa* was isolated from limestone formations in Cala Saint VincenG in Mallorca Island, so possibly subjected to some extent to saltiness [62]. Otherwise, *Verr. dirinae* is worth a mention as even if described as a parasite of the lichen *Dirina massiliensis* [63], it has been isolated from other substrates/salted niches such as wood and marine macroalgae [64,65]. Moreover, the presence of some BF has been suggested to be in relation with the biodeterioration degree of tombstones [66].

4.3. Marble Deterioration Degree Selects the Resident Fungal Community

The instrumental stone investigations performed (namely XRD, light reflection microscopy, and SEM) on the two most deteriorated monuments evidenced the presence of gypsum and chloride crystals (mainly NaCl), the latter of which was found in high amounts. Their presence on marble stone is strictly related to the microclimatic condition, salt solubility and surface physical conditions (e.g., porosity, roughness). The severe alteration of the decayed marble has been shown by SEM and by MIP pore size distribution, where yields were up to 10 times higher than the new unaffected control specimen. These values can be explained with the time of outdoor exposition (1876 RR, 1890 AV, and 1900 FGM their years of built) and even more by microclimatic conditions. Previous studies showed that the most damaged marble pieces in the cemetery of Cagliari were all characterized by low values of solar radiation, high humidity and a reduced thermoforesis, favoring sulfatation and weathering along with the surfaces' colonization [32,33,67,68]. Indeed, the nearby and dense vegetation, in addition to the low exposition to the dominant winds (Libeccio and Scirocco from south and Mistral from north-west), favored the stagnation of marine air and its penetration into damaged stones. Sea spray acts as a marine inoculum migrating between grains by capillarity; here the salt wet-dry cycles gradually and increasingly alter the materials' structure [69–72]. Each crevice and intergrain space can become a salty micro-niche available for specialized microorganisms; along with the salt physical action, the selected community can contribute to stone structural destabilization [73–76]. The exclusive isolation of the halophilic fungi from sand disaggregated monuments only, highlighted as salt-attacked monuments constitute a suitable habitat for selected osmophilic/osmotolerant microorganisms [75,77,78]. However, this phenomenon seems to be progressive since *Neoph. triangularis*, *N. bulbillosa*, and *Verr. dirinae* were isolated from the RR monument (8% MIP porosity) only, while *S. thailandicus* and *K. petricola* from the AV monument (5% MIP porosity). On this basis, further studies are needed to investigate the possible use of the halophilic and the most halotolerant black fungal strains (e.g., *K. petricola* CCFEE 5776) as possible indicators of stone degradation in the early steps of salt-mediated decoesion or, by contrast, the use of low salt tolerant BF (e.g., *C. uncinatum*) as indicators of stone integrity in coastal environments.

4.4. Metabolic Assay, Ecological Traits, and Detrimental Potential of the Investigated Species

Knowledge on the relationship between microorganisms and colonized substrate drives the identification of the major conservation threats and may address the control strategies designed to face biodeterioration [79]. Plate assays were useful to outline some important metabolic traits in BF. Acid production has been recorded in half of the studied strains. Even if not leading to a complete dissolution of carbonates in the culture media, BF showed a detrimental potential well beyond their assessed biomechanical action [7]. A local decrease in the pH has been previously reported for *K. petricola* A95 and found here, for example, in all the *Knufia* species and in two of the three *Neodevriesia* [80]. Further studies are needed to asses if this trait is limitedly present in some selected genera or not. In any case more methods and/or protocol adjustments are needed to detect even weak responses.

The amylases, pectinases, esterases and cellulases activities recorded showed that these fungi can feed on vegetal sources and debris. According to the conceptual model for MCF proposed by Chertov and colleagues [81], the availability of organic nutrients is the dominant factor limiting their development on stone surfaces in European temperate and Mediterranean climates, while growth becomes very intensive in the presence of water. The metabolic features highlighted here support the well-known yet rarely applied practice of removing the organic matter from monuments to prevent fungal growth [82]. The production of esterases and caseinases activities suggests as these strains may also feed on wall paintings where oils and proteinaceous binders were used as already reported [83–86]. Furthermore, the inhibition halo (competition for space) produced by two *Knufia* species (namely *K. petricola* and *K. mediterranea*) against a cladosporiacean plate contaminant, provided evidence that some BF species in peculiar conditions (PSAM test) can compete, more than expected [87–89], against high sporulating fungi.

From a conservative point of view, the ability to metabolize carboxymethylcellulose (CMC) and Tween 20 (LIP) should be careful evaluated in restoration practice for their possible drawbacks. These chemicals, indeed, are commonly used as thickening agents and surfactants, respectively [90,91], and their residues on the treated artifact could lead to a tertiary bioreceptivity if not thoroughly removed.

4.5. Tolerance to Biocides

Despite the significant impact of the green revolution in the biocides market [92], traditional biocidal products are still widely used [93]. This is due to the consolidated protocols used on different materials, the wide knowledge on the pros and cons of their use, and the non-negligible economic aspects.

In situ trials are often used in restoration practice to customize treatments as they confidently reproduce, in a short time, the interaction of biocides with biota and materials, and provide a more reliable indication of the in situ persistence of the biocide [94–97]. However, several factors can negatively affect the efficiency of any restoration protocol. The metabolic state of the organism is important for their responsiveness to biocides treatments. Metabolically active colonies are, indeed, more sensitive to chemical and physical stresses [98]. BF, being poikilohydric organisms, withstand the unfavorable conditions in a dormant state mainly induced by desiccation (anhydrobiosis) [12,99]. In addition, fungi growing in stone cracks and fissures may be not easily reached by biocides applied on a monument surface during restoration treatments. Even the typical clump-like BF colonies, ensuring the optimal surface-volume ratio, minimizes the direct exposure to external stressors, including biocides; moreover, external cells in the colony shield the inner layers from external injuries (shadow effect) [100]. Conversely ADT plate assays have an undeniable advantage as they reproduce the best conditions to record the sensitivity to biocides of single metabolically active cells in a medium that cannot seize the biocide as stones frequently do. For these reasons, the tolerance/resistance recorded in the most responsive condition should warn about their possible survival after treatments, as well as their spread.

The used biocides have different formulations, mechanisms of action, and thus different outcomes. BZC (CAS No 8001-54-5) is a quaternary ammonium compound (QAC) routinely used for chemical antisepsis and disinfection and widely applied in cultural heritage preservation treatments. According to the concentration used, BZC acts as a highly active detergent causing cell membrane permeability alterations, irreversible damage to the barrier function and proteins denaturation [101]. Benzalkonium chloride is also the only active compound in the PREV patent, albeit in a lower concentration (50%) than in BZC (90%). Despite the lower concentration of QAC salt in PREV, the inhibition halos produced were frequently larger than BZC, such as for *Sal. thailandicus* CCFEE 5723, *K. petricola* CCFEE 5776, and *S. tyrrhenica* CCFEE 5935. This apparent incongruence, already evidenced in previous studies [36], has been explained with the possible different composition in terms of n-alkyl groups length and/or degree of C-C saturation in the two chemicals [102], and by additives, not specified in the composition of biocides but included in the patent of the commercial formulas [21,103]. The other two biocides tested have instead intracellular targets. As a matter of fact, *N*,*N*-dimethyl-*N*′-phenyl-*N*′-(fluoro-dichloromethylthio)-sulphamide (CAS No 1085-98-9), the active principle of LICH, is supposed to affect a number of enzymes by reacting with –SH bonds [104]. While the fungicidal and antimicrobial action played by IPBC (3-iodo-2-propynyl butyl carbamate, CAS No 55406-53-6) and OIT (2-*n*-octyl-4-isothiazolin-3-one; CAS No 26530-20-1), the active principles of BioR, are probably due to iodine toxicity, and inhibition of cell respiration/ATP production, respectively [105,106].

The achieved results evidenced an increasing biocidal power, in terms of inhibition halos and number of strains completely inhibited, as follows: BZC-PREV, LICH, and BioR. The thick multilayered and strongly melanized cell wall typical of these fungi serve as an efficient barrier for toxic substances with positive influence on resistance. Moreover, the presence of halos of total and partial inhibition recorded frequently when PREV, LICH, and BioR were applied (Figure 6C), could be explained by a certain ability to repair sub-lethal damages as resistance mechanism to biocides. Several strains, such as *E. bonariae*, *A. pullulans*, *E. oligosperma*, *Verr. dirinae*, and secondly *K. petricola*, and *C. uncinatum*, exhibited low responsiveness even to the most powerful biocide. This fact should warn against the use of extremely powerful biocides as a quick remedy in controlling whatever biological patina as the drastic decrease in competitors favors the spreading of the ones who survived. In the meanwhile, a stronger biocidal action is often backed to a higher toxicity for the operator and the environment. Conversely, the widespread use of ammonium quaternary salts (e.g., benzalkonium chloride), should be reduced to oppose the rapid recolonizations by serving as nutrients for micro- and macro-organisms [107]. The wide use of sub-inhibitory concentrations of benzalkonium chloride (and QACs), it has been suggested, may be responsible for a significant decrease in microorganisms' sensitivity to biocides, driving the evolution of polyextremotolerant fungi towards the enhancement of their stress tolerance [108]. For this reason, the rotation of disinfectants is often recommended in practice, in order to kill the resistant biota [109]. In this context plate assays could be useful to test multiple chemicals. The ability to cope with different biocidal products shown by these strains also make them promising candidates as new reference organisms in material testing procedures [110].

5. Conclusions

This research gave an assessment of the main BF traits with respect to thermal and salt growth preferences, such as on the possible relation between the recorded fungal traits and the substrate. Plate assays resulted a fast, easy, cost-effective, and useful tool for such aims. To deepen the BF detrimental/colonizing potential, more plate tests should be designed or improved, and physical analyses were fundamental in proving the relation between halophilic strains and highly deteriorated marble surfaces. We also assessed the detrimental potential of the investigated species through metabolic assays, which were also useful to determine the conditions favoring their growth and substances that should be avoided to prevent tertiary bioreceptivity. We also evaluated the sensitivity to a few traditional biocides through ADT tests, which were useful to identify the most recalcitrant BF species.

Supplementary Materials: The following supporting information can be downloaded at: https://www.mdpi.com/article/10.3390/app12042038/s1, Table S1: Thermal preferences recorded within the range 0–40 °C, Table S2: Salt tolerance recorded within the range 0–20% NaCl, Table S3: ADT results, Table S4: ADT two-way ANOVA Tuckey test row analysis.

Author Contributions: Conceptualization, D.I.; methodology, D.I., F.B. and P.M.; data curation, D.I., P.M. and F.B.; writing—original draft preparation, D.I.; writing—review and editing, D.I., L.Z., P.M., F.B. and G.C.; All authors have read and agreed to the published version of the manuscript.

Funding: The Ordine Nazionale dei Biologi for funding a scholarship for the project "Fungal deteriogens and stone monuments".

Institutional Review Board Statement: No humans or animals are involved in this study.

Informed Consent Statement: No humans are involved in this study.

Data Availability Statement: All data resulting from this research are fully reported here or as Supplementary Material.

Acknowledgments: The authors wish to thank the Superintendence of Heritage Landscape, Historical, Artistic and Ethno-Anthropological Heritage for the metropolitan city of Cagliari and for the provinces of Oristano, Medio Campidano and, Municipality Cemeteries Direction of Cagliari for collaboration and sampling permission. Gianfranco Carcangiu is acknowledged for support in elaboration of XRD analysis.

Conflicts of Interest: The authors declare no conflict of interest.

References

1. Caneva, G.; Ceschin, S. Ecology of biodeterioration. In *Plant Biology for Cultural Heritage*; Caneva, G., Nugari, M.P., Salvadori, O., Eds.; The Getty Conservation Institute: Los Angeles, CA, USA, 2009; pp. 35–58.
2. Miller, A.Z.; Sanmartín, P.; Pereira-Pardo, L.; Dionísio, A.; Sáiz-Jiménez, C.; Macedo, M.F.; Prieto, B. Bioreceptivity of building stones: A review. *Sci. Total Environ.* **2012**, *426*, 1–12. [CrossRef]
3. Sanmartín, P.; Miller, A.Z.; Prieto, B.; Viles, H.A. Revisiting and reanalysing the concept of bioreceptivity 25 years on. *Sci. Total Environ.* **2021**, *770*, 145314. [CrossRef]
4. Tonon, C.; Favero-Longo, S.E.; Matteucci, E.; Piervittori, R.; Croveri, P.; Appolonia, L.; Meirano, V.; Serino, M.; Elia, D. Microenvironmental features drive the distribution of lichens in the House of the Ancient Hunt, Pompeii, Italy. *Int. Biodeterior. Biodegrad.* **2019**, *136*, 71–81. [CrossRef]
5. Isola, D.; Selbmann, L.; Meloni, P.; Maracci, E.; Onofri, S.; Zucconi, L. Detrimental rock black fungi and biocides: A study on the Monumental Cemetery of Cagliari. In *Science and Technology for the Conservation of Cultural Heritage*; Rogerio-Candelera, M.A., Lazzari, M., Cano, E., Eds.; CRC Press: London, UK, 2013; pp. 83–86.
6. Onofri, S.; Zucconi, L.; Isola, D.; Selbmann, L. Rock-inhabiting fungi and their role in deterioration of stone monuments in the Mediterranean area. *Plant. Biosyst.* **2014**, *148*, 384–391. [CrossRef]
7. Tonon, C.; Breitenbach, R.; Voigt, O.; Turci, F.; Gorbushina, A.A.; Favero-Longo, S.E. Hyphal morphology and substrate porosity -rather than melanization-drive penetration of black fungi into carbonate substrates. *J. Cult. Herit.* **2021**, *48*, 244–253. [CrossRef]
8. Selbmann, L.; Isola, D.; Zucconi, L.; Onofri, S. Resistance to UV-B induced DNA damage in extreme-tolerant cryptoendolithic Antarctic fungi: Detection by PCR assays. *Fungal Biol.* **2011**, *115*, 937–944. [CrossRef]
9. Selbmann, L.; Grube, M.; Onofri, S.; Isola, D.; Zucconi, L. Antarctic epilithic lichens as niches for black meristematic fungi. *Biology* **2013**, *2*, 784–797. [CrossRef]
10. Selbmann, L.; Isola, D.; Egidi, E.; Zucconi, L.; Gueidan, C.; de Hoog, G.S.; Onofri, S. Mountain tips as reservoirs for new rock-fungal entities: *Saxomyces* gen.nov. and four new species from Alps. *Fungal Divers.* **2014**, *65*, 167–182. [CrossRef]
11. Marvasi, M.; Donnarumma, F.; Frandi, A.; Mastromei, G.; Sterflinger, K.; Tiano, P.; Perito, B. Black microcolonial fungi as deteriogens of two famous marble statues in Florence, Italy. *Int. Biodeterior. Biodegrad.* **2012**, *68*, 36–44. [CrossRef]
12. Zakharova, K.; Tesei, D.; Marzban, G.; Dijksterhuis, J.; Wyatt, T.; Sterflinger, K. Microcolonial fungi on rocks: A life in constant drought? *Mycopathologia* **2013**, *175*, 537–547. [CrossRef]
13. Dornieden, T.; Gorbushina, A.A.; Krumbein, W.E. Biodecay of cultural heritage as a space/time related ecological situation—An evaluation of a series of studies. *Int. Biodeterior. Biodegrad.* **2000**, *46*, 261–270. [CrossRef]
14. Favero-Longo, S.E.; Gazzano, C.; Girlanda, M.; Castelli, D.; Tretiach, M.; Baiocchi, C.; Piervittori, R. Physical and chemical deterioration of silicate and carbonate rocks by meristematic microcolonial fungi and endolithic lichens (Chaetothyriomycetidae). *Geomicrobiol. J.* **2011**, *28*, 732–744. [CrossRef]
15. Gorbushina, A.A.; Krumbein, W.E.; Hamann, C.H.; Panina, L.; Soukharjevski, S.; Wollenzien, U. Role of black fungi in color change and biodeterioration of antique marbles. *Geomicrobiol. J.* **1993**, *11*, 205–222. [CrossRef]
16. Sert, H.B.; Sümbül, H.; Sterflinger, K. Microcolonial fungi from antique marbles in Perge/Side/Termessos (Antalya/Turkey). *Antonie van Leeuwenhoek* **2007**, *91*, 217–227. [CrossRef]
17. Toreno, G.; Isola, D.; Meloni, P.; Carcangiu, G.; Selbmann, L.; Onofri, S.; Caneva, G.; Zucconi, L. Biological colonization on stone monuments: A new low impact cleaning method. *J. Cult. Herit.* **2018**, *30*, 100–109. [CrossRef]
18. Caneva, G.; Isola, D.; Lee, H.J.; Chung, Y.J. Biological risk for hypogea: Shared data from Etruscan tombs in Italy and ancient tombs of the Baekje dynasty in Republic of Korea. *Appl. Sci.* **2020**, *10*, 6104. [CrossRef]
19. Caneva, G.; Bartoli, F.; Savo, V.; Futagami, Y.; Strona, G. Combining statistical tools and ecological assessments in the study of biodeterioration patterns of stone temples in Angkor (Cambodia). *Sci. Rep.* **2016**, *6*, 32601. [CrossRef]
20. Pinna, D. *Coping with Biological Growth on Stone Heritage Objects: Methods, Products, Applications, and Perspectives*; Apple Academic Press CRC Press: Oakville, ON, Canada, 2017; 360p.
21. Caneva, G.; Nugari, M.P.; Pinna, D.; Salvadori, O. *Il Controllo del Degrado Biologico: I Biocidi nel Restauro dei Materiali Lapidei*; Nardini: Firenze, Italy, 1996; pp. 1–200.

22. Nugari, M.P.; Salvadori, O. Biocides and treatment of stone: Limitation and future prospects. In *Art, Biology and Conservation: Biodeterioration of Works of Art*; The Met Fifth Avenue: New York, NY, USA, 2003; pp. 518–535.
23. Young, M.E.; Alakomi, H.L.; Fortune, I.; Gorbushina, A.A.; Krumbein, W.E.; Maxwell, I.; McCullagh, C.; Robertson, P.; Saarela, M.; Valero, J.; et al. Development of a biocidal treatment regime to inhibit biological growths on cultural heritage: BIODAM. *Environ. Geol.* **2008**, *56*, 631–641. [CrossRef]
24. Cappitelli, F.; Nosanchuk, J.D.; Casadevall, A.; Toniolo, L.; Brusetti, L.; Florio, S.; Principi, P.; Borin, S.; Sorlini, C. Synthetic consolidants attacked by melanin-producing fungi: Case study of the biodeterioration of Milan (Italy) Cathedral marble treated with acrylics. *Appl. Environ. Microbiol.* **2007**, *73*, 271–277. [CrossRef]
25. Cappitelli, F.; Villa, F.; Sanmartín, P. Interactions of microorganisms and synthetic polymers in cultural heritage conservation. *Int. Biodeterior. Biodegrad.* **2021**, *163*, 105282. [CrossRef]
26. De Leo, F.; Urzì, C. Fungal colonization on treated and untreated stone surfaces. In *Molecular Biology and Cultural Heritage*; Saiz-Jimenez, C., Ed.; Swets & Zeitlinger BV: Lisse, The Netherlands, 2003; pp. 213–218.
27. Sterflinger, K. Fungi: Their role in deterioration of cultural heritage. *Fungal Biol. Rev.* **2010**, *24*, 47–55. [CrossRef]
28. Nai, C.; Wong, H.Y.; Pannenbecker, A.; Broughton, W.J.; Benoit, I.; de Vries, R.P.; Gueidan, C.; Gorbushina, A.A. Nutritional physiology of a rock-inhabiting, model microcolonial fungus from an ancestral lineage of the Chaetothyriales (Ascomycetes). *Fungal Genet. Biol.* **2013**, *56*, 54–66. [CrossRef]
29. Voigt, O.; Knabe, N.; Nitsche, S.; Erdmann, E.A.; Schumacher, J.; Gorbushina, A.A. An advanced genetic toolkit for exploring the biology of the rock-inhabiting black fungus *Knufia petricola*. *Sci. Rep.* **2020**, *10*, 22021. [CrossRef]
30. Trovão, J.; Portugal, A. Current knowledge on the fungal degradation abilities profiled through biodeteriorative plate essays. *Appl. Sci.* **2021**, *11*, 4196. [CrossRef]
31. Isola, D.; Zucconi, L.; Onofri, S.; Caneva, G.; De Hoog, G.S.; Selbmann, L. Extremotolerant rock inhabiting black fungi from Italian monumental sites. *Fungal Divers.* **2016**, *76*, 75–96. [CrossRef]
32. Meloni, P.; Vacca, G.; Massidda, L.; Carcangiu, G.; Mameli, P.; Cocco, O.; Toreno, G. Vulnerabilità dei beni culturali: Sistemi informatici finalizzati alla conservazione programmata e preventiva di monumenti in marmo. In *Pensare la Prevenzione: Manufatti, Usi, Ambienti*; Ricerche, A., Ed.; Parco Scientifico Tecnologico di Venezia: Marghera-Venezia, Italy, 2010; pp. 385–392.
33. Meloni, P.; Vacca, G.; Massidda, L.; Carcangiu, G.; Mameli, P.; Toreno, G. Architettura in un Sistema GIS per la conservazione programmata delle opere all'aperto del cimitero storico di Bonaria in Cagliari (Sardegna). In Proceedings of the Atti del IV Convegno di Monitoraggio e Conservazione preventiva dei Beni Culturali, Fac. di Ingegneria, Cassino, Italy, 27–28 May 2010; pp. 76–82.
34. Meloni, P.; Toreno, G.; Carcangiu, G.; Cocco, O.; Deiana, M.; Murru, A. Sinergie per la conservazione degli spazi della memoria: L'esperienza didattica dei cantieri pilota del cimitero monumentale di Bonaria in Cagliari. In *Atti 2° Congresso Specialistico Internazionale sui Cimiteri Monumentali: Conoscenza, Conservazione e Restyling*; Cicop Italia—Centro Internazionale per la Conservazione del Patrimonio Architettonico: Firenze, Italy, 2013; pp. 166–171.
35. Isola, D.; Bartoli, F.; Langone, S.; Ceschin, S.; Zucconi, L.; Caneva, G. Plant DNA Barcode as a Tool for Root Identification in Hypogea: The Case of the Etruscan Tombs of Tarquinia (Central Italy). *Plants* **2021**, *10*, 1138. [CrossRef]
36. Isola, D.; Zucconi, L.; Cecchini, A.; Caneva, G. Dark-pigmented biodeteriogenic fungi in Etruscan hypogeal tombs: New data on their culture-dependent diversity, favouring conditions, and resistance to biocidal treatments. *Fungal Biol.* **2021**, *125*, 609–620. [CrossRef]
37. Gražulis, S.; Daškevič, A.; Merkys, A.; Chateigner, D.; Lutterotti, L.; Quiros, M.; Serebryanaya, N.R.; Moeck, P.; Downs, R.T.; Le Bail, A. Crystallography Open Database (COD): An open-access collection of crystal structures and platform for world-wide collaboration. *Nucleic Acids Res.* **2012**, *40*, D420–D427. [CrossRef]
38. Czaja-Szewczak, B. Burial tunics from Naqlun. In *Christianity and Monasticism in the Fayoum Oasis*; Gabra, G., Ed.; Essays of the 2004 International Symposium of the Saint Mark Foundation and the Saint Shenouda the Archimandrite Coptic Society in Honor of Martin Krause; American University in Cairo Press: Cairo, Egypt, 2005; pp. 133–142.
39. Hammer, Ø.; Harper, D.A.; Ryan, P.D. PAST: Paleontological statistics software package for education and data analysis. *Palaeontol. Electron.* **2001**, *4*, 9. Available online: http://palaeo-electronica.org/2001_1/past/issue1_01.htm (accessed on 9 December 2021).
40. Czachura, P.; Owczarek-Kościelniak, M.; Piątek, M. *Salinomyces polonicus*: A moderately halophilic kin of the most extremely halotolerant fungus *Hortaea werneckii*. *Fungal Biol.* **2021**, *125*, 459–468. [CrossRef]
41. Isola, D.; Scano, A.; Orrù, G.; Prenafeta-Boldú, F.X.; Zucconi, L. Hydrocarbon-contaminated sites: Is there something more than *Exophiala xenobiotica*? New insights into black fungal diversity using the long cold incubation method. *J. Fungi* **2021**, *7*, 817. [CrossRef]
42. Sterflinger, K.; Little, B.; Pinar, G.; Pinzari, F.; de los Rios, A.; Gu, J.D. Future directions and challenges in biodeterioration research on historic materials and cultural properties. *Int. Biodeterior. Biodegrad.* **2018**, *129*, 10–12. [CrossRef]
43. Martin-Sanchez, P.M.; Miller, A.Z.; Saiz-Jimenez, C. Lascaux Cave: An example of fragile ecological balance in subterranean environments. In *Microbial Life in Cave Systems*; Engel, A.S., Ed.; De Gruyter: Berlin, Germany; Boston, MA, USA, 2015; pp. 279–301. [CrossRef]
44. Urzì, C.; De Leo, F.; Krakova, L.; Pangallo, D.; Bruno, L. Effects of biocide treatments on the biofilm community in Domitilla's catacombs in Rome. *Sci. Total Environ.* **2016**, *572*, 252–262. [CrossRef]

45. Mazzoli, R.; Giuffrida, M.G.; Pessione, E. Back to the past: "Find the guilty bug—Microorganisms involved in the biodeterioration of archeological and historical artifacts". *Appl. Microbiol. Biotechnol.* **2018**, *102*, 6393–6407. [CrossRef]
46. Jurado, V.; Gonzalez-Pimentel, J.L.; Hermosin, B.; Saiz-Jimenez, C. Biodeterioration of Salón de Reinos, Museo Nacional del Prado, Madrid, Spain. *Appl. Sci.* **2021**, *11*, 8858. [CrossRef]
47. Ma, W.; Wu, F.; Tian, T.; He, D.; Zhang, Q.; Gu, J.-D.; Duan, Y.; Ma, D.; Wang, W.; Feng, H. Fungal diversity and its contribution to the biodeterioration of mural paintings in two 1700-Year-Old Tombs of China. *Int. Biodeterior. Biodegrad.* **2020**, *152*, 104972. [CrossRef]
48. Savković, Ž.; Unković, N.; Stupar, M.; Franković, M.; Jovanović, M.; Erić, S.; Šarić, K.; Stanković, S.; Dimkić, I.; Vukojević, J.; et al. Diversity and biodeteriorative potential of fungal dwellers on ancient stone stela. *Int. Biodeterior. Biodegrad.* **2016**, *115*, 212–223. [CrossRef]
49. Trovão, J.; Tiago, I.; Catarino, L.; Gil, F.; Portugal, A. In vitro analyses of fungi and dolomitic limestone interactions: Bioreceptivity and biodeterioration assessment. *Int. Biodeterior. Biodegrad.* **2020**, *155*, 105107. [CrossRef]
50. Unković, N.; Erić, S.; Šarić, K.; Stupar, M.; Savković, Ž.; Stanković, S.; Dimkić, I.; Vukojević, J.; Grbić, M.L. Biogenesis of secondary mycogenic minerals related to wall paintings deterioration process. *Micron* **2017**, *100*, 1–9. [CrossRef]
51. Sun, W.; Su, L.; Yang, S.; Sun, J.; Liu, B.; Fu, R.; Wu, B.; Liu, X.; Cai, L.; Guo, L.; et al. Unveiling the hidden diversity of rock-inhabiting fungi: Chaetothyriales from China. *J. Fungi* **2020**, *6*, 187. [CrossRef]
52. Sazanova, K.V.; Zelenskaya, M.S.; Vlasov, A.D.; Bobir, S.Y.; Yakkonen, K.L.; Vlasov, D.Y. Microorganisms in superficial deposits on the stone monuments in Saint Petersburg. *Microorganisms* **2022**, *10*, 316. [CrossRef]
53. Selbmann, L.; Onofri, S.; Zucconi, L.; Isola, D.; Rottigni, M.; Ghiglione, C.; Piazza, P.; Alvaro, M.C.; Schiaparelli, S. Distributional records of Antarctic fungi based on strains preserved in the Culture Collection of Fungi from Extreme Environments (CCFEE) Mycological Section associated with the Italian National Antarctic Museum (MNA). *MycoKeys* **2015**, *10*, 57. [CrossRef]
54. Gostinčar, C.; Turk, M.; Zajc, J.; Gunde-Cimerman, N. Fifty *Aureobasidium pullulans* genomes reveal a recombining polyextremotolerant generalist. *Environ. Microbiol.* **2019**, *21*, 3638–3652. [CrossRef]
55. Isola, D.; Selbmann, L.; de Hoog, G.S.; Fenice, M.; Onofri, S.; Prenafeta-Boldú, F.X.; Zucconi, L. Isolation and screening of black fungi as degraders of volatile aromatic hydrocarbons. *Mycopathologia* **2013**, *175*, 369–379. [CrossRef]
56. Prenafeta-Boldú, F.X.; Armjio-Medina, C.; Isola, D. Black fungi in the built environment—The good, the bad, and the ugly. In *Viruses, Bacteria, and Fungi in the Built Environment. Designing Healthy Indoor Environments*; Pacheco-Torgal, F., Ivanov, V., Falkinham, J.O., Eds.; Woodhead Publishing: Duxford, UK, 2021; pp. 65–99.
57. Gostinčar, C.; Stajich, J.E.; Zupancic, J.; Zalar, P.; Gunde-Cimerman, N. Genomic evidence for intraspecific hybridization in a clonal and extremely halotolerant yeast. *BMC Genom.* **2018**, *19*, 364. [CrossRef]
58. Gostinčar, C.; Zajc, J.; Lenassi, M.; Plemenitas, A.; de Hoog, S.; Al Hatmi, A.M.S.; Gunde-Cimerman, N. Fungi between extremotolerance and opportunistic pathogenicity on humans. *Fungal Divers.* **2018**, *93*, 195–213. [CrossRef]
59. Gostinčar, C.; Zalar, P.; Gunde-Cimerman, N. No need for speed: Slow development of fungi in extreme environments. *Fungal Biol. Rev.* **2022**, *39*, 1–14. [CrossRef]
60. Kogej, T.; Ramos, J.; Plemenitas, A.; Gunde-Cimerman, N. The halophilic fungus *Hortaea werneckii* and the halotolerant fungus Aureobasidium pullulans maintain low intracellular cation concentrations in hypersaline environments. *Appl. Environ. Microbiol.* **2005**, *71*, 6600–6605. [CrossRef]
61. Butinar, L.; Sonjak, S.; Zalar, P.; Plemenitaš, A.; Gunde-Cimerman, N. Melanized halophilic fungi are eukaryotic members of microbial communities in hypersaline waters of solar salterns. *Botanica Marina* **2005**, *48*, 73–79. [CrossRef]
62. Egidi, E.; De Hoog, G.S.; Isola, D.; Onofri, S.; Quaedvlieg, W.; De Vries, M.; Verkley, G.J.M.; Stielow, J.B.; Zucconi, L.; Selbmann, L. Phylogeny and taxonomy of meristematic rock-inhabiting black fungi in the Dothideomycetes based on multi-locus phylogenies. *Fungal Divers.* **2014**, *65*, 127–165. [CrossRef]
63. Crous, P.W.; Braun, U.; Schubert, K.; Groenewald, J.Z. Delimiting *Cladosporium* from morphologically similar genera. *Stud. Mycol.* **2007**, *58*, 33–56. [CrossRef]
64. Piñar, G.; Dalnodar, D.; Voitl, C.; Reschreiter, H.; Sterflinger, K. Biodeterioration risk threatens the 3100 year old staircase of Hallstatt (Austria): Possible involvement of halophilic microorganisms. *PLoS ONE* **2016**, *11*, e0148279. [CrossRef]
65. Gnavi, G.; Garzoli, L.; Poli, A.; Prigione, V.; Burgaud, G.; Varese, G.C. The culturable mycobiota of Flabellia petiolata: First survey of marine fungi associated to a Mediterranean green alga. *PLoS ONE* **2017**, *12*, e0175941. [CrossRef]
66. Owczarek-Kościelniak, M.; Krzewicka, B.; Piątek, J.; Kołodziejczyk, Ł.M.; Kapusta, P. Is there a link between the biological colonization of the gravestone and its deterioration? *Int. Biodeterior. Biodegrad.* **2020**, *148*, 104879. [CrossRef]
67. Casti, M.; Meloni, P.; Pia, G.; Palomba, M. Differential damage in the semi-confined Munazio Ireneo cubicle in Cagliari (Sardinia): A correlation between damage and microclimate. *Environ. Earth Sci.* **2017**, *76*, 529. [CrossRef]
68. Murru, A.; Freire-Lista, D.M.; Fort, R.; Varas-Muriel, M.J.; Meloni, P. Evaluation of post-thermal shock effects in Carrara marble and Santa Caterina di Pittinuri limestone. *Constr. Build. Mater.* **2018**, *186*, 1200–1211. [CrossRef]
69. Silva, B.; Rivas, T.; Prieto, B. Relation between type of soluble salt and decay forms in granitic coastal churches in Galicia (NW Spain). In Proceedings of the Origin, Mechanisms and Effects of Salts on Degradation of Monuments in Marine and Continental Environments, Bari, Italy, 25–27 March 1996; Zezza, Ed.; pp. 181–190.
70. Stryszewska, T. The change in selected properties of ceramic materials obtained from ceramic brick treated by the sulphate and chloride ions. *Constr. Build. Mater.* **2014**, *66*, 268–274. [CrossRef]

71. Vázquez-Nion, D.; Troiano, F.; Sanmartín, P.; Valagussa, C.; Cappitelli, F.; Prieto, B. Secondary bioreceptivity of granite: Effect of salt weathering on subaerial biofilm growth. *Mater. Struct.* **2018**, *51*, 158. [CrossRef]
72. Zanardini, E.; May, E.; Purdy, K.J.; Murrell, J.C. Nutrient cycling potential within microbial communities on culturally important stoneworks. *Environ. Microbiol. Rep.* **2019**, *11*, 147–154. [CrossRef]
73. Saiz-Jimenez, C.; Laiz, L. Occurrence of halotolerant/halophilic bacterial communities in deteriorated monuments. *Int. Biodeterior. Biodegrad.* **2000**, *46*, 319–326. [CrossRef]
74. Piñar, G.; Ripka, K.; Weber, J.; Sterflinger, K. The micro-biota of a sub-surface monument the medival chapel of St. Virgil (Vienna, Austria). *Int. Biodeterior. Biodegrad.* **2009**, *63*, 851–859. [CrossRef]
75. Adamiak, J.; Otlewska, A.; Gutarowska, B. Halophilic microbial communities in deteriorated buildings. *World J. Microbiol. Biotechnol.* **2015**, *31*, 1489–1499. [CrossRef]
76. Adamiak, J.; Bonifay, V.; Otlewska, A.; Sunner, J.A.; Beech, I.B.; Stryszewska, T.; Kańka, S.; Oracz, J.; Żyżelewicz, D.; Gutarowska, B. Untargeted metabolomics approach in halophiles: Understanding the biodeterioration process of building materials. *Front. Microbiol.* **2017**, *8*, 2448. [CrossRef]
77. Piñar, G.; Ettenauer, J.; Sterflinger, K. La vie en rose: A review of the rosy discoloration of subsurface monuments. In *The Conservation of Subterranean Cultural Heritage*; Saiz-Jimenez, C., Ed.; CRC Press/Balkema: Leiden, The Netherlands, 2014; pp. 113–124.
78. Otlewska, A.; Adamiak, J.; Stryszewska, T.; Kanka, S.; Gutarowska, B. Factors determining biodiversity of halophilic microorganisms on historic masonry buildings. *Microbes Environ.* **2017**, *32*, 164–173. [CrossRef]
79. Sanmartín, P.; DeAraujo, A.; Vasanthakumar, A. Melding the old with the new: Trends in methods used to identify, monitor, and control microorganisms on cultural heritage materials. *Microb Ecol.* **2018**, *76*, 64–80. [CrossRef]
80. Gerrits, R.; Pokharel, R.; Breitenbach, R.; Radnik, J.; Feldmann, I.; Schuessler, J.A.; von Blanckenburg, F.; Gorbushina, A.A.; Schott, J. How the rock-inhabiting fungus K. petricola A95 enhances olivine dissolution through attachment. *Geochim. Cosmochim. Acta* **2020**, *282*, 76–97. [CrossRef]
81. Chertov, O.; Gorbushina, A.; Deventer, B. A model for microcolonial fungi growth on rock surfaces. *Ecol. Modell* **2004**, *177*, 415–426. [CrossRef]
82. Salvadori, O.; Municchia, A.C. The role of fungi and lichens in the biodeterioration of stone monuments. *Open Conf. Proc. J.* **2016**, *7*, 39–54. [CrossRef]
83. He, D.; Wu, F.; Ma, W.; Zhang, Y.; Gu, J.D.; Duan, Y.; Xu, R.; Feng, H.; Wang, W.; Li, S.W. Insights into the bacterial and fungal communities and microbiome that causes a microbe outbreak on ancient wall paintings in the Maijishan Grottoes. *Int. Biodeterior. Biodegrad.* **2021**, *163*, 105250. [CrossRef]
84. Zucconi, L.; Gagliardi, M.; Isola, D.; Onofri, S.; Andaloro, M.C.; Pelosi, C.; Pogliani, P.; Selbmann, L. Biodeterioration agents dwelling in or on the wall paintings of the Holy Saviour's cave (Vallerano, Italy). *Int. Biodeterior. Biodegrad.* **2012**, *70*, 40–46. [CrossRef]
85. Zucconi, L.; Canini, F.; Isola, D.; Caneva, G. Fungi affecting wall paintings: A meta-analysis of their diversity. *Appl. Sci.* 2022; *under review*.
86. Suphaphimol, N.; Suwannarach, N.; Purahong, W.; Jaikang, C.; Pengpat, K.; Semakul, N.; Yimklan, S.; Jongjitngam, S.; Jindasu, S.; Thiangtham, S.; et al. Identification of microorganisms dwelling on the 19th century Lanna mural paintings from Northern Thailand using culture-dependent and-independent approaches. *Biology* **2022**, *11*, 228. [CrossRef]
87. Selbmann, L.; Isola, D.; Fenice, M.; Zucconi, L.; Sterflinger, K.; Onofri, S. Potential extinction of Antarctic endemic fungal species as a consequence of global warming. *Sci. Total Environ.* **2012**, *438*, 127–134. [CrossRef] [PubMed]
88. Quan, Y.; Muggia, L.; Moreno, L.F.; Wang, M.; Al-Hatmi, A.M.; da Silva Menezes, N.; Shi, D.; Deng, S.; Ahmed, S.; Hyde, K.D.; et al. A re-evaluation of the Chaetothyriales using criteria of comparative biology. *Fungal Divers.* **2020**, *103*, 47–85. [CrossRef]
89. Quan, Y.; van den Ende, B.G.; Shi, D.; Prenafeta-Boldú, F.X.; Liu, Z.; Al-Hatmi, A.M.; Ahmed, S.A.; Verweij, P.E.; Kang, Y.; de Hoog, S. A comparison of isolation methods for black fungi degrading aromatic toxins. *Mycopathologia* **2019**, *184*, 653–660. [CrossRef]
90. Baglioni, P.; Berti, D.; Bonini, M.; Carretti, E.; Dei, L.; Fratini, E.; Giorgi, R. Micelle, microemulsions, and gels for the conservation of cultural heritage. *Adv. Colloid Interface Sci.* **2014**, *205*, 361–371. [CrossRef]
91. Gagliano Candela, R.; Maggi, F.; Lazzara, G.; Rosselli, S.; Bruno, M. The essential oil of *Thymbra capitata* and its application as a biocide on stone and derived surfaces. *Plants* **2019**, *8*, 300. [CrossRef]
92. Fidanza, M.R.; Caneva, G. Natural biocides for the conservation of stone cultural heritage: A review. *J. Cultl. Herit.* **2019**, *38*, 271–286. [CrossRef]
93. Lo Schiavo, S.; De Leo, F.; Urzì, C. Present and future perspectives for biocides and antifouling products for stone-built cultural heritage: Ionic liquids as a challenging alternative. *Appl. Sci.* **2020**, *10*, 6568. [CrossRef]
94. Pinna, D.; Salvadori, B.; Galeotti, M. Monitoring the performance of innovative and traditional biocides mixed with consolidants and water-repellents for the prevention of biological growth on stone. *Sci. Total Environ.* **2012**, *423*, 132–141. [CrossRef]
95. Salvadori, O.; Charola, A.E. *Methods to Prevent Biocolonization and Recolonization: An Overview of Current Research for Architectural and Archaeological Heritage*; Smithsonian Institution Scholarly Press: Washington, DC, USA, 2011; pp. 37–50.
96. Urzì, C.; De Leo, F.; Galletta, M.; Salamone, P.; Balzarotti, R. Efficiency of biocide in "in situ" and "in vitro" treatment. Study case of the "Templete de Mudejar", Guadalupe, Spain. In *Proceedings of the 9th International Congress on Deterioration and Conservation of Stone Venice June 19–24*; Elsevier Science BV: Amsterdam, The Netherlands, 2000; pp. 531–539.

97. Delgado Rodrigues, J.; Valero Congil, J.; Wakefield, R.; Brechet, E.; Larrañaga, I. Monitoraggio della biocolonizzazione e valutazione dell'efficacia di un biocida. In *Arkos: Scienza e Restauro Dell'architettura*; Nardini: Firenze, Italy, 2004; Volume 5, pp. 52–58.
98. Tesei, D. Black fungi research: Out-of-this-world implications. *Encyclopedia* **2022**, *2*, 212–229. [CrossRef]
99. Onofri, S.; Selbmann, L.; Barreca, D.; Isola, D.; Zucconi, L. Do fungi survive under actual space conditions? Searching for evidence in favor of the lithopansperimia. *Plant. Biosyst.* **2009**, *143*, S85–S87. [CrossRef]
100. Onofri, S.; de la Torre, R.; de Vera, J.P.; Ott, S.; Zucconi, L.; Selbmann, L.; Scalzi, G.; Venkateswaran, K.J.; Rabbow, E.; Sánchez Iñigo, F.J.; et al. Survival of rock-colonizing organisms after 1.5 years in outer space. *Astrobiol* **2012**, *12*, 508–516. [CrossRef]
101. Acero, F.J.; Carbú, M.; El-Akhal, M.R.; Garrido, C.; González-Rodríguez, V.E.; Cantoral, J.M. Development of proteomics-based fungicides: New strategies for environmentally friendly control of fungal plant diseases. *Int. J. Mol. Sci.* **2011**, *12*, 795–816. [CrossRef] [PubMed]
102. Gilbert, P.; Moore, L.E. Cationic antiseptics: Diversity of action under a common epithet. *J. Appl. Microbiol.* **2005**, *99*, 703–715. [CrossRef]
103. Caneva, G.; Nugari, M.P.; Salvadori, O. *Plant. Biology for Cultural Heritage: Biodeterioration and Conservation*; Getty Publications: Los Angeles, CA, USA, 2008.
104. Corbett, J.R.; Wright, K.; Baillie, A.C. *The Biochemical Mode of Action of Pesticides*, 2nd ed.; Academic Press: London, UK, 1984.
105. Juergensen, L.; Busnarda, J.; Caux, P.Y.; Kent, R. Fate, behavior, and aquatic toxicity of the fungicide IPCB in the Canadian environment. *Environ. Toxicol.* **2000**, *15*, 201–213. [CrossRef]
106. Williams, T.M. The mechanism of action of isothiazolone biocides. *Power Plant. Chem.* **2007**, *9*, 14–22.
107. Scheerer, S.; Ortega-Morales, O.; Gaylarde, C. Microbial deterioration of stone monuments—An updated overview. *Adv. Appl. Microbiol.* **2009**, *66*, 97–139. [CrossRef]
108. Gostinčar, C.; Grube, M.; Gunde-Cimerman, N. Evolution of fungal pathogens in domestic environments? *Fungal Biol.* **2011**, *115*, 1008–1018. [CrossRef]
109. Langsrud, S.; Sundheim, G.; Borgmann-Strahsen, R. Intrinsic and acquired resistance to quaternary ammonium compounds in food-related *Pseudomonas* spp. *J. Appl. Microbiol.* **2003**, *95*, 874–882. [CrossRef]
110. Ruibal, C.; Selbmann, L.; Avci, S.; Martin-Sanchez, P.M.; Gorbushina, A.A. Roof-inhabiting cousins of rock-inhabiting fungi: Novel melanized microcolonial fungal species from photocatalytically reactive subaerial surfaces. *Life* **2018**, *8*, 30. [CrossRef]

applied sciences

MDPI

Article

Investigations of the Surface of Heritage Objects and Green Bioremediation: Case Study of Artefacts from Maramureş, Romania

Dorina Camelia Ilieş [1], Nicolaie Hodor [2], Liliana Indrie [3], Paula Dejeu [4], Alexandru Ilieş [1], Adina Albu [3], Tudor Caciora [1,*], Marin Ilieş [5], Lucian Barbu-Tudoran [6,7] and Vasile Grama [1]

1 Department of Geography, Tourism and Territorial Planning, Faculty of Geography, Tourism and Sport, University of Oradea, 1 Universitatii Street, 410087 Oradea, Romania; dilies@uoradea.ro (D.C.I.); ailies@uoradea.ro (A.I.); vgrama@uoradea.ro (V.G.)
2 Faculty of Geography, "Babes-Bolyai" University, 5–6 Clinicilor, 400090 Cluj Napoca, Romania; nicolaie.hodor@ubbcluj.ro
3 Department of Textile, Leather and Industrial Management, Faculty of Energy Engineering and Industrial Management, University of Oradea, B. St. Delavrancea Str. No. 4, 410058 Oradea, Romania; lindrie@uoradea.ro (L.I.); aalbu@uoradea.ro (A.A.)
4 Medical Laboratory Service M.D., Bethany Medical Clinic, 410004 Oradea, Romania; office@betania-centrumedical.ro
5 Faculty of Geography Extension, "Babes-Bolyai" University, 6 Avram Iancu Street, 435500 Sighetu Marmatiei, Romania; marin.ilies@ubbcluj.ro
6 Electron Microscopy Laboratory 'Prof. C. Craciun', Faculty of Biology and Geology, "Babes-Bolyai" University, 5–7 Clinicilor Str., 400006 Cluj-Napoca, Romania; lucian.barbu@itim-cj.ro
7 Electron Microscopy Integrated Laboratory, National Institute for R&D of Isotopic and Molecular Technologies, 67–103 Donat Str., 400293 Cluj-Napoca, Romania
* Correspondence: caciora.tudoriulian@student.uoradea.ro; Tel.: +40-740941144

Citation: Ilieş, D.C.; Hodor, N.; Indrie, L.; Dejeu, P.; Ilieş, A.; Albu, A.; Caciora, T.; Ilieş, M.; Barbu-Tudoran, L.; Grama, V. Investigations of the Surface of Heritage Objects and Green Bioremediation: Case Study of Artefacts from Maramureş, Romania. *Appl. Sci.* **2021**, *11*, 6643. https://doi.org/10.3390/app11146643

Academic Editors: Filomena De Leo and Daniela Isola

Received: 2 June 2021
Accepted: 14 July 2021
Published: 20 July 2021

Abstract: Old textiles are important elements of thecultural heritage. As a result of their composition mostly of natural elements old textiles are extremely prone to physical and chemical degradation due to fungal action. The treatments usually applied for the cleaning of heritage textiles target the use of synthetic fungicides, which are potentially harmful to both human health and the environment. Numerous studies highlight as an alternative to the use of conventional antifungals, the employment of essential oils and plant extracts, which are environmentally friendly and which have no adverse effects on human health. Against this background the present study aims to test six essential oils (*Lavandula angustifolia, Citrus limon, Mentha piperita, Marjoram, Melaleuca alternifolia, Origanum vulgare*) to establish their inhibitory effects against fungi identified on an old piece of traditional Romanian clothing from Maramureş. For the study, the types of fungi present on the objects was determined primarily through the open plates technique and microscopic identification. After identification, the essential oils were applied to the delimited surfaces, and their effects observed up to 32 days after application. The results show that these essential oils have a strong inhibitory effect on such fungal genera as *Penicillinum* sp., *Cladosporium* sp., *Aspergillus* spp., *Candida guillermondii, Botrys* sp., *Mucor* sp., having no observable side-effects on the physical properties of the materials concerned. The antimicrobial effects that essential oils and plant extracts have in the short term must be tested in future to ensure the enhanced preservation of heritage textiles and the health integrity of the restorers and visitors who view them in museums, collections or exhibitions.

Keywords: cultural heritage; materials; fungi; essential oils; antifungal; inhibitory effects

1. Introduction

The traditional folk costume from Maramureş is an important component of the Romanian cultural heritage. Certain elements of the peasant's traditional garments are two thousand years old, as evidenced by the scenes on Trajan's Column in Rome, Italy

(113 AC), those on the Monument to Adamclisi, *Tropaeum Traiani* (108–109 AC), from Dobruja, Romania and some other important archaeological discoveries from Romania. The preservation and conservation of popular traditional cloths is of great importance, especially now, when the modernisation of society and globalisation generate a growing standardization and lack of differentiation in the appearance of outfits [1,2], most especially those for men. While Maramureş celebration clothes still retain their former appearance, recently some materials and colours have come to be produced by relatively advanced and modernised techniques. The traditional objects are those most subject to degradation, due to the numerous insects, moths, fungi, bacteria and mechanical forces, among others, to which they are subjected over the years. At the same time traditional methods of preservation and conservation can be lost. Therefore, the preservation, in good condition, of the old components of traditional garments has acquired increasing importance, for the use, the identity and the pride of the future inhabitants of the region/country. Arguably, Maramureș has become an identity brand, symbolising the art and the creative power, as well as the youth, vigour and elegance of the Romanian inhabitants of Maramureș.

Under investigation is an aged garment object which is mainly made up of natural sheep's wool; for the popular motifs that adorn it materials as sheepskin, cotton yarn, pearly beads and hemp yarn also are used. All of these materials are organic, undergoing chemical processing to give them a long life, stability and elasticity, if they are stored and used properly. Historically, the causes of damage to natural materials vary over time. Biological damage is a factor that makes the leather and furs particularly vulnerable. Due to the acidity of the skin under conditions of high humidity, the development potential for colonies of microorganisms is very high, with them adhering to the substrate along with dust particles and other environmental substances. Furs and wool can be attacked by insects such as moth larvae, beetles and others, and specific microclimatic conditions, characterised by conditions of high humidity, low intensity of light and others, favour their multiplication. Bio-damage can lead, in time, to the complete destruction of the object. Physical and chemical damage is caused by natural factors, with chemical damage being faster than physical damage, with the former's effects being profound and irreversible [3]. Deterioration can be caused by natural factors [4]. Temperature levels can pose problems that are difficult to solve in relation to the conservation of collections if favouring the development of moulds (along with high humidity), along with insects and rodents, and accelerating chemical damage to materials. The amount of damage incurred is increased when the materials are present in a relative humidity below 50%, under which conditions they tend to become dry and brittle, and their physical strength decreases. At high humidity or on contact with aqueous solutions, the collagen present is destroyed, with negative effects occurring in terms of destruction of the structure and an increase in the hydrophilicity of the fibres involved. Light, due to photochemical effects, can also cause significant degradation, in terms of the weakening of the resistance of the material substrate, friability, chromatic changes, rupture of collagen fibres and others. Ultraviolet light should be removed without affecting the lighting concerned. The presence of dust promotes biological attacks on the skin and furs. Damage to the objects due to anthropogenic causes can also occur, as a result of improper storage and exposure (e.g., materials being placed directly on top of one another, wrinkled, placed in contact with metal objects, fixed with needles and/or nails, the improper affixing of labels among others) [5,6].

Under certain environmental conditions the colonization of microorganisms is possible with the combination of factors including: temperature, humidity fluctuation, natural or artificial lighting (favorable premises for the installation and evolution of microorganisms are improper exposure to sunlight and weather [7–9], dust content and carbon dioxide high values or when nutrients are favorable); these can induce very frequent biodeterioration, alteration processes [10–12]. Wool fibers are most susceptible to attack by bacteria and fungi, especially in conditions where moisture can be accentuated due to the hygroscopic properties of wool. The microorganisms most frequently [13] mentioned in degradation of wool and other protein fibers are bacteria which belongs to the genus:

Bacillus, Proteus, Actinomycetes (Streptomyces sp.*)* and micro-fungi: *Aspergillus, Fusarium and Trichoderma, Penicillium.*

The biodeteriogens represented by fungi can lead to the deterioration of the cultural heritage organic composition materials, with their hyphae penetrating the substrate material favouring the germination of spores. The possible physical damage is likely to be colouring/discoloration, the smell of mould, fissures, fragment detachments, fragility and the variation of dyeing properties, among others [13–17]. Possible chemical damage and changes can be wrought by fungal-derived carboxylic acids, including oxalic, citric, succinic, formic, malic, acetic, fumaric and others. In addition, chemical changes can also be reflected in aesthetical damages to the cultural materials concerned, such as in the form of the discoloration and deterioration of the surfaces involved, leading to the appearance of stains that may alter the original colour of the garment [18–21]. The presence of organic residues (e.g., glue, dirt, dust) may accelerate the processes of degradation, with the aesthetical changes and chromatic alteration taking place. At the same time, they can lead to the loss of strength and elongation specific to the material, oxidation, discoloration or coloration, due to the pigments involved and the modification of the molecular structures present [22].

Remediation can be realised using toxic materials (e.g., ethylene oxide [EtO], gamma rays, etc.), but the use of such materials can significantly affect the biodiversity and ecological systems concerned. Therefore, specialists in the field try to apply environmentally and ecologically friendly biocides instead. The use of antifungal natural extracts (i.e., essential oils [EOs]) has always been a viable alternative to the use of harmful chemicals, with, for example, plant extracts (*Allium ursinum* and *Ocimum basilicum*) having been tested as successful antifungals [23].

EOs (e.g., lemon, spearmint, rosemary, fennel, marjoram, pin, eucalyptus, etc.) can successfully serve as part of a green bioremediation procedure to be applied in the sustainable conservation of artworks. Having been tested against microbial colonies isolated from different substrates using green potential strategies, they have been found to act as a sound alternative to traditional procedures in terms of their selective action, human safety and impacts on artworks [20,24–41]. The antifungal activity of *Origanum vulgare, Rosmarinus officinalis* and *Lavandula angustifolia* EOs has been tested, with positive results having been achieved against fungi isolated on stone (*Bipolaris* sp.) and against *Aspergillus* sp., *Penicillium* sp., and *Trichoderma* sp. on different objects by Stupar et al. [42], Savković et al. [43] and Bayramoğlu [44], among others. Some relevant studies in the field [45–50] confirm that thymus EO has been found to have good antimicrobial activity against *Bacillus, Staphylococcus, Fusarium* and *Aspergillus* spp., in comparison to the effects of EO with commercial biocide. Radwan et al. [49] point out the inhibitory effect of thyme, clove and cinnamon EOs on *Candida albicans* and different finds of mould. The EOs of eucalyptus and lavender tested with good results as natural preservatives for leather [51], with oregano EO being successfully used as a bactericidal agent in the leather industry by Bayramoğlu et al. [26,44]. Combinations of the EOs [52], *Thymus vulgaris* and *Pimpinella anisum* and methanol extracts, have shown outstanding antibacterial properties against the pathogenic bacteria, *Staphylococcus aureus, Bacillus cereus, Escherichia coli, Proteus vulgaris, Proteus mirabilis, Salmonella typhi, Salmonella typhimurium, Klebsiella pneumoniae* and *Pseudomonas aeruginosa*. EOs obtained from *Matricaria chamomilla* have been tested [53] as environmentally friendly in cultural heritage environments, against *Aspergillus* spp.; similarly, Mahilrajan et al. [54] pinpoint in the same study, the intense antifungal activity of camphor oil, in comparison with other tested oils. Pepa et al. [55] point out that EOs from southern Italy, derived from *Origanum*, tested with the minimal inhibitory concentrations, *in vitro*, with evident antibacterial and antifungal activity. EOs like *Origanum vulgare* and *Thymus vulgaris* have also been applied by Palla et al. [56] to emphasise the green remediation in the biodeterioration of cultural heritage artworks, which is induced by fungal colonisation (e.g., by that of *Aspergillus flavus*). The above-mentioned authors concluded that valid green conservation strategies (with no negative effects for human health and no environmental pollution) could possibly replace traditional biocides.

In Romania, recently, an EO isolated from thyme *(Thymus vulgaris)* was assayed for antifungal activity against *Candida albicans* and *Aspergillus niger* on sheepskins, with it being shown to have outstanding antifungal proprieties [57–59]. Moreover, the studies by Niculescu et al. [60,61] and Marcu et al. [62] have emphasised the possibility of using antifungal ecological materials to treat cultural heritage objects made of natural leather.

Following the above, this research aimed to investigate the microbial load of an old heritage object, made of natural animal sheepskin, which metabolic activity could generate, thus accelerating the biodegradation process, and the loading of the aerosol with biological particles, like spores, toxins and allergens, as well as other harmful substances. Such action can cause problems regarding the health of workers and users and the application of natural biodegraders. Consequently, the antifungal properties of some EOs of *Lavandula angustifolia, Citrus limon, Mentha piperita, Marjoram, Melaleuca alternifolia, Origanum vulgare* were tested, with the results of their applications being recorded at certain time intervals.

2. Materials and Methods

2.1. Investigated Object

The object investigated was a handmade, short coat for young men, lined with sheepskin from the Mara Valley, Maramureș, Romania (Figure 1) [63–66], about 80 to 100 years old. The men's garment was created to cover the torso, with it being worn most often on important holidays or on Sundays.

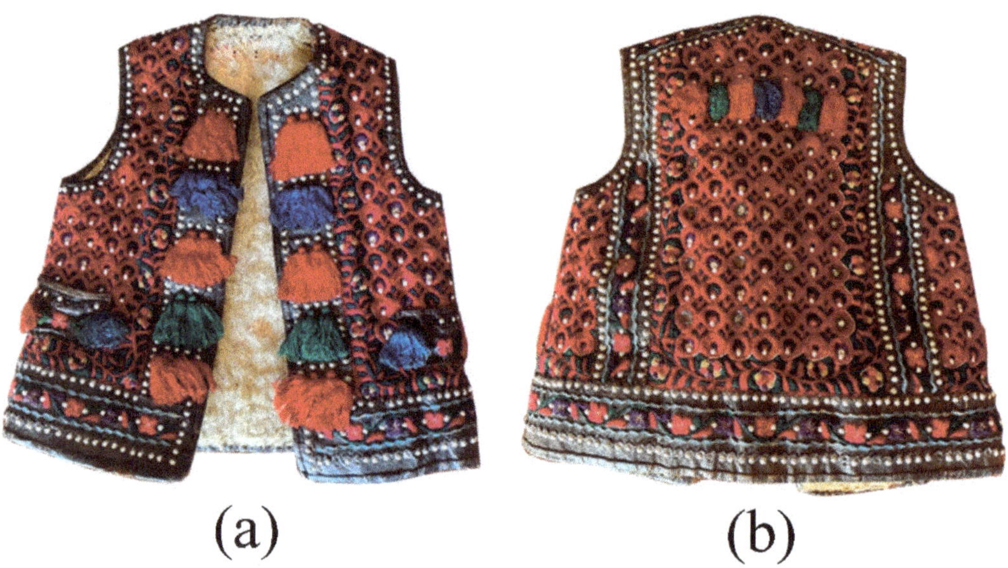

Figure 1. Traditional Romanian men's coat; (**a**)—front view; (**b**)—back view.

The coat was made of hemp thread (also called *'pânză de tort'* in Romanian). The coat's lining consisted of sewn tanned lambskin, of curly wool. On all its edges and at the seams, an intensely dark-hued strip (*'cipcă'*) or hem of red leather was applied, conveying a sense of perfect regularity and symmetry. On the inside, there was a large pocket on the left side, facilitating access by the right hand. The coat was decorated by hand-sewing (similar to embroidery), with coloured thread, and by applying decorative elements made of tanned leather, over the whole bridge of the coat, consisting of the tree of life, rose flowers called *'ruji'*, stitched flowers encircling mirrors and pseudo zig-zag patterning. In the middle of some rose flowers without mirrors, special metal buttons ('*bumbi*') had been placed, rounded and extended, with some fasteners facing inwards, which helped to affix them to the fabric structure, by means of their being bent. Near the free edge of the wings and on the back of the coat, close to the shoulders are several ornamental tassels (bell-shaped tassels) arranged symmetrically, in red, green and blue colours.

In the last half century, other examples of such fur coats (from Mara) have begun to be worn in other parts of Maramureș, becoming an identity brand [67–70], symbolising the art and creative power of the youth, as well as the vigour and elegance of the Romanians in Maramureș.

2.2. Analytical Methods

The procedure followed in the current study required the following materials: a delimiting frame; sterile swabs on a wooden rod; and six EOs with 100% purity from the Young Living Essential Oils, used for their antifungal properties [71–79]. The types of EOs employed were: *Mentha piperita, Lavandula angustifolia, Citrus limon, Melaleuca alternifolia, Marjoram* and *Origanum vulgare*. The products used consisted of: *Mentha piperita*, batch 103519, expiry date 7 August 2021; *Lavandula angustifolia*, batch 87491, expiry date 7 August 2021; *Citrus limon*, batch 103517, expiry date 7 August 2021; *Tea Tree*, batch 89075, expiry date October 2021; *Marjoram*, batch 86942, expiry date July 2022; and *Oregano*, batch 89094, expiry date September 2022. The material resources used consisted of: Sabouraud sterile culture medium for the yeast and mould isolation; an incubator ICT 18/FALC with a temperature range between 5 and 80 °C; a microbiological hood; glass slides and slides for the microscopic identification; microbiological handles; an optical microscope Micros Austria with binocular head series BIM-105B; an API yeast identification kit ®® 20C AUX; and a densitometer.

Six different areas on the face of the traditional coat were examined, as shown in Figure 2. The working areas were delimited by means of a metal dial with a size of 25 cm^2. Small fragments (millimetres) of fur and sheepskin were removed with tweezers previously sterilized by the incandescent method for testing of the fungi concerned; so as to observe the effects that they might have had on the individual strands of material. The tests were conducted both before and after applying the EOs to each delimited area, although the primary samples were considered as blank references.

After delimiting the surfaces on which the in-situ procedures were to be performed, the samples of the surface were taken. The samples were extracted using sterile swabs from all six areas of the sheepskin coat, both before and after the application of the Eos (see Table 1). Ten drops (the equivalent to 300 μL) of the corresponding EO were applied directly to the tested material surface in the centre of each delimited area [75]. After the application of the EOs, three samples were taken for each of the six different work areas delimited on the area of sheepskin to which the EOs were applied. The three samples were taken at pre-set intervals, with the first sample being taken 30 min after the administration of the EOs, the second test being administered after 24 h and the third after 48 h. Table 1 also shows the type of material examined and the type of EO applied to each of the delimited areas. The EO used for each type of material concerned was selected in accordance with the recommendations made in the literature, especially that of Palla et al. [56], which state that certain EOs tend to be more effective than others when applied to certain types of material, such as those which constitute heritage objects.

It must be noted that the air in the room in which the sheepskin coat was stored and examined was not circulated, and neither was the room ventilated for 48 h during the entire examination, so as to avoid contamination with other types of spores from the atmosphere that could, otherwise, have been introduced by currents of air moving through the room. The temperature was maintained between 22 °C and 24 °C, while the humidity was kept between 52% and 55%.

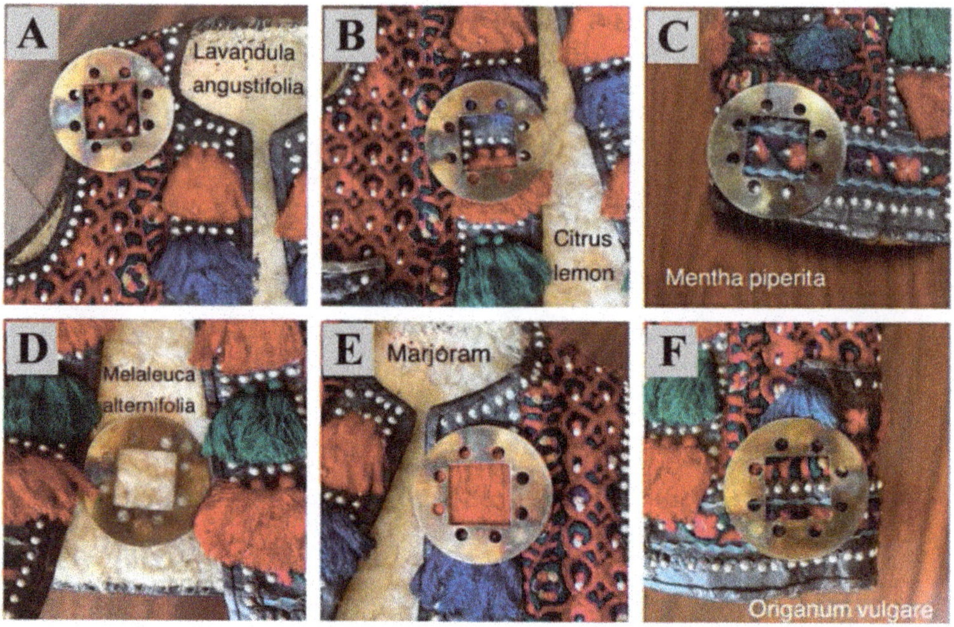

Figure 2. Delimitation of the six work areas on the front of the coat, indicating the type of essential oil applied to each (**A**)—*Lavandula angustifolia;* (**B**)—*Cistrus lemon;* (**C**)—*Mentha piperita;* (**D**)—*Melaleuca alternifolia;* (**E**)—*Marjoram;* (**F**)—*Origanum vulgare.*

Table 1. Areas of the sheepskin coat examined, and to which essential oils applied.

Examined Area (Figure 2)	Applied Essential Oil	Material Type
A	*Lavandula angustifolia* (lavender)	Cotton yarn
B	*Citrus limon* (limon)	Leather-cotton yarn mix
C	*Mentha piperita* (mint)	Leather
D	*Melaleuca alternifolia* (tea tree)	Wool
E	*Marjoram* (marjoram)	Cotton yarn
F	*Origanum vulgare* (oregano)	Pearly beads

With the EOs having been applied directly to the examined surfaces, all the samples were seeded on sterile Sabouraud culture media, with the plates used being incubated for 30 days at a temperature of 28 °C. The third stage of the procedure involved isolating and identifying the fungi developed on culture media. The plates were evaluated daily to observe their evolution in terms of the changes in the appearance of the colonies concerned, with the evaluation undertaken sequentially being that of shape, texture, consistency, diameter, colour and contour.

The moulds present were identified after evaluating the macroscopic and microscopic characteristics involved, and the yeast species was determined, using the API ®® 20 °C AUX6 identification kit, after examination of the 19 biochemical assimilation reactions that took place. The technique of working on the latter involved preparing the yeast suspension solution, calibrated at 2 McFarland, using a densitometer, and thereafter pipetting 100 μL into each microdwell of the gallery. After incubating the API gallery at 29 °C ± 2 °C for 72 h, the results were read, with a final result being obtained after comparing each well with the negative control. The final result was a 7-digit numerical profile, which was decoded using Apiweb™ computer software, so as to identify the corresponding yeast species.

3. Results

Fungal colonies were developed rapidly on Sabouraud media, from samples taken before the application of EOs. At 72 h of incubation, the first fungal colonies could be macroscopically visualised. After 10 days of incubation, the plates were completely invaded. Figure 3 shows the plates on the seventh day of incubation; each image indicates the examined area of the sheepskin coat. The high degree of fungal contamination of the coat can easily be observed.

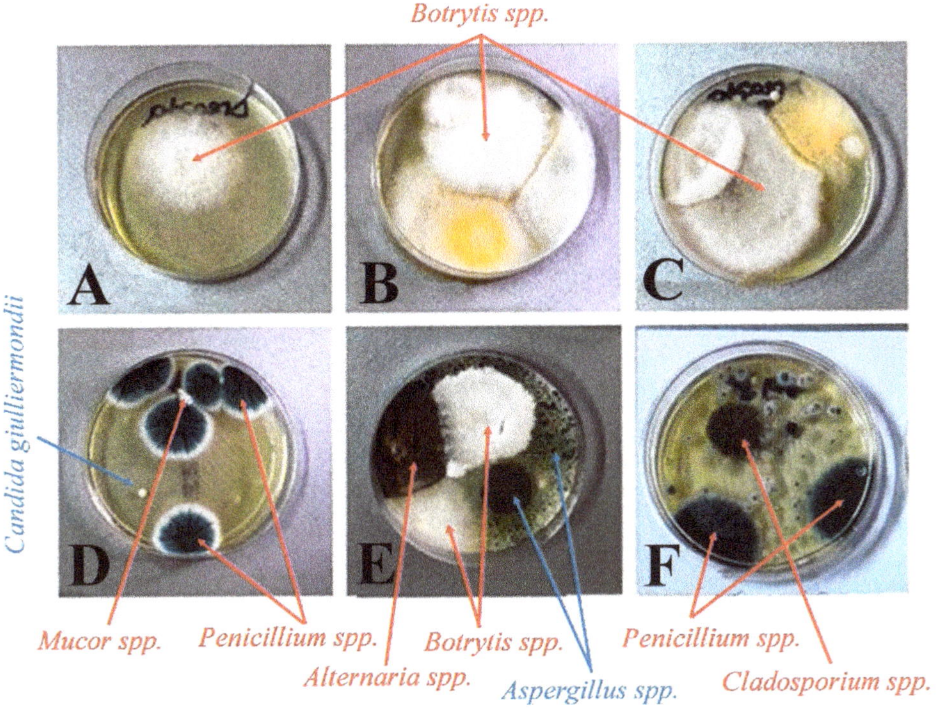

Figure 3. Plates seeded with primary samples (before application of essential oils). Reading taken after 7 days of incubation (**A**)—*Lavandula angustifolia*; (**B**)—*Cistrus lemon*; (**C**)—*Mentha piperita*; (**D**)—*Melaleuca alternifolia*; (**E**)—*Marjoram*; (**F**)—*Origanum vulgare*.

In the above-mentioned way, seven different types of colonies were identified [80,81], with six being from the mould class (*Alternaria* sp., *Aspergillus* sp., *Botrytis* sp., *Cladosporium* sp., *Mucor* sp. and *Penicillium* sp.) and one from the yeast class (*Candida guilliermondii*) (Table 2).

As previously stated the number of samples taken after applying the EOs was 3 sets of 6, taken 30 min, 24 h and 48 h after applying the corresponding EO. The plates were monitored for 30 days in order to evaluate the inhibitory effects of the EOs. Figure 4 shows the plates, as read on day 14 of the incubation. The inhibitory effects of the EOs are evident. The only plate on which a fungal colony developed is the one corresponding to the inner area of the sheepskin coat, present in the sample taken 48 h after applying the EO of *Melaleuca alternifolia* (tea tree). The isolated fungal colony consisted of *Cladosporium* sp., which became macroscopically visible on day 5 of incubation. It was observed that this type of mould was not isolated from the primary sample, as is detailed in the discussion section below.

Table 2. Identified fungi and the effect of essential oils.

Microscopic Image	Genus Name	The Examined Area of the Sheepskin Coat from Which It Was Isolated (Figure 2)	Applied Essential Oils—Effect
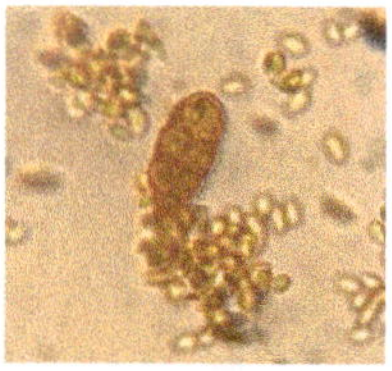	*Alternaria* sp.	Area E	*Marjoram*—inhibitory
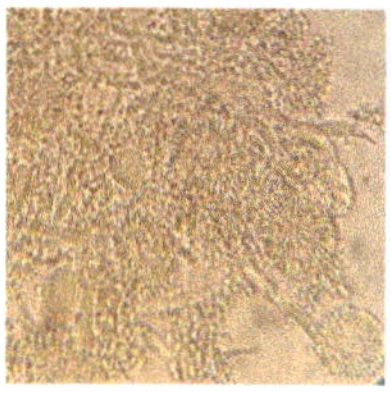	*Aspergillus* sp.	Area E	*Marjoram*—inhibitory
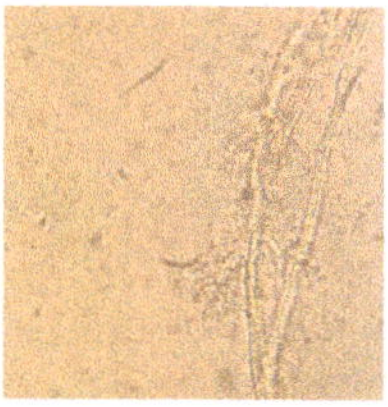	*Botrytis* sp.	Area A Area B Area C Area E	*Lavandula angustifolia*—inhibitory *Citrus limon*—inhibitory *Mentha piperita*—inhibitory *Marjoram*—inhibitory
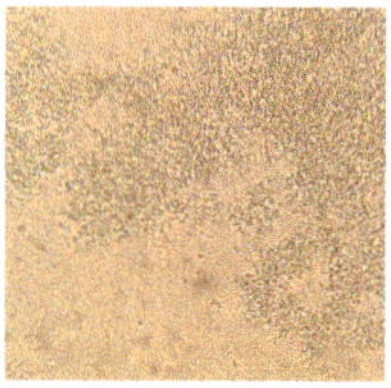	*Candida guilliermondii* Profile API: 6702377	Area D	*Melaleuca alternifolia*—inhibitory
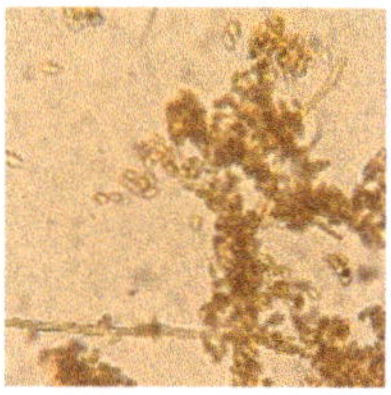	*Cladosporium* sp.	Area F	*Origanum vulgare*—inhibitory
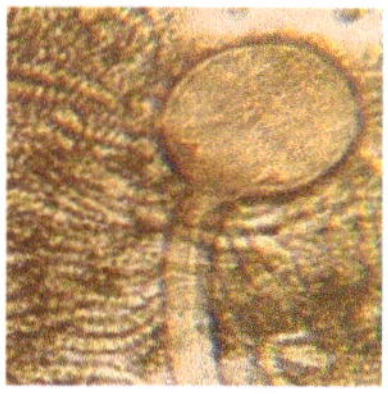	*Mucor* sp.	Area D	*Melaleuca alternifolia*—inhibitory

Table 2. *Cont.*

Microscopic Image	Genus Name	The Examined Area of the Sheepskin Coat from Which It Was Isolated (Figure 2)	Applied Essential Oils—Effect
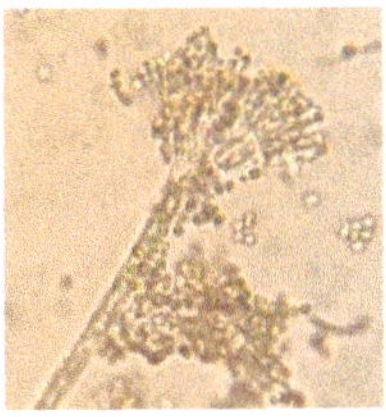	*Penicillium* sp.	Area D Area F	*Melaleuca alternifolia*—inhibitory *Origanum vulgare*—inhibitory

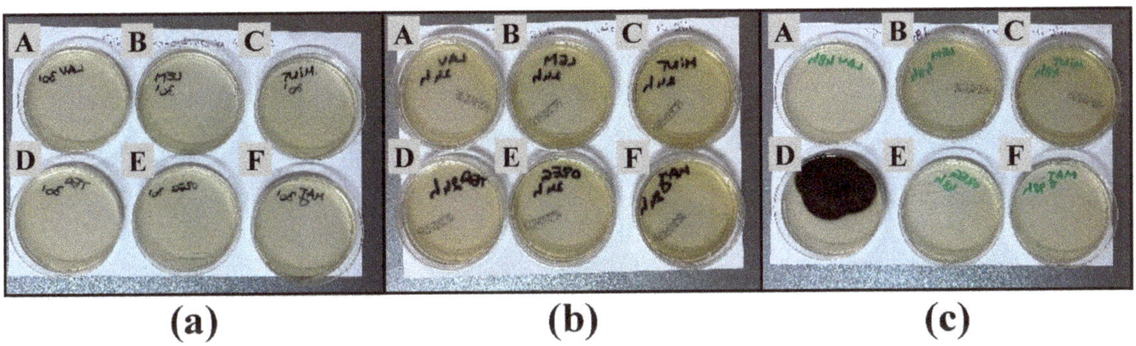

Figure 4. The plates read on day 14 of incubation: the samples at (**a**) 30 min, (**b**) 24 h and (**c**) 48 h.

Figure 5a–c show the plates, as read on day 30 of the incubation. The developed colonies became macroscopically visible after between 20 days to 25 days of incubation.

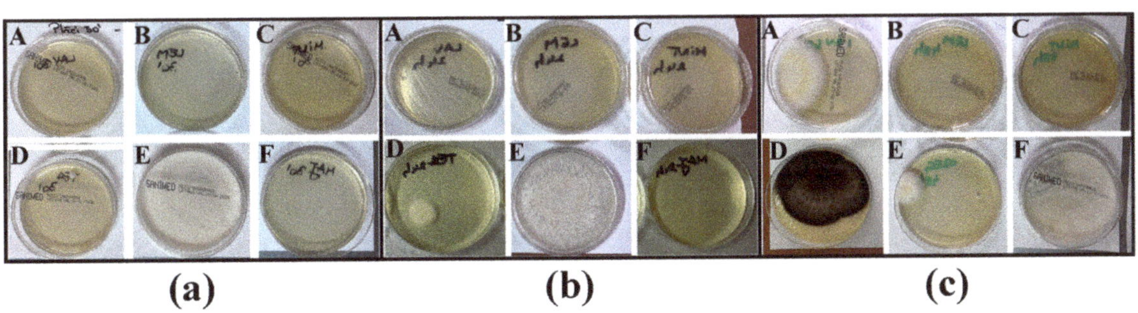

Figure 5. The plates read on day 30 of incubation: samples at (**a**) 30 min, (**b**) 24 h and (**c**) 48 h.

The results, obtained after the samples from the examined areas were examinedat 30 min, 24 h and 48 h following the application of the EOs, are presented in Table 3.

The existing difference was due to the duration of the effect that they had on certain types of fungi as shown on Table 4. It is evident that some EOs (e.g., *Lavandula angustifolia, Melaleuca alternifolia, Marjoram, Origanum vulgare*) inhibited the action of some fungal genera up to a period of 22 days, whereas on others the effect lasted for a period of at least 30 to 32 days (Table 4).

Table 3. Interpretation of the seeded plates, with samples taken 30 min, 24 h and 48 h after applying the essential oil.

Samples Taken	Examined Area	Essential Oil	Fungal Colony Growth
30 min after application of the essential oils	Area A	*Lavandula angustifolia*	Absent
	Area B	*Citrus limon*	Absent
	Area C	*Mentha piperita*	Absent
	Area D	*Melaleuca alternifolia*	*Mucor* sp.
	Area E	*Marjoram*	Absent
	Area F	*Origanum vulgare*	*Botrytis* sp.
24 h after the application of the essential oils	Area A	*Lavandula angustifolia*	Absent
	Area B	*Citrus limon*	Absent
	Area C	*Mentha piperita*	Absent
	Area D	*Melaleuca alternifolia*	*Mucor* sp.
	Area E	*Marjoram*	Absent
	Area F	*Origanum vulgare*	*Botrytis* sp.
48 h after the application of the essential oils	Area A	*Lavandula angustifolia*	*Botrytis* sp.
	Area B	*Citrus limon*	Absent
	Area C	*Mentha piperita*	Absent
	Area D	*Melaleuca alternifolia*	*Cladosporium* sp.
	Area E	*Marjoram*	*Botrytis* sp.
	Area F	*Origanum vulgare*	*Cladosporium* sp. *Botrytis* sp.

Table 4. Fungal species inhibited by tested essential oils and the tested duration of their inhibitory effect.

Essential Oil Used	Inhibited Fungal Species	Duration of the Inhibitory Effect from the Moment of Application
Lavandula angustifolia	*Botrytis* sp.	Up to 22 days
Citrus limon	*Botrytis* sp.	Minimum of 32 days
Mentha piperita	*Botrytis* sp.	Minimum of 32 days
Melaleuca alternifolia	*Candida guilliermondii* *Mucor* sp. *Penicillium* sp.	Minimum of 32 days Up to 22 days Minimum of 30 days
Marjoram	*Alternaria* sp. *Aspergillus* sp. *Botrytis* sp.	Minimum of 30 days Minimum of 30 days Up to 22 days
Origanum vulgare	*Cladosporium* sp. *Botrytis* sp. *Penicillium* sp.	Minimum of 30 days Up to 22 days Minimum of 30 days

4. Discussion

After the tests were performed on the delimited surfaces of the 80- to 100-year-old clothing object, the conclusion was drawn that all the EOs had a strong inhibitory effect on the fungi present, even if they were applied on different materials (cotton yard, leather, wool, pearly beads). This is due to the fungicidal effect of EOs, which persists in some cases over 30 days after application. Similar results were obtained by different authors [82–85] following the application of these extracts on various materials. Most EOs (except for *Citrus limon* and *Mentha piperita*) were observed to have relatively little effect on *Botrytis* sp., which might indicate some resistance of this type of fungus to the action of the substances concerned.

It was revealed that the genus *Cladosporium* sp. developed at 5 days of incubation on the culture medium shown with the sample taken 48 h after the application of the EO, without the species involved having been isolated from the primary sample (taken before the application of the EO). The sample concerned was taken from the wool threads of the inner region of the coat. Woollen garments seem to be easily contaminated with fungal spores, with the degree of fungal contamination being most likely to differ from layer to layer—superficial, intermediate and basal. The spores that are present in the lower layers may rise to the surface over time, and, if the EO is applied superficially or in insufficient quantity, its inhibitory effect is diminished. Moreover, in the case of studies like the present, the environmental conditions present in the room (temperature, humidity, air currents) must be taken into account, as well as the sampling procedure itself. All such factors form a set of features that can interfere with the final results.

The fungal action identified on the garment investigated could be pathological for humans, especially for the personnel handling the garment. Although the fungi of the *Aspergillus* spp. family are generally harmless, they can cause various pathologies in those with compromised immune systems, underlying lung disease or asthma, who inhale the relevant fungal spores. The pathologies resulting from infections with such a fungus, which usually affect the respiratory system, have greatly varying signs and severity. In some (such as asthmatics), the spores trigger an allergic reaction, whereas others tend to develop mild to severe lung infections. The most serious form of aspergillosis, invasive aspergillosis, occurs when the infection spreads through the blood vessels, with the signs and symptoms of the infection varying, depending on the type of pathology that the patients concerned develop, and with it not being contagious. Daily exposure to *Aspergillus* is rarely a problem for people with healthy immune systems [86–88]. Fungi from the *Candida guilliermondii* family appear to be one of the most common pathogens in localised infections of the nails (onychomycosis) and/or of the skin on the toes. In patients with low immunity, infections with such a fungus can be a potential cause of fungal infections in the blood, namely fungemia, especially in patients with haematological malignancies, organ transplants and central venous catheters. This type of fungus is one of the most common opportunistic agents in severely immunocompromised patients. In such cases, depending on the associated pathologies, fungemia can prove fatal [89]. Fungal infections of the *Cladosporium* spp. family either aggravate the symptoms of an existing asthma, or are associated with atopy located mainly in the upper respiratory tract (allergic rhinitis/sinusitis) and less often in the skin (superficial or deep skin lesions). The impact of patients' symptoms (dyspnoea, sneezing, stuffy nose, pruritus, nasal/ocular secretions) was significantly correlated with spore concentrations. In patients with weakened immune systems, infections with this type of fungus can even cause disseminated infections, called fungemia [90–92]. Fungi of the genus *Penicillium* spp., which are occasionally the cause of infection in humans, can be isolated from patients with keratitis, endophthalmitis, otomycosis, necrotising esophagitis, pneumonia, endocarditis, peritonitis and urinary tract infections. Most *Penicillium* infections occur in immunosuppressed hosts, and the mechanism of infection with this type of fungus is either by means of inhalation (most commonly resulting in fungemia) or post-traumatic [93,94].

5. Conclusions

The clothing objects that make up the cultural heritage are prone to deterioration, due to fungal action, especially if they are stored or exposed in environments that lack careful adjustment of the internal microclimate parameters. In addition to the harmful effects that fungi can have on heritage objects, people (restorers, museographers, collectors, visitors) who come in contact with such infested objects can develop specific conditions. Thus, treating historical textiles in the most responsible way possible becomes a dual purpose issue; on the one hand, it is necessary to preserve the materials under the best possible conditions, and, on the other hand, it is necessary to ensure the integrity of the health of those who are interested in them. The current study demonstrates that EOs are one of the most viable solutions to help ensure responsible treatment given that their antifungal properties can exceed 30 days of use, and with the products being both ecofriendly and accessible in terms of cost. The six EOs (*Lavandula angustifolia, Citrus limon, Mentha piperita, Melaleuca alternifolia, Marjoram, Origanum vulgare*) applied on different types of materials served to inhibit the presence of six different fungal colonies (*Alternaria* sp., *Aspergillus* sp., *Botrytis* sp., *Cladosporium* sp., *Mucor* sp. and *Penicillium* sp.) and a class of yeast (*Candida guilliermondii*) identified as being on the traditional coat. Simultaneously, the EOs had a strong antifungal effect on several materials (cotton, leather, wool) that formed part of the object investigated. Thus, all of the products studied were proven to be effective on certain surfaces and against certain types of fungi, in terms of both their cleaning and their protection effects. Prospects for the future must include the use of natural antifungals for the treatment of heritage objects; but at the same time, the tests must also aim at determining the possible negative effects that they may have on fragile objects.

Author Contributions: Conceptualisation: D.C.I. and N.H.; Methodology: P.D. and L.I.; Software: L.B.-T. and A.A.; Validation: P.D. and A.A.; Formal analysis: P.D., L.I. and L.B.-T.; Investigation: P.D., A.I. and M.I.; Resources: L.I. and A.A.; Data curation: P.D., D.C.I., N.H. and T.C.; Writing—original draft preparation: D.C.I., P.D., M.I. and T.C.; Writing—review and editing: D.C.I., T.C., A.I., M.I. and V.G. All authors have read and agreed to the published version of the manuscript.

Funding: This research was funded by grant PN-III-P1-1.2-PCCDI-2017-0686.

Institutional Review Board Statement: Not applicable.

Informed Consent Statement: Not applicable.

Data Availability Statement: The data presented in this study may be obtained on request from the corresponding author.

Acknowledgments: The research undertaken was made possible by the equal scientific involvement of all the authors concerned. The authors wish to thank the anonymous reviewers for their thoughtful suggestions and comments made, and to acknowledge the support of the grant PN-III-P1-1.2-PCCDI-2017-0686. The investigations and analyses involved were done at the Medical Laboratory Service M.D., Bethany Medical Clinic Oradea, University of Medicine and Pharmacy of University of Oradea and at the National Institute of Research-Development for Isotopic and Molecular Technologies, Cluj-Napoca.

Conflicts of Interest: The authors declare no conflict of interest.

References

1. Albu, A.V.; Caciora, T.; Berdenov, Z.; Ilies, D.C.; Sturzu, B.; Sopota, D.; Herman, G.V.; Ilies, A.; Kecse, G.; Ghergheles, C.G. Digitalization of garment in the context of circular economy. *Ind. Text.* **2021**, *72*, 102–107. [CrossRef]
2. Ilies, D.C.; Herman, G.V.; Caciora, T.; Ilies, A.; Indrie, L.; Wendt, J.A.; Axinte, A.; Diombera, M.; Lite, C.; Berdenov, Z.; et al. Considerations Regarding the Research for the Conservation of Heritage Textiles in Romania. In *Waste in Textile and Leather Sectors*; Körlü, A., Ed.; IntechOpen: Rijeka, Croatia, 2020; Chapter 10; pp. 88–93.
3. Cappitelli, F.; Cattò, C.; Villa, F. The Control of Cultural Heritage Microbial Deterioration. *Microorganisms* **2020**, *8*, 1542. [CrossRef]
4. Indrie, L.; Oana, D.; Ilies, M.; Ilies, D.C.; Lincu, A.; Ilies, A.; Baias, Ș.; Herman, G.; Onet, A.; Costea, M.; et al. Indoor air quality of museums and conservation of textiles art works. Case study: Salacea Museum House, Romania. *Ind. Text.* **2019**, *70*, 88–93. [CrossRef]
5. Creangă, D.M. Cauzele degradarii bunurilor de patrimoniu pe suport de piele. *Codrul Cosm.* **2004**, *10*, 153–157.
6. Di Carlo, E.; Chisesi, R.; Barresi, G.; Barbaro, S.; Lombardo, G.; Rotolo, V.; Sebastianelli, M.; Travagliato, G.; Palla, F. Fungi and Bacteria in Indoor Cultural Heritage Environments: Microbial-related Risks for Artworks and Human Health. *Environ. Ecol. Res.* **2016**, *4*, 257–264. [CrossRef]
7. Jain, P.C. *Microbial Degradation of Grains, Oil Seeds, Textiles, Wood, Corrosion of Metals and Bioleaching of Mineral Ores*; Department of Applied Microbiology & Biotechnology, Dr. Harisingh Gour University: Sagar, India, 2008; Available online: http://awarticles.s3.amazonaws.com/PCJain2008MicrobialDegradationOfGrainsOilSeeds.pdf (accessed on 1 September 2020).
8. De Leoa, F.; Urzì, C. Microfungi from Deteriorated Materials of Cultural Heritage. In *Fungi from Different Substrates*; Misra, J.K., Tewari, J.P., Deshmukh, S.K., Vagvolgyi, S., Eds.; Taylor and Francis Press: Abingdon, UK, 2015; pp. 144–158.
9. Kavkler, K.; Gunde-Cimerman, N.; Zalar, P.; Demsar, A. Fungal contamination of textile objects preserved in Slovene museums and religious institutions. *Int. Biodeter. Biodegr.* **2015**, *97*, 51–59. [CrossRef]
10. Caneva, G.; Nugari, M.P.; Salvadori, O. *La biologia nel Restauro*; Nardini Editore: Firenze, Italy, 1994.
11. Sterflinger, K.; Piñar, G. Microbial deterioration of cultural heritage and works of art—Tilting at windmills? *Appl. Microbiol. Biotechnol.* **2013**, *97*, 9637–9646. [CrossRef]
12. Valentin, N. Microbial contamination in museum collections: Organic materials. In *Molecular Biology and Cultural Heritage*; Saiz Jimenez, C., Ed.; Swets & Zeitlinger: Lisse, The Netherlands, 2003; pp. 85–91.
13. Caneva, G.; Nugari, M.P.; Salvadori, O. *Biology in the Conservation of Works of Art*; Sintesi Grafica s.r.l.: Roma, Italy, 1991.
14. Szostak-Kotowa, J. Biodeterioration of textiles. *Int. Biodeterior. Biodegrad.* **2004**, *53*, 165–170. [CrossRef]

15. Mukerji, K.G.; Garg, K.L.; Mishra, A.K. Fungi in deterioration of museum objects. In Proceedings of the 3rd International conference on Biodeterioration of Cultural Property, Bangkok, Thailand, 4–7 July 1995; pp. 226–241.
16. Kowalik, R. Microbiodeterioration of Library Materials. *Restaurator* **1980**, *4*, 99–114. [CrossRef]
17. Montegut, D.; Indictor, N.; Koestler, R.J. Fungal deterioration of cellulosic textiles: A review. *Int. Biodeterior. Biodegrad.* **1991**, *28*, 209–226. [CrossRef]
18. Arif Hossain, M. *Investigating Bio Deterioration of Cultural Heritage: Detection and Identification of Microbial Communities and Biocides Application*; University of Évora: Évora, Portugal, October 2017; Available online: https://www.academia.edu/32181292/Investigating_bio_deterioration_of_Cultural_Heritage_Detection_and_identification_of_microbial_com-408_munities_and_biocides_application (accessed on 10 September 2020).
19. Sterflinger, K. Fungi as geologic agents. *Geomicrobiol. J.* **2000**, *17*, 97–124. [CrossRef]
20. Vukojević, J.; Grbić, M.L. Moulds on paintings in Serbian fine art museums. *Afr. J. Microbiol. Res.* **2010**, *4*, 1453–1456.
21. Manente, S.; Micheluz, A.; Ganzerla, R.; Ravagnan, G.; Gambaro, A. Chemical and biological characterization of paper: A case study using a proposed methodological approach. *Int. Biodeter. Biodegr.* **2012**, *74*, 99–108. [CrossRef]
22. Di Carlo, E.; Barresi, G.; Palla, F. Biodeterioration. In *Biotechnology and Conservation of Cultural Heritage*; Palla, F., Barresi, G., Eds.; Springer: New York, NY, USA, 2017; pp. 1–32.
23. Fierascu, I.; Dima, R.; Fierascu, R.C. Natural Extracts for preventing Artefacts Biodeterioration. In Proceedings of the 17 International Conference on Cultural Heritage and New Technologies (CHNT 17, 2012), Vienna, Austria, 5–7 November 2012; Fischer Ausserer, K., Ed.; Museen der Stadt Wien—Stadtarchäologie: Vienna, Austria, 2012.
24. Pattnaik, S.; Subramanyam, V.R.; Kole, C. Antibacterial and antifungal activity of ten essential oils in vitro. *Microbios* **1996**, *86*, 237–246.
25. Abad, M.J.; Ansuategui, M.; Bermejo, P. Active antifungal substances from natural sources. *ARKIVOC* **2007**, *7*, 116–145.
26. Bayramoğlu, E.E.; Gülümser, G.; Karaboz, I. The Investigation of Antibacterial Activities of Some Essential Oils in Wet Blue Leather. *Int. J. Nat. Eng. Scien.* **2008**, *2*, 33–36.
27. Guiamet, P.S.; Gómez de Saravia, S.G.; Arenas, P.; Pérez, M.L.; De la Paz, J.; Borrego, S.F. Natural products isolated from plants used in biodeterioration control. *Pharmacologyonline* **2006**, *3*, 537–544.
28. Sasso, S.; Scrano, L.; Ventrella, E.; Bonomo, M.G.; Crescenzi, A.; Salzano, G.; Bufo, S.A. Natural biocides to prevent the microbial growth on cultural heritage. In Proceedings of the Conference Built Heritage 2013—Monitoring Conservation and Management, Milan, Italy, 18–20 November 2013; pp. 1035–1042.
29. Orlita, A. Microbial biodeterioration of leather and its control: A review. *Int. Biodeter. Biodegr.* **2004**, *53*, 157–163. [CrossRef]
30. Al-Hussaini, R.; Mahasneh, A.M. Microbial growth and quorum sensing antagonist activities of herbal plants extracts. *Molecules* **2009**, *14*, 3425–3435. [CrossRef]
31. Abdel-Maksoud, G.; El-Amin, A.-R.; Afifi, F. Insecticidal activity of Cinnamomum cassia extractions against the common Egyptian mummies' insect pest (Dermestes maculatus). *Int. J. Conserv. Sci.* **2014**, *5*, 355–368.
32. Barresi, G.; Carlo, E.; Trapani, M.R.; Parisi, M.G.; Chillè, C.; Mulè, M.F.; Cammarata, M.; Palla, F. Marine organisms as source of bioactive molecules applied in restoration projects. *Herit. Sci.* **2015**, *3*, 17–20. [CrossRef]
33. Borrego, S.; Valdés, O.; Vivar, I.; Lavin, P.; Guiamet, P.; Battistoni, P.; Gómez de Saravia, S.; Borges, P. Essential oils of plants as biocides against microorganisms isolated from Cuban and Argentine Documentary heritage. *ISRN Microbiol.* **2012**, 826786. [CrossRef] [PubMed]
34. Warscheid, T. Integrated concepts for the protection of cultural artifacts against biodeterioration. In *Of Microbes and Art, the Role of Microbial Communities in the Deterioration and Protection of Cultural Heritage*; Ciferri, O., Tiano, P., Mastromei, G., Eds.; Kluwer Academic/Plenum Publishers: New York, NY, USA, 2000; pp. 185–201.
35. Calvo, M.A.; Arosemena, E.L.; Shiva, C.; Adelantado, C. Antimicrobial activity of plant natural extracts and essential oils. In *Science against Microbial Pathogens: Communicating Current Research and Technological Advances*; Méndez-Villas, A., Ed.; Formatex Research Center: Badajoz, Spain, 2001; pp. 1179–1185.
36. Rotolo, V.; Barresi, G.; Di Carlo, E.; Giordano, A.; Lombardo, G.; Crimi, E.; Costa, E.; Bruno, M.; Palla, F. Plant extracts as green potential strategies to control the biodeterioration of Cultural Heritage. *Int. J. Conserv. Sci.* **2016**, *7*, 839–846.
37. Ćirković, I.; Jovalekić, M.; Jegorović, B. In vitro antibacterial activity of garlic and synergism between garlic and antibacterial drugs. *Arch. Biol. Sci.* **2012**, *64*, 1369–1375. [CrossRef]
38. de la Paz, J.; Larionova, M.; Maceira, M.A.; Borrego, S.F.; Echevarría, E. Control of biodeterioration using a fraction isolated from leaves of Ricinus communis linn. *Pharmacologyonline* **2006**, *3*, 462–466.
39. Guiamet, P.S.; de la Paz, N.J.; Arenas, P.M.; Gómez de Saravia, S.G. Differential sensitivity of Bacillus sp. isolated from archive materials to plant extracts. *Pharmacologyonline* **2008**, *3*, 649–658.
40. Afifi, H.A.M. Comparative efficacy of some plant extracts against fungal deterioration of stucco ornaments in the Mihrab of Mostafa Pasha Ribate, Cairo, Egypt. *Am. J. Biochem. Mol. Biol.* **2012**, *2*, 40–47. [CrossRef]
41. Bacci, L.; Lima, J.K.A.; Araùjo, A.P.A.; Blank, A.F.; Silva, I.M.A.; Santos, A.A.; Santos, A.C.C.; Alves, P.B.; Picanço, M.C. Toxicity, behavior impairment, and repellence of essential oils from pepper-rosmarin and patchouli to termites. *Entomol. Exp. Appl.* **2015**, *156*, 66–76. [CrossRef]

42. Stupar, M.; Grbić, M.L.J.; Džamić, A.; Unkovia, N.; Ristić, M.; Jelikić, A.; Vukojević, J. Antifungal activity of selected essential oils and biocide benzalkonium chloride against the fungi isolated from cultural heritage objects. *S. Afr. J. Bot.* **2014**, *93*, 118–124. [CrossRef]
43. Savković, Z.D.; Stupar, M.Č.; Grbić, L.M.V.; Vukojević, J.B. Comparison of anti-Aspergillus activity of Origanum vulgare L. essential oil and commercial biocide based on silver ions and hydrogen peroxide. *Acta Bot. Croat.* **2016**, *75*, 121–128. [CrossRef]
44. Bayramoglu, E. Unique biocide for the leather industry. *J. Am. Leather Chem. Assoc.* **2007**, *102*, 347–352.
45. Casiglia, S.; Bruno, M.; Scandolera, E.; Senatore, F.; Senatore, F. Influence of harvesting time on composition of essential oil of Thymus capitatus (L.) Hoffmanns. & Link. growing wild in northern Sicily and its activity on microorganisms affecting historical art crafts. *Arab. J. Chem.* **2019**, *12*, 2704–2712. [CrossRef]
46. Soković, M.; Vukojević, J.; Marin, P.; Brkić, D.; Vajs, V.; Griensven, L.J.L.D. Chemical composition of essential oils of Thymus and Mentha species and their antifungal activities. *Molecules* **2009**, *14*, 238–249. [CrossRef]
47. Walentowska, J.; Foksowicz-Flaczyk, J. Thyme essential oil for antimicrobial protection of natural textiles. *Int. Biodeter. Biodegr.* **2013**, *84*, 407–411. [CrossRef]
48. Othman, M.; Saada, H.; Matsuda, Y. Antifungal activity of some plant extracts and essential oils against fungi-infested organic archaeological artefacts. *Archaemometry* **2020**, *62*, 187–199. [CrossRef]
49. Radwan, I.A.; Abed, A.H.; Abeer, M.R.; Ibrahim, R.A.; Abdallah, A.S. Effect of thyme, clove and cinnamon essential oils on Candida albicans and moulds isolated from different sources. *Am. J. Anim. Vet. Sci.* **2014**, *9*, 303–314. [CrossRef]
50. Stević, T.; Berić, T.; Šavikin, K.; Soković, M.; Gođevac, D.; Dimkić, I.; Stanković, S. Antifungal activity of selected essential oils against fungi isolated from medicinal plan. *Ind. Crop. Prod.* **2014**, *55*, 116–122. [CrossRef]
51. Širvaitytė, J.; Šiugždaitė, J.; Valeika, V. Application of commercial essential oils of Eucalyptus and Lavender as natural preservative for leather tanning industry. *Rev. Chim.* **2011**, *62*, 884–893.
52. Al-Bayati, F.A. Synergistic antibacterial activity between Thymus vulgaris and Pimpinella anisum essential oils and methanol extracts. *J. Ethnopharmacol.* **2007**, *116*, 403–406. [CrossRef]
53. EL-Hefny, M.; Abo Elgat, W.A.; Al-Huqail, A.A.; Ali, H.M. Essential and Recovery Oils from Matricaria chamomilla Flowers as Environmentally Friendly Fungicides Against Four Fungi Isolated from Cultural Heritage Objects. *Processes* **2019**, *7*, 809. [CrossRef]
54. Mahilrajan, S.; Nandakumar, J.; Kailayalingam, R.; Manoharan, N.A.; SriVijeindran, S. Screening the antifungal activity of essential oils against decay fungi from palmyrah leaf handicrafts. *Biol. Res.* **2014**, *47*, 35. [CrossRef]
55. Pepa, T.D.; Elshafie, H.S.; Capasso, R.; De Feo, V.; Camele, I.; Nazzaro, F.; Scognamiglio, M.R.; Caputo, L. Antimicrobial and Phytotoxic Activity of Origanum heracleoticum and O. majorana Essential Oils Growing in Cilento (Southern Italy). *Molecules* **2019**, *24*, 2576. [CrossRef]
56. Palla, F.; Bruno, M.; Mercurio, F.; Tantillo, A.; Rotolo, V. Essential Oils as Natural Biocides in Conservation of Cultural Heritage. *Molecules* **2020**, *25*, 730. [CrossRef] [PubMed]
57. Chirila, C.; Berechet, M.D.; Deselnicu, V. Thyme essential oil as natural leather preservative against fungi. *Int. Conf. Adv. Mater. Syst.* **2016**, 132268064. [CrossRef]
58. Deselnicu, V.; Maier, S.S.; Albu, L.; Buruntea, N. Antimicrobial and Antifungal Leathers for Increasing the Health and the Comfort of the Individuals. In Proceedings of the CORTEP 2007, Iasi, Romania, 18–21 October 2007.
59. Berechet, M.D.; Chirila, C.; Deselnicu, V. Antifungal Activity of Thyme Essential Oil on Woolen Sheepskins. In Proceedings of the 6th International Conference on Advanced Materials and Systems—ICAMS 2016, Bucharest, Romania, 20–23 October 2016; pp. 203–208. [CrossRef]
60. Niculescu, O.; Leca, M.; Moldovan, Z.; Deselnicu, D.C. Obtaining and characterizing of a product with antifungal properties based on essential oils and natural waxes for finishing natural leathers. *Rev. Chim.* **2015**, *66*, 1733–1736.
61. Niculescu, O.; Gaidau, C.; Simon, D.; Berechet, M.D. The Study on the Possibility of Using Ecological Materials with Antifungal Properties for Treating Natural Leathers. *Rev. Chim.* **2020**, *71*, 445–449. [CrossRef]
62. Marcu, F.; Ilieș, D.C.; Wendt, A.J.; Indrie, L.; Ilieș, A.; Burta, L.; Caciora, T.; Herman, G.V.; Todoran, A.; Baias, S.; et al. Investigations regarding the biodegradation of cultural heritage. Case study of traditional embroidered peasant shirt (Maramures, Romania). *Rom. Biotechnol. Lett.* **2020**, *25*, 1362–1368. [CrossRef]
63. Ilies, A.; Hurley, P.D.; Ilies, D.C.; Baias, S. Tourist animation—A chance adding value to traditional heritage: Case study in the Land of Maramures (Romania). *Rev. Etnogr. Si Folc.* **2017**, *1–2*, 131–151.
64. Ilies, A.; Grama, V. The external western Balkan border of the European Union and its borderland: Premises for building functional transborder territorial systems. *Ann. Anal. Za Istrske Mediter. Studije-Ser. Hist. Et Sociol.* **2010**, *20*, 457–468.
65. Wei, Q.F.; Wang, X.Q.; Mather, R.R.; Fotheringham, A.F. New Approaches to Characterisation of Textile Materials Using Environmental Scanning Electron Microscope. *Fibres Text. East. Eur.* **2004**, *12*, 79–83.
66. Ilieș, D.C.; Indrie, L.; Ilieș, A.; Marcu, F.; Axinte, A.; Burtă, L.; Herman, G.V.; Atasoy, E.; Baidog, A.; Iovan, C.; et al. Investigations of aged textiles using scanning electron microscopy. *J. Environ. Biol.* **2020**, *41*, 499–504. [CrossRef]
67. Dezsi, S.; Rusu, R.; Ilies, M.; Ilies, G.; Badarau, A.S.; Rosian, G. The role of rural tourism in the social and economic revitalisation of Lapus land (Maramures County, Romania). *Proc. Pap. Geoconference Ecol. Econ. Educ. Legis.* **2014**, *2*, 783–790.
68. Deac, L.A.; Gozner, M.; Sambou, A. Ethnographic museums in the rural areas of Crișana region, Romania—keepers of local heritage, tradition and lifestyle. *GeoJournal Tour. Geosites* **2019**, *27*, 1251–1260. [CrossRef]

69. Wendt, J.A.; Ilieş, A.; Wiskulski, T.; Ilieş, M. Wooden orthodox churches in the Maramureş Region—opportunities and threats of tourism. *Ekonom. Prob. Tur.* **2018**, *43*, 133–140. [CrossRef]
70. Indrie, L.; Zlatev, Z.; Ilies, D.C.; Sturza, A.; Dochia, M.; Gozner, M.; Herman, G.; Caciora, T. Implementation of image processing techniques as a tool for form analysis of Romanian folk elements. *Ind. Text.* **2020**, *71*, 492–498. [CrossRef]
71. Adam, K.; Sivropoulu, A.; Kokkini, S.; Lanaras, T.; Arsenakis, M. Antifungal activities of Origanum vulgare subsp. hirtum, Mentha spicata, Lavandula angustifolia and Salvia fruticosa essential oils against human pathogenic fungi. *J. Agric. Food Chem.* **1998**, *46*, 1739–1745. [CrossRef]
72. Daferera, D.J.; Ziogas, B.N.; Polissiou, M.G. The effectiveness of plant essential oils on the growth of Botrytis cinerea, Fusarium sp. and Clavibacter michiganensis subsp. michiganensis. *Crop Prot.* **2003**, *22*, 39–44. [CrossRef]
73. Behmanesh, F.; Pasha, H.; Sefidgar, A.A.; Taghizadeh, M.; Moghadamnia, A.A.; Adib Rad, H.; Shirkhani, L. Antifungal Effect of Lavender Essential Oil (Lavandula angustifolia) and Clotrimazole on Candida albicans: An in Vitro Study. *Scientifica* **2015**, 1–5. [CrossRef] [PubMed]
74. Ghoorchibeigi, M.; Larijani, K.; Aberoomand Azar, P.; Zare, K.; Mehregan, I. Chemical composition and radical scavenging activity of Citrus limon peel essential oil Orient. *J. Chem.* **2017**, *33*, 458–461. [CrossRef]
75. Griffin, S.G.; Markham, J.L.; Leach, D.N. An agar dilution method for the determination of the minimum inhibitory concentration of essential oils. *J. Essent. Oil Res.* **1999**, *12*, 249–255. [CrossRef]
76. Lee, T.K.; Nguyen, T.T.H.; Le, T.T.T. Antifungal activity of the tea tree essential oils (Melaleuca alternofolia) against pathogenic fungi. *Int. J. Adv. Res.* **2019**, *7*, 1239–1248. [CrossRef]
77. Rushton, R.T.; Davis, N.W.; Page, J.C.; Durkin, C.A. The effect of tea tree oil extract on the growth of fungi. *Low. Extrem.* **1997**, *4*, 113–116.
78. Nenoff, P.; Haustein, U.F.; Brandt, W. Antifungal activity of the essential oil of Melaleuca alternifolia (tea tree oil) against pathogenic fungi in vitro. *Ski. Pharmacol.* **1996**, *9*, 388–394. [CrossRef]
79. Carson, C.F.; Riley, T.V. Antimicrobial activity of the major components of the essential oil of Melaleuca alternifolia. *J. Appl. Bacteriol.* **1995**, *78*, 264–269. [CrossRef]
80. Procop, G.W.; Church, D.L.; Hall, G.S.; Janda, W.M.; Koneman, E.W.; Schreckenberger, P.C.; Woods, G.L. *Koneman's Color Atlas and Textbook of Diagnostic Microbiology*, 7th ed.; Wolters Kluwer: Alpen, The Netherlands, 2016.
81. Buiuc, D.; Negut, M. *Tratat de Microbiologie Clinică*, 3rd ed.; Editura Medicală: Bucuresti, Romania, 2009.
82. Sil, A.; Pramanik, K.; Samantaray, P.; Firoz, M.; Yadav, V. Essential oils: A boon towards eco-friendly management of phytopathogenic fungi. *J. Entomol. Zool. Stud.* **2020**, *8*, 1884–1891.
83. Bielak, E.; Marcinkowska, E.; Sygula-Cholewinska, J. The Durability of Antimicrobial Effect of Leathers Finished with Oregano Oil. *J. Am. Leather Chem. Assoc.* **2017**, *112*, 377–386.
84. Özcan, M.M.; Starovic, M.; Aleksic, G.; Figueredo, G.; Juhaimi, F.A.; Chalchat, J.-C. Chemical Composition and Antifungal Activity of Lavender (Lavandula stoechas) Oil. *Nat. Prod. Commun.* **2018**, *13*, 895–898. [CrossRef]
85. Smigielski, K.; Prusinowska, R.; Stobiecka, A.; Kunicka-Styczyñska, A.; Gruska, R. Biological Properties and Chemical Composition of Essential Oils from Flowers and Aerial Parts of Lavender (Lavandula angustifolia). *J. Essent. Oil Bear. Plants* **2018**, *21*, 1303–1314. [CrossRef]
86. Bazaz, R.; Denning, D.W. Aspergillosis: Causes, types and treatment. *Pharm. J.* **2019**, *303*, 7927. [CrossRef]
87. Paul, D.; Paul, K. Aspergillosis: An overview. *IJPSR* **2018**, *9*, 5032–5049. [CrossRef]
88. Pop, O.L.; Judea Pusta, C.T.J.; Buhas, C.L.; Judea, A.S.; Huniadi, A.; Jurca, C.; Sandor, M.; Negrutiu, B.M.; Buhas, B.A.; Nikin, Z.; et al. Anaplastic Lymphoma Kinase (ALK) Overexpression in Lung Cancer Biopsies—An 18 month study in north western Romania. *Rev. Chim.* **2019**, *70*, 2690–2693. [CrossRef]
89. Girmenia, C.; Pizzarelli, G.; Cristini, F.; Barchiesi, F.; Spreghini, E.; Scalise, G.; Martino, P. Candida guilliermondii Fungemia in Patients with Hematologic Malignancies. *J. Clin. Microbiol.* **2006**, *44*, 2458–2464. [CrossRef] [PubMed]
90. Katotomichelakis, M.; Nikolaidis, C.; Makris, M.; Proimos, E.; Constantinidis, T.C.; Papadakis, C.E.; Danielides, V. Alternaria and Cladosporium calendar of Western Thrace: Relationship with allergic rhinitis symptoms. *Laryngoscope* **2016**, *126*, E51–E56. [CrossRef]
91. Sandoval-Denis, M.; Sutton, D.A.; Martin-Vicente, A.; Cano-Lira, J.F.; Wiederhold, N.; Guarro, J.; Genéa, J. Cladosporium Species Recovered from Clinical Samples in the United States. *J. Clin. Microbiol.* **2015**, *53*, 2990–3000. [CrossRef]
92. Huniadi, A.; Sorian, A.; Maghiar, A.; Mocuta, D.; Antal, L.; Pop, O.L.; Judea Pusta, C.T.; Buhas, C.L.; Pascalau, A.; Sandor, M. 6 (2, 3-Dichlorodiphenyl)-1, 2, 4-Triazine-3, 5-Diamine Use in Pregnancy and Body Stalk Anomaly—A Possible Association? *Rev. Chim.* **2019**, *7*, 2656–2659. [CrossRef]
93. Visagie, C.M.; Hirooka, Y.; Tanney, J.B.; Whitfield, E.; Mwange, K.; Meijer, M.; Amend, A.S.; Seifert, K.A.; Samson, R.A. Aspergillus, Penicillium and Talaromyces isolated from house dust samples collected around the world. *Stud. Mycol.* **2014**, *78*, 63–139. [CrossRef] [PubMed]
94. Hu, Y.; Zhang, J.; Li, X.; Yang, Y.; Zhang, Y.; Ma, J.; Xi, L. Penicillium marneffei Infection: An Emerging Disease in Mainland China. *Mycopathologia* **2013**, *175*, 57–67. [CrossRef] [PubMed]

applied sciences

MDPI

Article

Identification of Fungal Community Associated with Deterioration of Optical Observation Instruments of Museums in Northern Vietnam

Cao Cuong Ngo [1,2,†], Quang Huy Nguyen [3,†], Thu Hoai Nguyen [2], Ngoc Tung Quach [1,4], Pravin Dudhagara [5], Thi Hanh Nguyen Vu [1,4], Thi Thanh Xuan Le [1], Thi Thu Hang Le [3], Thi Thu Hong Do [2], Van Duc Nguyen [6], Nam Trung Nguyen [1,4] and Quyet-Tien Phi [1,4,*]

1 Institute of Biotechnology, Vietnam Academy of Science and Technology, Hanoi 100000, Vietnam; cuongnc@vrtc.org.vn (C.C.N.); qn.tung@ibt.ac.vn (N.T.Q.); hagiangyeu@yahoo.com (T.H.N.V.); xuankhanhan@gmail.com (T.T.X.L.); nam@ibt.ac.vn (N.T.N.)

2 Vietnam-Russia Tropical Centre, Hanoi 100000, Vietnam; thuhoaicnsh@gmail.com (T.H.N.); hongdt1009@gmail.com (T.T.H.D.)

3 LMI DRISA, Department of Life Sciences, University of Science and Technology of Hanoi, Vietnam Academy of Science and Technology, Hanoi 100000, Vietnam; nguyen-quang.huy@usth.edu.vn (Q.H.N.); le-thi-thu.hang@usth.edu.vn (T.T.H.L.)

4 Graduate University of Science and Technology, Vietnam Academy of Science and Technology, Hanoi 100000, Vietnam

5 Department of Biosciences (UGC-SAP-II & DST-FIST-I), Veer Narmad South Gujarat University, Surat 395007, India; pravindudhagara@vnsgu.ac.in

6 General Department of Technology, Ministry of Defense, Hanoi 100000, Vietnam; Ductckt1963@gmail.com

* Correspondence: tienpq@ibt.ac.vn or qtien.phi@gmail.com

† These authors have contributed equally to this work and share the first authorship.

Citation: Ngo, C.C.; Nguyen, Q.H.; Nguyen, T.H.; Quach, N.T.; Dudhagara, P.; Vu, T.H.N.; Le, T.T.X.; Le, T.T.H.; Do, T.T.H.; Nguyen, V.D.; et al. Identification of Fungal Community Associated with Deterioration of Optical Observation Instruments of Museums in Northern Vietnam. *Appl. Sci.* **2021**, *11*, 5351. https://doi.org/10.3390/app11125351

Academic Editor: Gyungsoon Park

Received: 5 May 2021
Accepted: 5 June 2021
Published: 9 June 2021

Abstract: Fungi are the most harmful microorganisms responsible for the deterioration of nonmetallic materials such as glass, polymers, and composites. To date, biological aspects of glass deterioration have been poorly investigated. The present study aimed to evaluate the diversity of the fungal community colonizing eyepieces of binoculars collected from museums of the northern provinces of Vietnam and the biodeterioration effects on accurate glass reproductions. A total of 40 isolates belonging to 14 genera were identified based on internal transcribed spacer (ITS) sequencing, morphological features, and maximum likelihood analysis. The most abundant fungal genera included *Aspergillus* (43.8%) and *Penicillium* (31.3%). Among those detected, *Byssochlamys, Curvularia, Phomopsis, Coprinellus, Perenniporia, Talaromyces, Pithomyces, Neopestalotiopsis, Trichoderma, Pleospora,* and *Humicola* were found for the first time. Of the 40 strains tested, 8 strains showed great organic acid production, and the extent of mycelium covered from 33.6 to 46.24%. Specifically, the highest extracellular polymeric substance production was observed in *Byssochlamys spectabilis* BXMA1-2 (14.96 g/L), *Aspergillus niger* BXMA5-2 (12.17 g/L), and *Aspergillus ochraceopetaliformis* BMLC1-2 (9.89 g/L). Glass biodeterioration experiments revealed that the light transmission through the fungal-treated glasses was decreased by 30–42.2% as compared to the nontreated glass. In addition, the main alterations resulted from hyphal fingerprints and spots, leading to apparent damage and biocorrosion.

Keywords: glass corrosion; biodeterioration; optical instruments; fungal community; *Aspergillus*; *Penicillium*; light transmission

1. Introduction

For many years, the deterioration and corrosion of glass due to physicochemical processes have been documented [1]. Moreover, the role of microorganisms in glass deterioration has been investigated in recent decades [2–5]. Many studies have proved that along with the physicochemical attack, the deterioration process is strongly stimulated by microbial contamination. Microorganisms can grow on glass materials and cause damage in situ, such as crack formation, pitting, etching, chipping, leaching, and discoloration [6]. Consequently, the glass will lose its transparency and quality as well as light transmission. Many studies have been conducted mainly focusing on the deterioration of stained-glass samples at historic churches in Germany [7], Spain [8,9], and Brazil [10], but little is known about the biodeterioration of modern glass materials of optical instruments.

Among microorganisms, fungi are very widely dispersed and grow in almost every environment on the planet. Fungi are the most common biological invaders responsible for the biodegradation and biocorrosion of glass materials [3,6,10]. Fungi are well-known producers of organic acids and therefore contribute to the biodegradation processes [4,11]. The presence of a tiny amount of substrates from water vapor in the air is enough to initiate the life of fungi on the surface of the glass. The inorganic composition and physical properties of the glass support fungal growth, and fungi can acquire the elements needed for growth from the glass itself [12].

Moreover, fungi can form spores highly resistant to adverse conditions and easily travel into the air, binding to all materials [13]. The biodegradation of glass caused by fungi often occurs slowly but continuously [6]. *Aspergillus, Cladosporium, Trichoderma, Penicillium, Chaetomium, Aureobasidium, Eurotium, Phoma, Scopulariopsis*, and *Rhizopus* are commonly reported on the surface of glass materials [2,6,10]. Experimental evidence showed that significant chemical and morphological changes in the surface layer were observed in a short period after inoculation with fungi [3,4].

The tropical climate, characterized by temperatures above 25 °C and humidity of 80–100%, imposes challenging conditions for preserving materials. Such environmental conditions are perfect for the growth of fungi and, therefore, facilitate the colonization of inorganic materials, including glass surfaces [2,7]. Especially in the rainy season, many of the nutrients come continuously from water vapor in the air, promoting biodeterioration and corrosion activities of fungi [10]. Fungus-mediated alteration of the glass is a slow and multistep process starting with the deposition of the organic material on the glass surface [14]. In suitable environmental conditions, the glass surface facilitates the growth of airborne fungal species. Colonization of fungi initiates biofilm formation, and slowly the mineralization process occurs on the glass surface, leading to glass decay. In many African, South-East Asian, and Latin American regions, high temperature and relative humidity provide perfect growth conditions for fungi attacking glass materials, especially optical instruments [15]. However, the risk and severity of damage to instruments vary widely within these regions. Moreover, a recent study showed that the biodeterioration of optical glass is induced by lubricant used in the operation of optical instruments [16]. So, timely cleaning of the glass surface is essential for the durability and service life of glass and optical instruments.

In this context, the present study aimed to evaluate the co-occurrence and diversity of culturable fungi isolated from eyepieces of binoculars in the northern provinces of Vietnam. Another objective of this work was to assess the biodeterioration of optical glass reproductions by isolated fungi under laboratory conditions. Understanding the fungal community growing on eyepieces of binoculars plays an important role in developing modern glasses resistant to harmful fungi. In addition, this finding will also provide crucial information for effective conservation solutions and prolong the use of optical devices in studied areas.

2. Materials and Methods

2.1. Materials

The binoculars (model 6nu5 8 × 30 M) used frequently for nature and bird watching were collected between June and September 2019 from a total of 3 three different museums located in northern Vietnam, including Muong Cultural Space Museum, Hanoi city; Museum of Biology, Phu Tho province; Thu Museum, Vinh Phuc province. In each museum, five binoculars used for around 4 years were collected. During 4 years of use, the exterior of binoculars was cleaned with ethanol solution every month. After that, five eyepieces contaminated with fungi were removed from five binoculars, placed in sterile plastic bags, and directly transported to the laboratory for fungal isolation.

2.2. Methods

2.2.1. Microscopical Investigations

The surface of fungus-contaminated eyepieces was primarily observed using an OPTIKA stereo microscope at a magnification of 60×. The figures were taken by using a digital camera via OPTIKA Vision Pro software. ImageJ v.1.51 software was used to analyze the figures to assess the level of fungal mycelium covering the glass surface. The colonization of fungi on the surface of eyepieces was then examined under a JEOL 5410 scanning electron microscope (SEM) (Japan). After that, the surface of eyepieces was cleaned twice with ethanol 70% to remove biomass and fungal byproducts (biofilms, biogenic crystals) entirely, and the biodeterioration patterns on the glass surface were analyzed.

2.2.2. Fungal Isolation

For each sample, the surface of eyepieces was swabbed using a sterile tube previously wetted in sterilized water and submerged into 1 mL sterile water containing 0.05% Tween 80, shaking at 200 rpm/min for 30 min. About 100 μL of the supernatant was then spread on a Czapek–Dox agar medium (30 g sucrose, 7.5 g peptone, 1 g K_2HPO_4, 0.5 g $MgSO_4{\cdot}7H_2O$, 0.5 g KCl, 0.01 g $FeSO_4$, pH 7.0) and incubated at 30 °C for 4 to 5 days. Fungal colonies were streaked to obtain pure strains for the analysis of morphological features. Fungal spore and spore chain were observed under a light microscope.

2.2.3. Identification and Analysis of Fungal Community

Fungal strains were incubated on the Czapek–Dox agar medium at 30 °C for 4 to 5 days, and then colonies were characterized by the morphological method. Subsequently, the pure cultures were used for genomic DNA extraction using Fungi/Yeast DNA Extraction Kit (Norgen, Canada). Molecular identification of fungal strains was carried out based on the analysis of internal transcribed spacer (ITS) region sequences. Specifically, the ITS regions were amplified using primer pairs ITS1F (5′-CTT GGT CAT TTA GAG GAA GTA A-3′) and ITS4 (5′-TCC TCC GCT TAT TGA TAT GC-3′) with PCR conditions described previously [17]. According to the manufacturer's protocol, the PCR amplicons were purified using the QIAQuick PCR Purification Kit (Qiagen, Germany). Finally, the purified products were sequenced using ABI PRISM 3100-Avant Genetic Analyzer (Applied Biosystems, USA). The obtained sequences were analyzed by using BioEdit v7.2.5, and a similarity search was performed using the BLASTn tool on GenBank (http://www.ncbi.nlm.nih.gov/) for fungal identification. Sequences were aligned using the ClustalX software v. 1.81 [18]. The clustering tree was constructed by maximum likelihood based on the analysis of the ITS sequences using MEGA7. The ITS sequences of fungal strains were deposited to GenBank.

2.2.4. Screening of Fungi for Significant Growth and pH Reduction

All fungal strains were incubated on MT1 medium (2.5 g glucose, 0.75 g $(NH_4)_2SO_4$, 1.0 g $MgSO_4{\cdot}7H_2O$, 1.0 g NaCl, 0.1 g $CaCl_2{\cdot}2H_2O$, 0.5 mL trace element (1.3 g $CuSO_4{\cdot}5H_2O$, 6.9 g $FeSO_4{\cdot}7H_2O$, 3.5 g $MnCl_2{\cdot}4H_2O$, 7.2 g $ZnSO_4{\cdot}7H_2O$, 0.5 g $NiCl_2{\cdot}6H_2O$), pH 7.2) and MT2 medium (2.0 g glucose, 2.0 g $NaNO_3$, 1.0 g K_2HPO_4, 0.5 g $MgSO_4$, 0.5 g KCl, 0.1 g $FeSO_4$, 0.1 g $CaCO_3$, pH 6.5) to evaluate their growth and pH reduction [14,19].

2.2.5. Glass Biodeterioration Experiments

The glass biodeterioration experiment was performed according to the guideline of the ISO 9022-11:2015 document (https://www.iso.org/standard/67535.html, accessed on April 2015). Briefly, fungal strains were separately incubated on potato dextrose agar (PDA) medium at 30 °C for two weeks to enable their growth and spore formation. Fungal spores were harvested to prepare a suspension of approximately 10^6 spores/mL in mineral salts medium (0.7 g KH_2PO_4, 0.7 g K_2HPO_4, 1 g NH_4NO_3, 0.7 g $MgSO_4{\cdot}7H_2O$, 0.005 g NaCl,

0.002 g $FeSO_4 \cdot 7H_2O$, 0.002 g $ZnSO_4 \cdot 7H_2O$, 0.001 g $MnSO_4 \cdot 7H_2O$, pH 7.0) supplemented with 0.05% Tween 80. Then, the spore suspension was spread on the previously sterilized (180 °C for 1 h) surface of the glass (10 cm × 20 cm) and incubated at 30 °C, and relative humidity was adjusted to 90% to stimulate the germination of spores into fungal mycelia. The glass samples used as the negative control were prepared in the same way, but no spores were inoculated. After 28 days of incubation, the coverage of fungal mycelia on the glass surface was evaluated in comparison with the untreated glasses using the ImageJ v.1.51 software [9,20].

In addition, the light transmittance through glass samples was analyzed by a spectrophotometer UV-2550 at wavelengths from 400 to 800 nm, both before and after the cleaning procedure. The light absorbance values at a given wavelength were determined.

Finally, the growth and colonization of fungi on glass surfaces during the biodeterioration experiment were also observed under the SEM JEOL 5410 (Japan). The alteration of the glass surface was evaluated right after the cleaning procedure to assess the corrosion patterns.

2.2.6. Assessment of Exopolysaccharide (EPS) Production

Fungal plugs (1 cm × 1 cm) grown on Czapek–Dox agar medium for 72 h at 28 °C were transferred into a 250 mL flask containing 50 mL of Czapek–Dox broth. The cultures were then incubated on a rotary shaker at 150 rpm for two days at 28 °C. Then, 4 mL of prepared fungal inoculant was transferred into 96 mL of modified minimal Czapek–Dox medium containing 5 g/L of glucose as a sole carbon source for 7 days [21]. The EPS production in the broth culture of each fungus was measured as described by Jaroszuk-Ściseł et al. (2020) [22]. Briefly, crude EPS was collected by addition of 96% ethanol/supernatant (1:1 v/v) and then left for 24 h at 4 °C. The precipitate was harvested by centrifugation at 10,000 rpm for 15 min and freeze-dried immediately to yield a white powder. The powder was weighed, and the concentration in grams per liter of cultivation broth was calculated.

2.2.7. Statistical Analysis

All the data are expressed as mean ± standard deviation (SD) of three replicates. Values with different letters within a column are significantly different according to Fisher's least significant difference (LCD) test ($p < 0.05$).

3. Results

3.1. Visual Examination and Observation of Biodeterioration

The growth of fungi over the inner surface of binocular eyepieces and the extent of glass surface covered by hyphae can be observed in Figure 1a,b. As viewed under the light microscope, the coverage of fungal hyphae was extensive on the surface of 15 glass samples, with the extent of coverage ranging from 21 to 48% (Figure 1c,d), which corresponded to harmful grades 2 and 3 based on the ISO 9022-11 criteria (Table 1). The growth of fungi caused dense colonization on the surface of all glass samples (Figure 1e). After the cleaning procedure, imprints of fungal hyphae, including spots and stains, fingerprints, and etching, that covered the entire surface were detected by using SEM (Figure 1f). In addition, crack formation was not observed on any samples, indicating that the fungi did not bind strongly to the glass surface.

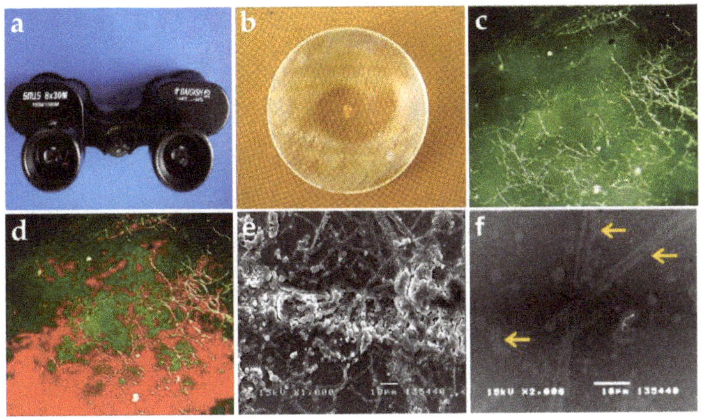

Figure 1. (**a**) The binocular model. (**b**) Binocular eyepieces contaminated with fungi. (**c**) Image captured using light microscopy at a magnification of 60×. (**d**) The estimation of hyphae covering the glass surface treated by the ImageJ tool. (**e**) SEM image exhibiting the fungal growth and colonization on the eyepieces (image at a magnification of 1000×, scale bar = 10 µm). (**f**) Yellow arrows indicate fingerprinting/etching of fungal hyphae on the surface of eyepieces after the cleaning procedure (image at a magnification of 2000×, scale bar = 10 µm).

Table 1. The extent of fungal hyphae on eyepiece samples.

Place	Glass Sample	Evaluation of the Extent of Fungal Growth *	
		Hyphal Surface Coverage (%)	Harmful Grade
Muong Cultural Space Museum, Hanoi	BXMA1	29 ± 1.2	2
	BXMA2	35 ± 1.5	3
	BXMA3	26 ± 1.3	2
	BXMA4	21 ± 1.2	2
	BXMA5	21 ± 1.5	2
Museum of Biology, Phu Tho	BTBB1	28 ± 1.5	2
	BTBB2	41 ± 1.7	3
	BTBB3	27 ± 1.6	2
	BTBB4	43 ± 2.1	3
	BTBB5	48 ± 1.4	3
Thu Museum, Vinh Phuc	BMLC1	41 ± 1.6	3
	BMLC2	36 ± 1.8	3
	BMLC3	34 ± 1.6	3
	BMLC4	25 ± 1.3	2
	BMLC5	43 ± 2.2	3

* ISO 9022-11 interpretation for the coverage of a glass surface with fungal hyphae: >0 to 10% (Grade 1)—restricted fungal growth; >10 to 30% (Grade 2)—intermittent spread fungal colonies (visible with the naked eye); >30 to 70% (Grade 3)—a substantial amount of fungal growth (easily visible); >70% (Grade 4)—massive fungal growth.

3.2. Distribution and Identification of the Isolated Fungi

A total of 186 fungal colonies were recovered from Muong Cultural Space Museum (n = 49), Museum of Biology (n = 73), and Thu Museum (n = 64) samples. Analysis of morphological and spore features of fungal isolates primarily classified them into 40 distinct clusters (Figure 2). Specifically, the most remarkable diversity and highest abundance of fungal species were found in Thu Museum samples (n = 18), followed by the Muong Cultural Space Museum samples (n = 16) and Museum of Biology samples (n = 6). Notably, at least three different fungal species were found in each glass sample, indicating co-contamination. Thus, the disinfection of fungal-contaminated optical equipment would face a significant challenge.

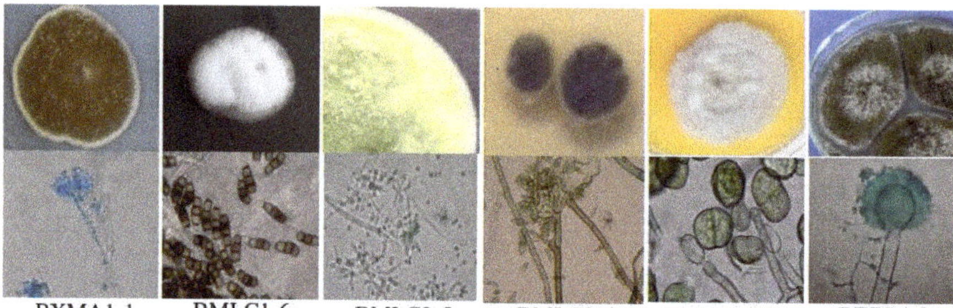

Figure 2. The colony and morphological characteristics of some fungi isolated from binocular eyepieces.

The molecular identification based on the ITS sequencing revealed high similarities (between 99 and 100%) to known fungal species belonging to 14 genera, including *Penicillium, Byssochlamys, Talaromyces, Cladosporium, Phomopsis, Neopestalotiopsis, Trichoderma, Coprinellus, Perenniporia, Curvularia, Pithomyces, Pseudopithomyces,* and *Aspergillus* (Figure S1). *Aspergillus* (43.8%) and *Penicillium* (31.3%) were the two predominant fungal genera, followed by species of the genus *Curvularia* (4%). The comparative sequence analyses combined with the clustering tree analysis further allowed us to assign 40 different fungal species (Figure S1). Fungal strains distributed throughout the tree in multiple single branches demonstrate high genetic diversity in both local and regional scales.

3.3. Evaluation of Growth, pH Changes, and Biodeterioration

To mimic the natural environment and evaluate the pH reduction of fungal isolates, two glucose mineral media were used in this study. All fungal strains secreted organic acids on two mineral salt media in the present study, leading to a significant pH reduction compared to the original pH value for almost all cases (Table 2). In the MT1 medium, a drastic decline in pH was caused by most fungi. Strong growth and pH reduction were observed for *Aspergillus fumigatus* BMLC1-1, *Aspergillus niger* BXMA5-2, *Cladosporium tenuissimum* BMLC2-7, *Penicillium decumbens* BTBB5-2, and *Perenniporia* cft *tephropora* BXMA4-4, suggesting that they might take advantage of other fungal species to develop on glass surfaces.

According to ISO 9022-11, the glass corrosion of each fungal strain was evaluated in the present study. It was seen that fungal species were growing on the surface of all glass samples, albeit to varying degrees. The extent of mycelium covered from 2.26 to 46.24%, compared to the noninoculated glass samples (Table 2). The most vigorous growth was observed for *Byssochlamys spectabilis* BXMA1-2 (46.2%), followed by *Curvularia* cft *veruculosa* BXMA2-2 (41.0%), *Penicillium ramusculum* BXMA4-2 (36.2%), *Aspergillus niger* BXMA5-2 (35.7%), *Penicillium oxalicum* BTBB2-2 (35.6%), *Humicola* cft *insolens* BMLC5-5 (35.3%), *Aspergillus ochraceopetaliformis* BMLC1-2 (33.6%), and *Talaromyces* cft *trachyspermus* BTBB1-3 (33.0%), which are highlighted in bold. These fungal species seem to have a good affinity for adhesion on the glass surface by developing mycelium networks in a relatively short time (28 days). Additionally, these strains also exhibited a significant reduction in pH in the two glucose mineral media tested. Based on the growth, pH reduction, and hyphal surface coverage, 8 fungal strains, including *B. spectabilis* BXMA1-2, *C.* cft *veruculosa* BXMA2-2, *P. ramusculum* BXMA4-2, *A. niger* BXMA5-2, *P. oxalicum* BTBB2-2, *H.* cft *insolens* BMLC5-5, *A. ochraceopetaliformis* BMLC1-2, *T.* cft *trachyspermus* BTBB1-3, were selected for further study (Table 2).

Table 2. Identification, the ability of growth and pH values observed in culture media, and percentage of hyphal surface coverage of isolated fungi.

Fungal Strain	GenBank Accession Number	The Changes in the pH of Media after Fungal Growth				
		MT1 (Initial pH 7.2)		MT2 (Initial pH 6.5)		Hyphal Surface Coverage (%)
		pH	Growth	pH	Growth	
Aspergillus fumigatus BMLC1-1	MW911781	2.53 ± 0.4	+++	3.17 ± 0.1	+++++	4.64 ± 1.4
***Aspergillus ochraceopetaliformis* BMLC1-2**	**MN394129**	**3.15 ± 0.1**	**++++**	**5.26 ± 0.6**	**+++**	**33.61 ± 3.7**
Aspergillus asperescens BMLC1-3	MZ292395	4.28 ± 0.3	++++	5.68 ± 0.1	++++	2.26 ± 1.1
Aspergillus sclerotiorum BMLC1-4	MN394130	4.34 ± 0.1	++++	5.66 ± 0.3	++++	3.87 ± 1.3
Penicillium lanoso BMLC1-5	MZ292396	2.78 ± 0.3	++++	6.26 ± 0.2	++++	15.28 ± 2.4
Pithomyces chartarum BMLC1-6	MN394131	3.18 ± 0.4	++++	6.46 ± 0.8	++++	32.33 ± 3.5
Neopestalotiopsis sp. BMLC1-7	MN394132	5.43 ± 0.5	++++	4.87 ± 0.6	+++	23.14 ± 2.8
Penicillium chermesinum BMLC2-1	MN394134	2.91 ± 0.1	++++	8.93 ± 0.5	+++++	20.28 ± 2.2
Penicillium roqueforti BMLC2-3	MZ292397	3.47 ± 0.2	+++	7.21 ± 0.9	++++	19.21 ± 2.4
Trichoderma koningiopsis BMLC2-5	MN394133	4.15 ± 0.4	++++	4.27 ± 0.6	+++	18.90 ± 2.3
Cladosporium tenuissimum BMLC2-6	MN394135	5.34 ± 0.6	+++	8.24 ± 0.7	++	36.43 ± 3.4
Cladosporium tenuissimum BMLC2-7	MN394136	2.22 ± 0.3	+++	2.68 ± 0.7	++++	27.65 ± 2.6
Penicillium toxicarium BMLC2-8	MN394137	4.19 ± 0.2	+++	6.91 ± 0.5	+++	11.18 ± 2.3
Aspergillus niger BMLC3-4	MW911782	2.56 ± 0.8	+++	6.23 ± 0.4	++++	21.53 ± 2.2
Pithomyces maydicus BMLC3-6	MN394138	5.95 ± 0.4	++++	5.35 ± 0.5	+++	12.15 ± 2.1
Byssochlamys cft *spectabilis* BMLC4-6	MN394139	4.28 ± 0.8	++++	5.68 ± 0.5	++++	14.99 ± 2.5
Pleospora herbarum BMLC4-7	MZ292398	6.21 ± 0.7	++++	6.43 ± 0.8	++++	10.91 ± 2.3
***Humicola* cft *insolens* BMLC5-5**	**MZ292399**	**3.35 ± 0.7**	**+++**	**5.35 ± 0.7**	**+++**	**35.26 ± 3.8**
Aspergillus sydowii BTBB1-1	MN396671	5.15 ± 0.6	++	5.53 ± 0.3	++	10.40 ± 2.7
Aspergillus flavus BTBB1-2	MN396672	5.21 ± 0.3	+++	5.85 ± 0.4	+++	4.25 ± 1.5
***Talaromyces* cft *trachyspermus* BTBB1-3**	**MW911783**	**3.78 ± 0.6**	**++++**	**5.79 ± 0.8**	**+++**	**36.95 ± 3.4**
***Penicillium oxalicum* BTBB2-2**	**MN396673**	**2.24 ± 0.1**	**++++**	**7.02 ± 0.9**	**+++++**	**35.95 ± 3.2**
Aspergillus cft *salwaensis* BTBB5-1	MN396674	5.47 ± 0.6	+++	6.85 ± 0.4	+++	34.07 ± 3.8
Penicillium decumbens BTBB5-2	MN396675	2.17 ± 0.5	++++	3.39 ± 0.9	+++++	21.58 ± 2.5
Penicillium brevissimum BXMA1-1	MH634479	4.9 ± 0.9	+	6.77 ± 0.3	+	24.18 ± 2.5
***Byssochlamys spectabilis* BXMA1-2**	**MH634480**	**2.68 ± 0.1**	**+++**	**6.23 ± 0.7**	**++++**	**46.24 ± 3.3**
Aspergillus sydowii BXMA1-3	MH634481	3.37 ± 0.2	+++	7.2 ± 0.2	++++	3.79 ± 1.2
Aspergillus tritici BXMA1-4	MH634482	3.18 ± 0.5	+++	7.11 ± 0.6	+++	20.63 ± 2.6
Aspergillus sydowii BXMA1-5	MH634483	2.6 ± 0.4	+++	7.47 ± 0.6	+++++	13.97 ± 2.9
Byssochlamys spectabilis BXMA2-1	MH634484	2.7 ± 0.1	++++	6.05 ± 0.2	++++	19.41 ± 2.2
***Curvularia* cft *veruculosa* BXMA2-2**	**MH634485**	**3.2 ± 0.6**	**++++**	**5.32 ± 0.3**	**++++**	**41.04 ± 3.1**
Phomopsis cft *tuberivora* BXMA2-3	MH634486	3.12 ± 0.4	++++	8.96 ± 0.2	+++++	7.86 ± 1.5
Coprinellus radians BXMA2-4	MH634487	3.63 ± 0.2	+++	8.32 ± 0.6	+++++	7.74 ± 1.6
Aspergillus flavus BXMA3-1	MH634488	3.12 ± 0.7	++	7.98 ± 0.6	++++	31.82 ± 3.6
Penicillium oxalicum BXMA3-2	MH634489	3.77 ± 0.2	++	7.4 ± 0.2	++++	20.22 ± 2.4
Curvularia lunata BXMA3-5	MH634490	2.91 ± 0.2	++++	6.32 ± 0.4	++++	27.30 ± 2.3
Penicillium brevissimum BXMA4-1	MH634491	2.68 ± 0.1	+++	7.28 ± 0.3	++++	31.533.6
***Penicillium ramusculum* BXMA4-2**	**MH634492**	**3.26 ± 0.7**	**+++**	**5.15 ± 0.9**	**++++**	**36.20 ± 3.2**
Perenniporia cft *tephropora* BXMA4-4	MH634493	2.87 ± 0.5	++++	4.5 ± 0.4	++	11.51 ± 2.8
***Aspergillus niger* BXMA5-2**	**MH634494**	**2.65 ± 0.9**	**+++**	**4.24 ± 0.7**	**+++++**	**35.66 ± 3.8**

Fair growth +; good growth ++; luxurious growth +++; heavy growth ++++; extreme growth +++++.

3.4. EPS Production by Fungal Strains

All fungi were selected and evaluated for their EPS production. Among these (Figure 3), the highest EPS production was reported for *Byssochlamys spectabilis* BXMA1-2. The yield of crude EPS and mycelial growth from the modified Czapek–Dox medium were 14.96 and 4.4 g/L, respectively. *Aspergillus niger* BXMA5-2 (12.17 g/L) and *Aspergillus ochraceopetaliformis* BMLC1-2 (9.89 g/L) were also found to produce high levels of EPS. *Talaromyces trachyspermus*, *Curvularia veruculosa*, *Talaromyces trachyspermus*, and *Humicola insolens* are for the first time reported as EPS producers in this present study. During cultivation, the cells formed pellets with high hairiness, leading to maximum biomass and EPS production.

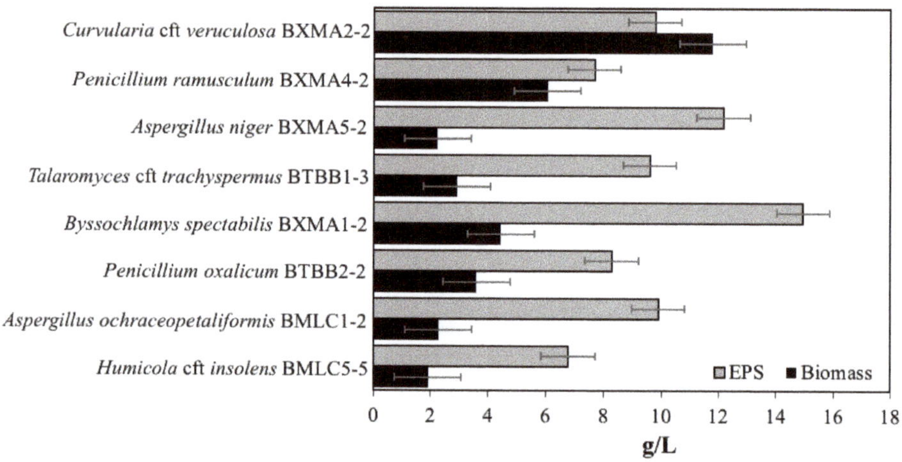

Figure 3. Exopolysaccharide (EPS) production by eight selected fungal isolates causing deterioration on a glass model.

3.5. Biodeterioration of the Glass Reproductions

The biodeteriogenic potential of the selected fungal strains was assessed by measuring light transmission through the glass in the visible light spectrum (wavelengths from 400 to 800 nm). After 28 days of incubation, the light transmission value through the fungal-inoculated glass samples significantly decreased by 30–42.2% compared to the noninoculated ones (Figure 4). The glass treated with *Byssochlamys spectabilis* BXMA1-2 exhibited the highest reduction in light transmission, followed by *Curvularia* cft *veruculosa* BXMA2-2, *Aspergillus niger* BXMA5-2, and *Penicillium ramusculum* BXMA4-2.

The glass surfaces showed some little fingerprints after 28 days, attributed to the presence of fungi (Figure 5a,b). This causes corrosion of the glass and loss of quality of the equipment. After the glass surface cleaning procedure, the ratio of light transmission through fungus-treated glasses was reduced by 0.5 to 17.8% compared with nontreated glasses (Table S1). After six months of incubation, the fungal strain produced a well-developed mycelia biofilm on the glass surfaces (Figure 5c). The phenomenon was observed on all fungus-treated glasses. The cleaning procedure showed precise etching and hyphal fingerprints (Figure 5d).

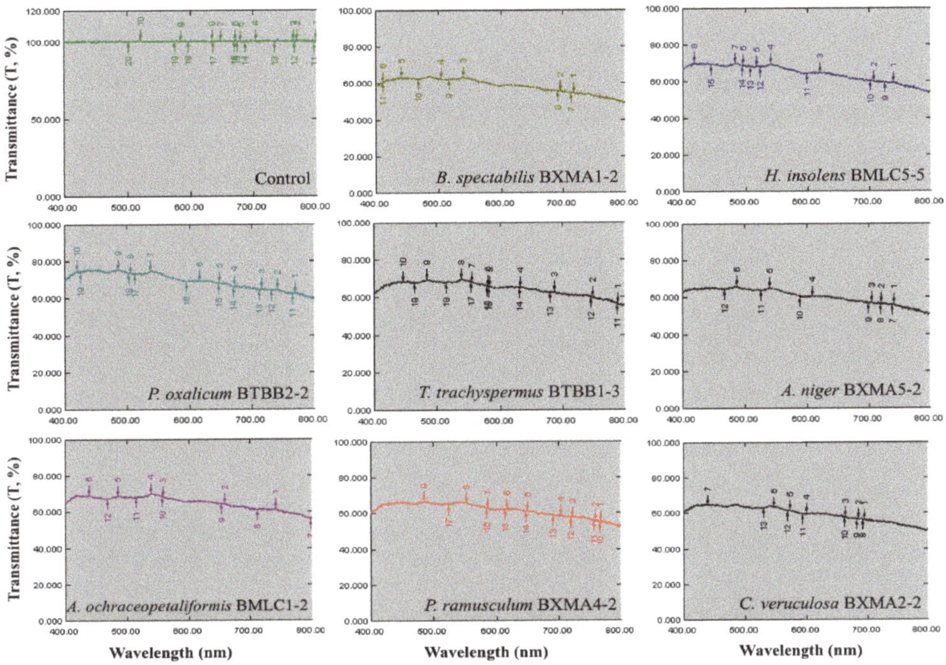

Figure 4. The effect of fungal growth on light transmission through fungus-inoculated glass samples in the visible light spectrum.

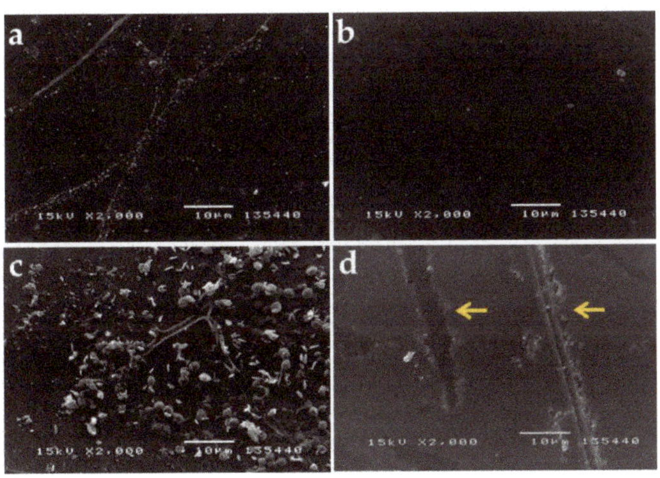

Figure 5. SEM micrographs of glass after 28 days of the test: (**a**) fungal growth on glass after 28 days; (**b**) sample photo taken after 28 days of the alcohol-purge test; (**c**) glass sample after six months of testing showing fungal spores; (**d**) glass sample after six months of testing showing the glass corrosion caused by fungi.

4. Discussion

The fungi could acquire the elements needed for growth from the deteriorating glass without an external supply of mineral nutrients [11]. Fungi have a great ability to metabolize a wide range of carbon sources and grow on every inorganic material easily [23]. Thus, the coinfection of different microorganisms and favorable environmental conditions play a significant role and could affect the biodeterioration and biocorrosion outcomes on the optical instruments.

The main component of glass is SiO_2 and other metal oxides, i.e., Na_2O, K_2O, CaO, Al_2O_3, B_2O, BaO, ZnO, As_2O_3, and La_2O_3. Elements such as Mg, Zn, Co, and Cr are also added to change the glass properties, including the magnification coefficient, mechanical strength, refractive index, color, and opacity transparency [24,25]. The glass surface of binoculars is also coated by antireflective layers, including the outer layer consisting of low-refractive-index dielectric materials that are environmentally resistant, such as MgF_2, Na_3AlF_6, CaF_2, and LiF_2. The inner layer is usually a dielectric with a high refractive index and good adhesion to the glass surface, including ZnO, TiO_2, CeF_3, and ThO_2 [26]. Moreover, a lipid layer is coated around the edges of eyepieces to prevent dirt particles from spreading and sticking to the inner glass surface and the optical system. This lipid layer consists of long-chain hydrocarbons, fatty acids, and some alkaline metals such as Li, Na, K, Ca, Ba, Al, Zn, and Pb, together with dust and other organic contaminants, providing the primary growth medium to initiate microbial growth. Such nutrients are sufficient for airborne fungal spores to germinate and develop into hyphae. Once the hyphal growth occurs on the glass surface, it quickly corrodes the inner surface of eyepieces.

Aspergillus, *Cladosporium*, and *Penicillium* are the most prevalent fungi genera commonly found in indoor and outdoor environments, especially glass materials [4,27,28]. Rodrigues et al. identified six fungal genera, among which *Penicillium* and *Cladosporium* were the two predominant fungal genera isolated from stained-glass windows in Portugal [4]. In addition, *Cladosporium* and *Phoma* were also reported from glass samples collected at two Catalonian churches [9]. *Aspergillus*, *Penicillium*, and *Alternaria* were found to contribute to the formation of pits in optical glass [16]. In the present study, *Aspergillus* and *Penicillium* were the two most dominant fungal genera, accounting for 75% of the total identified fungal strains. They are known as primary contaminants that harm historical and cultural heritages and cause glass corrosion [2,9,29]. Because of the ability to produce numerous spores easily dispersed in the air, *Aspergillus* and *Penicillium* are primary corrosive agents [30]. Earlier regional studies and reports on the frequent occurrence of these fungi in all areas of the northern provinces of Vietnam suggest the high load of these fungal spores. Therefore, chances of risk are increased for fungal contamination of optical equipment if no preventive measures are taken.

Interestingly, fungal genera Byssochlamys, Curvularia, Phomopsis, Coprinellus, Perenniporia, Talaromyces, Pithomyces, Neopestalotiopsis, Pleospora, and Humicola were identified for the first time as harmful agents on optical instruments. Byssochlamys, Curvularia, Phomopsis, Coprinellus, and Perenniporia were only found in Muong Cultural Space Museum, while Cladosporium, Humicola, Pithomyces, Neopestalotiopsis, Pleospora, and Trichoderma were only found in Thu Museum. In contrast, Talaromyces was only found in the Museum of Biology. The fungal community involved in the biodeterioration process of binocular eyepieces was more diverse than that of historical church window glasses [4].

The identified fungal species might secrete various organic acids, extracellular enzymes, pigments, and EPSs to colonize on glass materials; consequently, biofilms will be developed on the glass surface, accelerating biodegradation and biocorrosion processes [31–33]. Fungi are well known for their potential to overproduce and accumulate various organic acids [34,35]. Intrinsic abilities provide the fungi a competitive advantage over other organisms and the ability to proliferate on inorganic materials such as metal and glass [36]. As for the biodeterioration of glass, organic acids produced by fungi dissolve metal oxides and minerals on the glass surface, thus releasing nutrient ions for growth [19]. Organic acid production recorded here was in agreement with previous studies showing that *Aspergillus*, *Penicillium*, and *Cladosporium* are organic acid producers commonly found as the primary contaminants causing biocorrosion of glass materials [1,2,14]. As documented, the most common acids produced by fungi include formic, citric, acetic, oxalic, gluconic, and itaconic acids [34]. The magnitude of corrosion ranges from 0.27 to 0.03 mA cm^{-2} for 1% organic acids [37] and causes significant etching of glasses.

Fungi readily digest the organic materials such as oils from fingerprints and lens coatings and produce strong hydrofluoric acid as a waste product, which decreases the glass surface pH to lower than 3.0, causing permanent etching of the glass [38]. Thus, glass damage is even more potent in vivo because of the condensation and accumulation of a mixture of the above-mentioned organic acids.

Furthermore, organic acids move through the silica network and continue to dissolve metal oxides, leading to carbonate and sulfate salt production due to the absorption of CO_2 and SO_2 from the atmosphere [39]. Krumbein et al. demonstrated the formation of carbonate, sulfate, and silicate crystals due to fungal acids excreted during the biocorrosion process [11]. Altogether, these agents, in turn, etch and destroy the antireflective layers and the surface of glass permanently, resulting in a decreased degree of transparency and refractive index and a loss of image brightness and sharpness. Moreover, the acidic environment also favors fungal growth and, together with the hot, moist atmosphere conditions, causes physical–chemical corrosion, which speeds up glass deterioration.

In this study, the high level of EPSs recorded for *Aspergillus, Byssochlamys,* and *Penicillium* species was consistent with many previous studies [40–42]. The EPS composition varies depending on producer species. *Aspergillus* and *Penicillium* species produce EPSs composed of glucose, mannose, galactose, galactosamine, and acetate [43–45]. Due to excess sugars in EPSs, organic acid production occurs on glass surfaces, leading to glass corrosion. EPS production by the fungi is usually intracellular and secreted outside the cell to form biofilms, leading to permanent colonization of the fungi on the surface of materials [42]. However, the glass biodeterioration ability might differ between fungal species, and the identification of harmful fungi is crucial to protect glass material. Light scattering and absorption by the fungal mycelia on the glass surface could account for the rapid loss of light transmission and image quality. The etching of the glass by excretion of organic acids and other metabolic products results in grooves formed in the glass surface in contact with the hyphae. Furthermore, the fungi remove the glass surface layer and part of the glass matrix, leading to alteration of optical properties [2,20].

The high fungal biomass, colonization, EPS formation, and biofilm development are responsible for the etching and hyphal fingerprints on the glass. Furthermore, EPSs contain polysaccharides, uronic acids, and some enzymes, contributing to the corrosion process [46]. It is important to note that all fungal strains under investigation produce EPSs, contributing to chemical deterioration by acid and chelate production. Therefore, all these results demonstrate that all fungal strains could biodeteriorate glasses and that the degree of damage depended on fungal characteristics. The present study will provide baseline information about the significant fungal species responsible for the biodeterioration of optical instruments. It will be helpful in the development of fungus-resistant optical glass. A more detailed investigation is needed to determine the proper method of protection and preservation of such museum specimens.

5. Conclusions

Fungal biodeterioration of optical equipment represents a serious concern leading to notable research efforts over the last decades. In this study, we found that *Aspergillus* sp., *Penicillium* sp., and *Cladosporium* sp. were major fungi colonizing the binocular eyepieces at museums in northern Vietnam. Moreover, this is the first report that identifies *Byssochlamys spectabilis, Curvularia veruculosa, Humicola insolens,* and *Talaromyces lagena* as robust glass-deteriorating fungi. The growth of fungi resulted in dense colonization and biofilm formation on the glass surface, corresponding to the deterioration and corrosion of binocular eyepieces. Organic acid and EPS production were the main factors contributing to the leaching of elements from glass surfaces and severe physical, chemical, and aesthetical modifications. Thus, effective procedures for protecting optical observation equipment can be proposed based on an understanding of the fungal community. The proper cleaning of the optical lens is highly recommended to control fungal contamination as per the manufacturer's guidelines. However, for prolonged storage in museums, the optical devices must be kept under a controlled condition with unfavorable temperature and humidity for fungi. Finally, to stop the biodeterioration, an understanding of microbial communities is required to optimize biocides against fungi that reduce the biodeterioration of optical glasses.

Supplementary Materials: The following are available online at https://www.mdpi.com/article/10.3390/app11125351/s1, Figure S1: Clustering tree from the maximum likelihood analysis of all fungi isolated from binocular eyepieces using ITS sequences. *Cunninghamella elegans* CBS 160.28^{T} was selected as the outgroup taxon and 100% bootstrap values indicate the relevant internode. Identified strains are indicated in bold, Table S1: Effect of mycelium growing on glass plates after 28 days of alcohol-purge test as per ISO 9022-11 on light transmission at wavelengths of visible light range from 400 to 800 nm.

Author Contributions: Conceptualization, C.C.N., Q.H.N., and Q.-T.P.; methodology, T.H.N.V., T.T.X.L., T.T.H.D. and T.T.H.L.; investigation, V.D.N. and T.H.N.; writing—original draft preparation, C.C.N. and Q.H.N.; writing—review and editing, P.D., N.T.Q., and Q.-T.P.; visualization, N.T.N.; project administration, C.C.N. and T.H.N. All authors have read and agreed to the published version of the manuscript.

Funding: This research was funded by the Vietnam Academy of Science and Technology, grant number NVCC 08.08/20-20, and Vietnam-Russia Tropical Centre, grant number TCKT2020.

Institutional Review Board Statement: Not applicable.

Informed Consent Statement: Not applicable.

Data Availability Statement: Data sharing is not applicable to this article.

Acknowledgments: The authors would like to acknowledge the support of the VAST Culture Collection of Microorganisms, Institute of Biotechnology, Vietnam Academy of Science and Technology (www.vccm.vast.vn).

Conflicts of Interest: The authors declare no conflict of interest.

References

1. Schreiner, M. Glass of the past: The degradation and deterioration of medieval glass artifacts. *Microchim. Acta* **1991**, *104*, 255–264. [CrossRef]
2. Drewello, R.; Weissmann, R. Microbially influenced corrosion of glass. *Appl. Microbiol. Biotechnol.* **1997**, *47*, 337–346. [CrossRef]
3. Gorbushina, A.A.; Palinska, K.A. Biodeteriorative processes on glass: Experimental proof of the role of fungi and cyanobacteria. *Aerobiologia* **1999**, *15*, 183–192. [CrossRef]
4. Rodrigues, A.; Gutierrez-Patricio, S.; Miller, A.Z.; Saiz-Jimenez, C.; Wiley, R.; Nunes, D.; Vilarigues, M.; Macedo, M.F. Fungal biodeterioration of stained-glass windows. *Int. Biodeter. Biodegr.* **2014**, *90*, 152–160. [CrossRef]
5. Sterflinger, K.; Piñar, G. Microbial deterioration of cultural heritage and works of art-tilting at windmills? *Appl. Microbiol. Biotechnol.* **2013**, *97*, 9637–9646. [CrossRef]
6. Pinna, D. *Coping with Biological Growth on Stone Heritage Objects: Methods, Products, Applications, and Perspectives*; Apple Academic Press: Palm Bay, FL, USA, 2017.
7. Müller, E.; Drewello, U.; Drewello, R.; Weißmann, R.; Wuertz, S. In situ analysis of biofilms on historic window glass using confocal laser scanning microscopy. *J. Cult. Herit.* **2001**, *2*, 31–42. [CrossRef]
8. Carmona, N.; Laiz, L.; Gonzalez, J.M.; Garcia-Heras, M.; Villegas, M.A.; Saiz-Jimenez, C. Biodeterioration of historic stained glasses from the Cartuja de Miraflores (Spain). *Int. Biodeter. Biodegr.* **2006**, *58*, 155–161. [CrossRef]
9. Piñar, G.; Garcia-Valles, M.; Gimeno-Torrente, D.; Fernandez-Turiel, J.L.; Ettenauer, J.; Sterflinger, K. Microscopic, chemical, and molecular-biological investigation of the decayed medieval stained window glasses of two Catalonian churches. *Int. Biodeter. Biodegr.* **2013**, *84*, 388–400. [CrossRef] [PubMed]
10. Corrêa Pinto, A.M.; Palomar, T.; Alves, L.C.; da Silva, S.H.M.; Monteiro, R.C.; Macedo, M.F.; Vilarigues, M.G. Fungal biodeterioration of stained-glass windows in monuments from Belém do Pará (Brazil). *Int. Biodeter. Biodegr.* **2019**, *138*, 106–113. [CrossRef]
11. Krumbein, W.E.; Urzì, C.E.; Gehrmann, C. Biocorrosion and biodeterioration of antique and medieval glass. *Geomicrobiol. J.* **1991**, *9*, 139–160. [CrossRef]
12. Marvasi, M.; Vedovato, E.; Balsamo, C.; Macherelli, A.; Dei, L.; Mastromei, G.; Perito, B. Bacterial community analysis on the Mediaeval stained glass window "Natività" in the Florence Cathedral. *J. Cult. Herit.* **2009**, *10*, 124–133. [CrossRef]
13. Bindschedler, S.; Cailleau, G.; Verrecchia, E. Role of fungi in the biomineralization of calcite. *Minerals* **2016**, *6*, 41. [CrossRef]

14. Weaver, J.L.; DePriest, P.T.; Plymale, A.E.; Pearce, C.I.; Arey, B.; Koestler, R.J. Microbial interactions with silicate glasses. *NPJ Mater. Degrad.* **2021**, *5*, 11. [CrossRef]
15. Watkins, R.D. Mould in optical instruments. *Community Eye Health* **2003**, *16*, 28.
16. Bartosik, M.; Zakowska, Z.; Cedzińska, K.; Rozniakowski, K. Biodeterioration of optical glass induced by lubricants used in optical instruments technology. *Pol. J. Microbiol.* **2010**, *59*, 295–300. [CrossRef] [PubMed]
17. Ihrmark, K.; Bödeker, I.T.M.; Cruz-Martinez, K.; Friberg, H.; Kubartova, A.; Schenck, J.; Strid, Y.; Stenlid, J.; Brandström-Durling, M.; Clemmensen, K.E.; et al. New primers to amplify the fungal ITS2 region—evaluation by 454-sequencing of artificial and natural communities. *FEMS Microbiol. Ecol.* **2012**, *82*, 666–677. [CrossRef] [PubMed]
18. Thompson, J.D.; Gibson, T.J.; Higgins, D.G. Multiple sequence alignment using ClustalW and ClustalX. *Curr. Protoc. Bioinform.* **2003**, 2.3.1–2.3.22. [CrossRef]
19. Liaud, N.; Giniés, C.; Navarro, D.; Fabre, N.; Crapart, S.; Gimbert, I.H.; Levasseur, A.; Raouche, S.; Sigoillot, J.-C. Exploring fungal biodiversity: Organic acid production by 66 strains of filamentous fungi. *Fungal Biol. Biotechnol.* **2014**, *1*, 1. [CrossRef]
20. Garcia-Vallès, M.; Gimeno-Torrente, D.; Martínez-Manent, S.; Fernández-Turiel, J.L. Medieval stained glass in a Mediterranean climate: Typology, weathering and glass decay, and associated biomineralization processes and products. *Am. Mineral.* **2003**, *88*, 1996–2006. [CrossRef]
21. Vasil'chenko, L.G.; Khromonygina, V.V.; Karapetyan, K.N.; Vasilenko, O.V.; Rabinovich, M.L. Cellobiose dehydrogenase formation by filamentous fungus *Chaetomium* sp. INBI 2-26(-). *J. Biotechnol.* **2005**, *119*, 44–59. [CrossRef]
22. Jaroszuk-Ściseł, J.; Nowak, A.; Komaniecka, I.; Choma, A.; Jarosz-Wilkołazka, A.; Osińska-Jaroszuk, M.; Tyśkiewicz, R.; Wiater, A.; Rogalski, J. Differences in production, composition, and antioxidant activities of exopolymeric substances (EPS) obtained from cultures of endophytic *Fusarium culmorum* strains with different effects on cereals. *Molecules* **2020**, *25*, 616. [CrossRef]
23. Davis, S.; Chemello, C. Glass: Conservation and Preservation. In *Encyclopedia of Global Archaeology*; Smith, C., Ed.; Springer: New York, NY, USA, 2014; pp. 3047–3050.
24. Stábile, F.M.; Volzone, C.; Ortiga, J. Thermal evolution of Na_2O-K_2O-CaO-SiO_2-P_2O_5-Al_2O_3 glass system, and possible applications as biomedical devices. *Procedia Manuf. Sci.* **2015**, *8*, 332–337. [CrossRef]
25. Partyka, J.; Gasek, K.; Pasiut, K.; Gajek, M. Effect of addition of BaO on sintering of glass–ceramic materials from SiO_2–Al_2O_2–Na_2O–K_2O–CaO/MgO system. *J. Therm. Anal. Calorim.* **2016**, *125*, 1095–1103. [CrossRef]
26. Song, H.S.; Yoo, Y.J.; Lee, G.J.; Chang, K.S.; Song, Y.M. Optical design of porous ZnO/TiO_2 films for highly transparent glasses with broadband ultraviolet protection. *J. Nanomater.* **2017**, *2017*, 2738015. [CrossRef]
27. Johansson, W.; Peralta, A.; Jonson, B.; Anand, S.; Österlund, L.; Karlsson, S. Transparent TiO_2 and ZnO thin films on glass for UV protection of PV modules. *Front. Mater. Sci.* **2019**, *6*, 259. [CrossRef]
28. Frisvad, J.; Gravesen, S. *Penicillium* and *Aspergillus* from Danish homes and working places with indoor air problems: Identification and mycotoxin determination. *Health Implic. Fungi Indoor Environ.* **1994**, *2*, 281–290.
29. Guiamet, P.; Borrego, S.; Lavin, P.; Perdomo, I.; de Saravia, S.G. Biofouling and biodeterioration in materials stored at the Historical Archive of the Museum of La Plata, Argentine and at the National Archive of the Republic of Cuba. *Colloids Surf. B* **2011**, *85*, 229–234. [CrossRef]
30. Fog Nielsen, K. Mycotoxin production by indoor molds. *Fungal Genet. Biol.* **2003**, *39*, 103–117. [CrossRef]
31. Vadillo-Rodríguez, V.; Logan, B.E. Localized attraction correlates with bacterial adhesion to glass and metal oxide substrata. *Environ. Sci. Technol.* **2006**, *40*, 2983–2988. [CrossRef]
32. Borrego, S.; Guiamet, P.; Gómez de Saravia, S.; Batistini, P.; Garcia, M.; Lavin, P.; Perdomo, I. The quality of air at archives and the biodeterioration of photographs. *Int. Biodeter. Biodegrad.* **2010**, *64*, 139–145. [CrossRef]
33. Trovão, J.; Portugal, A. Current knowledge on the fungal degradation abilities profiled through biodeteriorative plate essays. *Appl. Sci.* **2021**, *11*, 4196. [CrossRef]
34. Magnuson, J.K.; Lasure, L.L. Organic acid production by filamentous fungi. In *Advances in Fungal Biotechnology for Industry, Agriculture, and Medicine*; Tkacz, J.S., Lange, L., Eds.; Springer US: Boston, MA, USA, 2004; pp. 307–340.
35. Goldberg, I.; Rokem, J.S. Organic and fatty acid production, microbial. In *Encyclopedia of Microbiology*, 4th ed.; Schmidt, T.M., Ed.; Academic Press: Cambridge, MA, USA, 2019; pp. 358–382. [CrossRef]
36. Naranjo-Ortiz, M.A.; Gabaldón, T. Fungal evolution: Major ecological adaptations and evolutionary transitions. *Biol. Rev.* **2019**, *94*, 1443–1476. [CrossRef]
37. Little, B.; Staehle, R.; Davis, R. Fungal influenced corrosion of post-tensioned cables. *Int. Biodeter. Biodegr.* **2001**, *47*, 71–77. [CrossRef]
38. Cordero, I. Fungus: How to prevent growth and remove it from optical components. *Community Eye Health* **2013**, *26*, 57.
39. Sand, W.; Bock, E. Biodeterioration of mineral materials by microorganisms—biogenic sulfuric and nitric acid corrosion of concrete and natural stone. *Geomicrobiol. J.* **1991**, *9*, 129–138. [CrossRef]
40. Gomoiu, I.; Catley, B.J. Properties of a kaolin-flocculating polymer elaborated by *Byssochlamys nivea*. *Enzyme Microb. Technol.* **1996**, *19*, 45–49. [CrossRef]
41. Kogan, G.; Matulová, M.; Michalková, E. Extracellular polysaccharides of *Penicillium vermiculatum*. *Z. Naturforsch. C. J. Biosci.* **2002**, *57*, 452–458. [CrossRef] [PubMed]
42. Mahapatra, S.; Banerjee, D. Fungal exopolysaccharide: Production, composition and applications. *Microbiol. Insights* **2013**, *6*, 1–16. [CrossRef] [PubMed]

43. Ruperez, P.; Leal, J.A. Extracellular galactosaminogalactan from *Aspergillus parasiticus*. *Trans. Brit. Mycol. Soc.* **1981**, *77*, 621–625. [CrossRef]
44. Sutherland, I.W. *Biotechnology of Microbial Exopolysaccharides*; Cambridge University Press: Cambridge, UK, 1990. [CrossRef]
45. Jansson, P.-E.; Lindberg, B. Structural studies of varianose. *Carbohydr. Res.* **1980**, *82*, 97–102. [CrossRef]
46. Roca, C.; Alves, V.D.; Freitas, F.; Reis, M.A.M. Exopolysaccharides enriched in rare sugars: Bacterial sources, production, and applications. *Front. Microbiol.* **2015**, *6*, 288. [CrossRef] [PubMed]

applied sciences

MDPI

Article

Diversity and Seasonal Dynamics of Airborne Fungi in Nerja Cave, Spain

Valme Jurado [1], Yolanda Del Rosal [2], Cristina Liñan [2,3], Tamara Martin-Pozas [4], Jose Luis Gonzalez-Pimentel [5] and Cesareo Saiz-Jimenez [1,*]

[1] Instituto de Recursos Naturales y Agrobiologia, IRNAS-CSIC, 41012 Sevilla, Spain; vjurado@irnase.csic.es
[2] Instituto de Investigacion Cueva de Nerja, 29787 Nerja, Spain; yolanda@cuevadenerja.es (Y.D.R.); cbaena@cuevadenerja.es (C.L.)
[3] Departamento de Ecologia y Geologia, Facultad de Ciencias, Universidad de Malaga, 29071 Malaga, Spain
[4] Museo Nacional de Ciencias Naturales, MNCN-CSIC, 28006 Madrid, Spain; tmpozas@mncn.csic.es
[5] Laboratorio Hercules, Universidade de Evora, 7000-809 Evora, Portugal; pimentel@irnas.csic.es
* Correspondence: saiz@irnase.csic.es

Abstract: Nerja Cave, Southern Spain, was revealed as an important biodiversity reservoir from which several novel species of *Aspergillus* were described. We carried out an aerobiological study in Nerja Cave to assess the origin of airborne fungi. This study quantified the fungi present in the air of ten representative halls covering the three sectors comprising the cave: Touristic Galleries, High Galleries, and New Galleries. Microclimatological monitoring allowed us to understand the dynamic of airborne fungi in two seasons of the year (winter and summer), corresponding to the strongest and the lowest cave ventilation, and to validate the influence that the transport of airborne fungi from outside may have on the cave itself. The data show that cold air enters in winter, as confirmed by the abundant presence of *Aspergillus* and *Penicillium* spores inside and outside the cave. In summer, the abundance of some fungi in the air of Nerja Cave, which are not detected outside, indicates a stagnation or low ventilation, and therefore, the concentration of fungal spores is maxima. The high occurrence of *Cladosporium* outside the cave and the scarce abundance inside support the cave stagnation in this season.

Keywords: aerobiology; airborne fungi; *Aspergillus*; *Penicillium*; *Parengyodontium*; entomopathogenic fungi

Citation: Jurado, V.; Del Rosal, Y.; Liñan, C.; Martin-Pozas, T.; Gonzalez-Pimentel, J.L.; Saiz-Jimenez, C. Diversity and Seasonal Dynamics of Airborne Fungi in Nerja Cave, Spain. *Appl. Sci.* **2021**, *11*, 6236. https://doi.org/10.3390/app11136236

Academic Editors: Filomena De Leo and Daniela Isola

Received: 16 June 2021
Accepted: 1 July 2021
Published: 5 July 2021

1. Introduction

One of the most important topics in the conservation of caves with Paleolithic paintings is the assessment of the microbial communities inhabiting speleothems, rocks, water pools and air. A wide diversity of approaches on cave fungi can be found in the literature [1–6] and aerobiological studies were introduced as a management tool to control airborne fungi in caves with paintings [7–10].

Microclimate control and monitoring of gases (CO_2, CH_4, ^{222}Rn, etc.) are used to investigate cave aerodynamic and ventilation or stagnation periods [11–17]. These studies, in combination with aerobiology, can provide clues for controlling airborne fungi and dispersion patterns in the caves [18,19].

The microbiology of Nerja Cave, Southern Spain, was studied regarding the impact of lighting on the development of phototrophic communities [20–24], but scarcely from an aerobiological point of view. A preliminary study carried out by Del Rosal et al. [25] was focused on the survey of pathogenic bacteria and the genus *Aspergillus*, very abundant in the cave air in summer. Further studies by Docampo et al. [26,27] using nonviable Hirst-type volumetric pollen traps located near the entrance and at the end of the visited area found that *Aspergillus/Penicillium* were the most abundant spore types, representing 50% of the total, followed by *Cladosporium*. However, the morphology of the spores, studied by microscopy, does not allow distinguishing between *Aspergillus* and *Penicillium* nor the identification of species.

The main objective of this aerobiological study was to assess the presence of airborne fungi in the cave to elucidate its origin. The identification of fungi was carried out by molecular methods. The study quantified the fungi present in the air from 10 representative halls covering the three sectors composing the cave: Touristic Galleries, High Galleries, and New Galleries. A sampling point outside the cave was used as a control of environmental fungi. This monitoring allowed to know the dynamic of airborne fungi in two seasons of the year (winter and summer), corresponding to the strongest and the lowest cave ventilation, and to validate the influence that the transport of airborne fungi from outside may have on the cave itself.

2. Materials and Methods

2.1. Site Description

Nerja Cave is one of the biggest and most important show caves in Spain, with a volume of about 300,000 m^3, almost 5 km of passages, and more than 400,000 visitors annually. It is also one of the most significant prehistoric sites in Europe, with more than 500 artistic representations. About a third of the cave, the Touristic Galleries, is open to the public since 1960. The other two sections, named High Galleries and New Galleries, only can be accessed by researchers.

The climate of the area is coastal Mediterranean, with an average precipitation slightly lower than 500 mm yr^{-1}. The external air average temperature ranges from about 13 °C to 26 °C (January and August, respectively), with an average annual value of 18.8 °C [28].

The cave is developed within fissured and karstified Triassic dolomitic marbles more than 500 m thick at the unsaturated zone of the Sierra Almijara carbonate massif. A significant portion of the dolomitic marbles is devoid of vegetation. The soil cover is scarce, with a thickness lower than 0.15 m, and the natural vegetation consists mainly of shrubs and pine trees. However, in the areas nearest to the touristic entrance, there are several gardens with many ornamental plants [29], and thus, the vegetation cover is much more developed over the Touristic Galleries than over the nonvisited sector.

The cavity develops between 123 and 191 m a.s.l. According to the outside orography, rock thickness above the cave varies from 4 to 50 m in the area open to tourism, and exceeds 80 m in the nonvisited area. The cavity has three natural entrances: an entrance with a semicircular form and about 12 m^2 of surface (E1 in Figure 1), which is used for the access and exit of visitors and two subcircular sinkholes (E2 and E3 in Figure 1).

Nerja Cave is connected to at least two other nearby cavities, Pintada Cave (PC)—a small cavity with a natural entrance located to 250 m a.s.l.—and another great and nonaccessible cave from the surface (GC), found recently with indirect methods [13]. Although there is no speleological connection between these three caves, they are effectively linked for air circulation patterns from the surface.

Four ventilation regimes (winter, summer, spring, and autumn) and two ventilation modes with opposite airflows direction (named DAF- and UAF-modes) were defined in Nerja Cave karstic system during the annual cycle [13]. In all the cases, the natural ventilation of the system is determined by variations in the differences of density between the external and internal air.

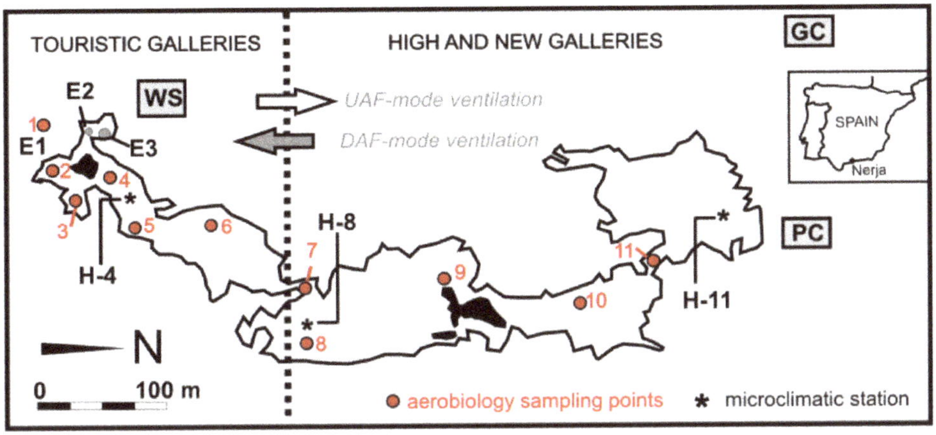

Figure 1. Cave map and location of the aerobiology sampling points and microclimatic stations. 1: Cave Exterior, 2: Entrance Hall, 3: Nativity Hall, 4: Cascade Hall, 5: Ghosts Hall, 6: Cataclysm Hall, 7: Kitchen Hall, 8: Heracles Hall, 9: Immensity Hall, 10: Lance Hall, 11: Mountain Hall. PC: Pintada Cave, GC: non accessible cave from surface, WS: weather station, E1: tourist entrance, E2 and E3: sinkholes. Location of PC, GC, and WS is approximate.

2.2. Microclimatology

Nerja Cave has a microclimatic monitoring network (Figure 1) consisting of dataloggers CR1000 (Campbell Scientific) with probes measuring air CO_2 concentration, air temperature, and relative humidity, amongst other parameters, at hourly intervals. Air temperature and relative humidity were measured with a probe HMP155A- Vaisala (measuring ranges: 0–100% and −80 to +60 °C; accuracy: ± 1.0% and ± 0.17 °C). Cave air CO_2 concentrations were measured using an infrared absorption sensor (GMM222-Vaisala; measuring range 0–5000 ppmv, accuracy ± 1.5%).

The ^{222}Rn concentration of cave air was registered every 30 min using a Radim 5WP radon monitor (GT-Analytic KG). The detectable activity is between 80 Bq/m^3 and 150 KBq/m^3 (for one-hour measurements with a statistical error ± 20%) (Table 1).

The environmental data of the external air were provided by the instrumental equipment placed in the weather station (WS in Figure 1) located a few meters from Nerja Cave entrances.

2.3. Aerobiology

Two sampling campaigns were carried out on 9 July 2016 and 12 January 2017. These two campaigns corresponded to the stronger ventilation (winter, UAF-mode ventilation) and lower ventilation (summer, DAF-mode ventilation) periods, previously described by Liñán et al. [13]. In each sampling, a total of 10 control sites were selected inside and one outside the cave. The sites sampled were: Touristic Galleries (Entrance Hall, Nativity Hall, Cascade Hall, Ghosts Hall, Cataclysm Hall), High Galleries (Kitchen Hall, Heracles Hall, Immensity Hall), and New Galleries (Lance Hall and Mountain Hall), in addition to the exterior air (Figure 1).

Samplings were performed with a Duo SAS (Surface Air System) Super 360 system (International pBI, Milan, Italy). This equipment is a type of suction impact collector that was widely used in aerobiological studies [8,10]. It allows the detection of a great diversity of cultivable fungi by filtering a preselected volume of air through two heads provided with a series of orifices, which impact upon Petri dishes.

At each sampling site, samples were taken in duplicate. The volume of filtered air was 100 L in 35 s. This volume was established because at higher volumes the number of spores was too high for an accurate count [8]. The culture medium used to promote fungal growth was dichloran rose bengal-chloramphenicol-agar (DRBCA, Merck, Darmstad, Germany).

The Petri dishes were incubated at 25 °C and after 5 days counted. After counting, the fungi were isolated, cultured and the DNA extracted.

The methodology was thoroughly described by Sanchez–Moral et al. [19] and Dominguez–Moñino et al. [30]. DNA was extracted from the mycelia of each fungal strain. Mycelia were scraped from the plates and transferred to a 1.5 mL Eppendorf tube containing 500 μl lysing buffer and 200 μl glass beads. The mixture was shaken in a cell disrupter (Fast Prep-24, Solon, OH, USA) at full speed for 3 min.

Fungal internal transcribed spacer (ITS) regions, including ITS1, 5.8S rDNA and ITS2, were amplified using the primers ITS1 (5′-TCCGTAGGTGAACCTGCGG-3′) and ITS4 (5′-TCCTCCGCTTATTGATATGC-3′). Amplification protocol comprised one cycle of denaturation (94 °C for 2 min), followed by 35 cycles of denaturation (94 °C for 1 min), and a terminal elongation step (72 °C for 5 min).

PCR products were analyzed by Macrogen Inc. (Amsterdam, The Netherlands) and sequenced using the same primer set. For phylogenetic identification of fungal strains, the sequences were compared using BLASTn algorithm with nonredundant databases of sequences deposited at the National Center for Biotechnology Information (http://www.ncbi.nlm.nih.gov/). The sequences are available from the GenBank database under accession numbers: MZ374012-MZ374055 and MZ375763-MZ375863.

3. Results

3.1. Microclimatology

Table 1 shows the main statistical parameters of the microclimatic data recorded outside (WS) and inside the cave: Cascade Hall (H-4), Heracles Hall (H-8), and Mountain Hall (H-11) for the period 2016–2017. The external air temperature ranged between 7.27 and 30.34 °C, with a mean value (*m*) of 18.93 °C. The relative humidity ranged from 29 to 94%, with a mean value of 67%.

Table 1. Microclimatic data registered in Nerja Cave, Spain, and the exterior atmosphere during 2016 and 2017.

	WS	H-4	H-8	H-11	WS	H-4	H-8	H-11
	Air Temperature (°C)				Relative Humidity (%)			
n	731	727	600	731	731	688	600	731
max	30.34	20.08	19.66	19.92	93.68	98.11	99.77	100.00
min	7.27	18.15	19.18	19.75	28.92	54.40	81.11	100.00
m	18.93	19.21	19.40	19.86	67.08	84.49	95.21	100.00
s	4.80	0.60	0.13	0.04	11.69	12.27	4.96	0.00
v (%)	25.33	3.10	0.69	0.19	17.43	14.53	5.21	0.00
range	22.88	1.93	0.48	0.16	65.00	43.71	18.66	0.00
	WS	**H-4**	**H-8**	**H-11**	**WS**	**H-4**	**H-8**	**H-11**
	Air CO_2 (ppmv)				Air ^{222}Rn (Bq/m^3)			
n	223	688	600	731	nd	1379	nd	nd
max	518	1483	1170	986	nd	622	nd	nd
min	436	489	512	499	nd	80	nd	nd
m	474	797	699	633	nd	213	nd	nd
s	19	226	158	111	nd	161	nd	nd
v (%)	4	28	23	18	nd	76	nd	nd
range	83	994	657	487	nd	542	nd	nd

Legends: exterior weather station (WS), Cascade Hall (H-4), Heracles Hall (H-8), Mountain Hall (H-11), *n*—number of measures, *max*-maximum value, *min*—minimum value, *m*—mean value, *s*—standard deviation, *v* (%)—variation coefficient, nd—not measured. Data "Air CO_2" registered in exterior correspond to year 2018, and "Air ^{222}Rn" corresponds to data registered between 2008 and 2013.

Inside the cave, the air temperature in H-4 hall ranged between 18.15 and 20.08 °C (m= 19.21 °C) and the relative humidity ranged between 54% and 98% (m = 85%). The air CO_2 content ranged from 489 to 1483 ppmv, with a mean value of 797 ppmv.

In the H-11 hall, the air temperature ranged between 19.75 and 19.92 °C (m = 19.86 °C) so the difference between the maximum and minimum temperatures was lower than 0.2 °C (Table 1). The air CO_2 content ranged from 499 to 986 ppmv, with a mean value of 633 ppmv. The air relative humidity was 100% during all the studied period.

In H-8 intermediate values of air temperature (m = 19.40 °C) and relative humidity (m = 95%) were recorded. In this case, the air CO_2 content ranged from 512 to 1170 ppmv, with a mean value of 699 ppmv. Therefore: (1) the air of the tourist sector is colder, less wet, and has a higher CO_2 content than in that of the nonvisited galleries of Nerja Cave, (2) the temperature and relative humidity variability is noticeably reduced in the nonvisited sector, located below greater thickness of rock and at greater distances from the cave entrances (E1, E2 and E3 in Figure 1), and (3) the variation coefficients (v in Table 1) in the external media are higher than in the cave, except for air CO_2.

Figure 2 shows the temporal evolution of the main environmental parameters relative to the external air and cave air during 2016 and 2017. In summer, the temperature of the external air (T_{ext}) is higher than that of the cave air (T_{int}), and the latter is denser than the exterior (Figure 2A,B). CO_2 concentrations in the Touristic Galleries (Figure 2C) reach maximum values, coinciding with a lower index of natural ventilation (daily average values of ^{222}Rn over the 400 Bq/m^3, Figure 2A) and a greater number of visitors (Figure 2B).

Air relative humidity is high in the cave, with values ranging between 95 and 100% (Figure 2D). When the external temperature markedly exceeds the cave temperature ($\Delta T > 5.0$ °C) the DAF-mode ventilation is activated (yellow orange bars in Figure 2). The cave air, denser than the external air, is blown out by the lowest entrances, and atmospheric air is sucked in by the highest entrances. During the day corresponding to the summer campaign (9 July 2016), ΔT was 7.53 °C (daily average value).

In winter, the external air is generally colder and denser than the cave air, although exceptionally short-term periods exist in which $T_{ext} > T_{int}$ (Figure 2 A,B). CO_2 concentrations inside the Touristic Galleries are minimal and much more similar to those of the external atmosphere (Figure 2C), given the more effective natural ventilation (daily average values of ^{222}Rn of 80 Bq/m^3) and lower number of visitors.

Cave air relative humidity is minimum in the Touristic Galleries and shows the same temporal evolution than the external air (Figure 2D). When $T_{ext} < T_{int}$, the external air, denser and colder than the internal air, easily enters Nerja Cave through the lowest entrances, moves through the Touristic Galleries, arrives at the High and New Galleries (the same day or 1–2 days later), and finally leaves the cave through Pintada Cave. During the day corresponding to the winter campaign (12 January 2017), the daily average value of ΔT was −5.30 °C.

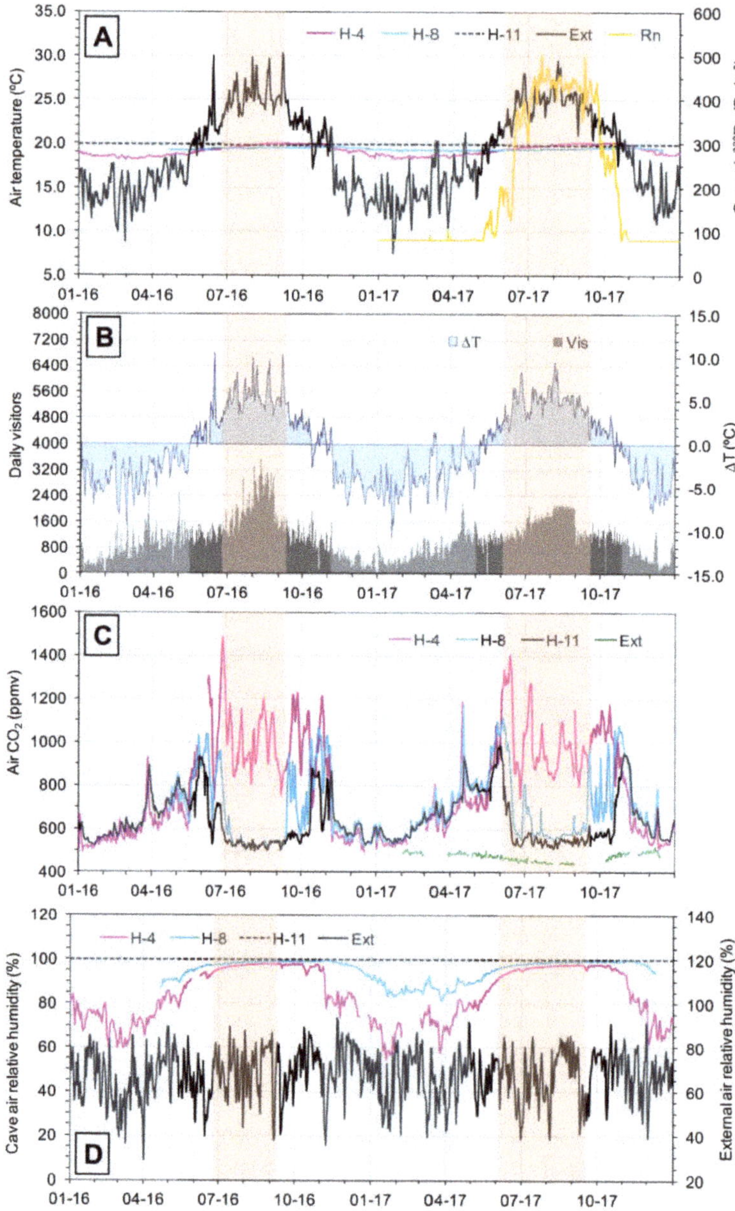

Figure 2. Daily evolution from 01/01/16 to 31/12/17 of (**A**) external and cave air temperatures in Nerja Cave versus air ^{222}Rn concentration in H-4 (average for the period 2008–2013); (**B**) daily visitors versus air temperature differences ($\Delta T = T_{ext} - T_{int}$); T_{ext}: external air temperature; T_{int}: internal air temperature (H-4); (**C**) air CO_2 content. Data of external air CO_2 correspond to year 2018 and (**D**) external and cave air relative humidity. Blue bars: winter ventilation period, UAF-mode. Yellow orange bars: summer ventilation period, DAF-mode.

3.2. Aerobiology

Two aerobiological samplings were performed to cover the aerodynamic periods in Nerja Cave (Figure 3): high ventilation, corresponding to the winter station (12 January 2017), and very low ventilation produced in the summer station (9 July 2016).

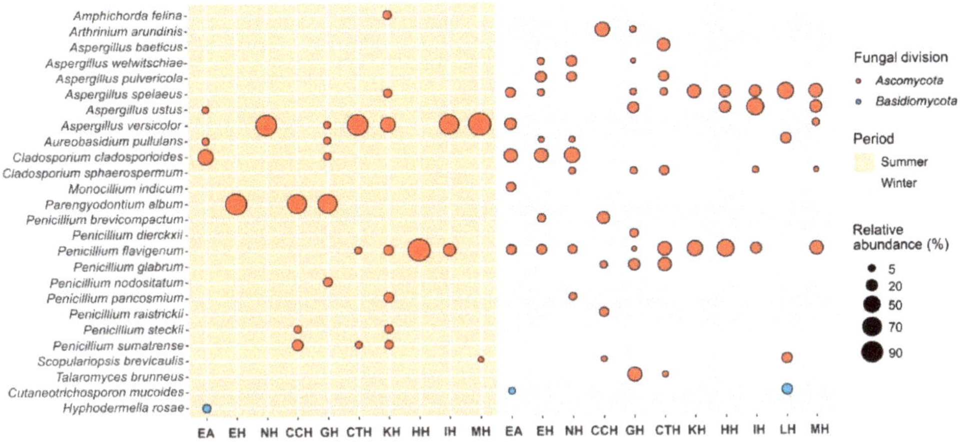

Figure 3. Bubble Plot. Most abundant fungi, as distributed in each sampling point (EA: Outside, EH: Entrance Hall, NH: Nativity Hall, CCH: Cascade Hall, GH: Ghosts Hall, CTH: Cataclysm Hall, KH: Kitchen Hall, HH: Heracles Hall, IH: Immensity Hall, LH: Lance Hall, MH: Mountain Hall) and period (summer and winter).

The concentration of fungal spores outside was relatively high in summer and near the half in winter. In general, the concentration of fungal spores in all the galleries was considerably lower in the summer period with respect to the higher ventilation period, as shown in Tables 2 and 3. Inside the cave, the fungal patterns were different depending on the galleries trail.

Two behaviors were distinguished: in the Touristic Galleries, fungal spore abundances increased between 2- and 5-times in winter, while the figures were greater (from 7- to 133-times in the High and New Galleries).

Remarkable were the low concentrations in the Lance Hall (Table 3), while in the Kitchen and Heracles halls, they were very high. Obviously these abundances should be related with definite sources and/or the aerodynamic of the halls.

3.2.1. Summer Campaign, 9 July 2016

The lower ventilation period (summer) is characterized by the abundant presence of spores of three genera, *Parengyodontium*, *Aspergillus*, and *Penicillium*. Other less abundant genera were *Cladosporium*, *Alternaria*, *Aureobasidium*, *Bipolaris*, and *Amphichorda* (Table 2). In total, 16 different species were retrieved inside the cave and 13 outside.

Parengyodontium album was only limited to the visited Touristic Galleries: Entrance, Cascade, and Ghosts Halls, where reached abundances between 70 and 90%. However, *Aspergillus versicolor*, also abundant, was distributed across the three galleries: Touristic (Nativity and Ghosts Halls), High (Kitchen and Immensity Halls), and New (Mountain Hall).

Other *Aspergillus* species was *A. spelaeus*, only identified in the Kitchen Hall. Six species of *Penicillium* were collected in the Touristic and High Galleries, from which the abundances of *P. flavigenum*, *P. sumatrense*, *P. pancosmium*, and *P. nodositatum* were above 10% in any of the halls and those of *P. citreosulfuratum* and *P. steckii* below 10%. Interestingly, *P. flavigenum* was the unique fungus in Heracles Hall and *A. versicolor* in Mountain Hall.

Outside, none of the most abundant fungi inside the cave could be identified, and only *Cladosporium cladosporioides* (43.06% outside and 5.0% inside) and *Aureobasidium pullulans* (5.56% outside and 5.0%) were represented in the Ghosts Hall, and *Alternaria tenuissima* in the Nativity Hall (6.67%).

Table 2. Total counts, identifications, and abundance of fungi isolated from Nerja Cave in 2016 summer campaign.

Sampling	CFU m^{-3} (SD) *	Identification (% Similarity)	Abundance (%)
Entrance Hall (EH)	170 (40)	*Cladosporium lebrasiae* (100%)	6.45
		Parengyodontium album (100%)	90.32
		Penicillium citreosulfuratum (100%)	3.23
Nativity Hall (NH)	80 (20)	*Alternaria tenuissima* (100%)	6.67
		Aspergillus versicolor (100%)	86.67
		Penicillium citreosulfuratum (100%)	6.67
Cascade Hall (CCH)	90 (20)	*Parengyodontium album* (100%)	70.59
		Phoma sp. (99%)	5.88
		Penicillium sumatrense (100%)	17.65
		Penicillium steckii (100%)	5.88
Ghosts Hall (GH)	100 (0)	*Aspergillus versicolor* (100%)	5.00
		Aureobasidium pullulans (100%)	5.00
		Bipolaris cynodontis (100%)	5.00
		Cladosporium cladosporioides (100%)	5.00
		Parengyodontium album (100%)	70.00
		Penicillium nodositatum (99%)	10.00
Cataclysm Hall (CTH)	110 (20)	*Aspergillus versicolor* (100%)	85.71
		Penicillium citreosulfuratum (100%)	4.76
		Penicillium flavigenum (100%)	4.76
		Penicillium sumatrense (100%)	4.76
Kitchen Hall (KH)	120 (10)	*Aspergillus spelaeus* (100%)	8.70
		Aspergillus versicolor (100%)	39.13
		Amphichorda felina (100%)	8.70
		Penicillium flavigenum (100%)	13.04
		Penicillium pancosmium (100%)	13.04
		Penicillium steckii (100%)	8.70
		Penicillium sumatrense (100%)	8.70
Heracles Hall (HH)	10 (0)	*Penicillium flavigenum* (100%)	100.00
Immensity Hall (IH)	20 (0)	*Aspergillus versicolor* (100%)	75.00
		Penicillium flavigenum (100%)	25.00
Lance Hall (LH)	0	No detected	-
Exterior air (EA)	390 (60)	*Tremateia chromolaenae* (99%)	5.56
		Victoriomyces antarcticus (100%)	2.78
		Alternaria tenuissima (100%)	1.39
		Aspergillus tubingensis (100%)	1.39
		Aspergillus melleus (100%)	5.56
		Aspergillus ustus (100%)	4.17
		Aureobasidium pullulans (100%)	5.56
		Botrytis cinerea (100%)	2.78
		Cladosporium cladosporioides (100%)	43.06
		Hyphodermella rosae (100%)	8.33
		Microascus croci (99%)	5.56
		Penicillium olsonii (99%)	5.56
		Purpureocillium lilacinum (100%)	8.33

* CFU: colony forming units; SD: standard deviation.

Table 3. Total counts, identifications, and abundance of fungi isolated from Nerja Cave in 2017 winter campaign.

Sampling	CFU m^{-3} (SD) *	Identification (% Similarity)	Abundance (%)
Entrance Hall (EH)	430 (10)	*Aspergillus welwitschiae* (99%)	5.13
		Aspergillus pulvericola (99%)	19.23
		Aspergillus spelaeus (99%)	5.13
		Aureobasidium pullulans (100%)	2.56
		Botrytis cinerea (100%)	1.28
		Cladosporium cladosporioides (100%)	33.33
		Penicillium brevicompactum (100%)	8.97
		Aspergillus europaeus (100%)	2.56
		Penicillium flavigenum (100%)	14.10
		Penicillium rubens (100%)	7.69
Nativity Hall (NH)	360 (30)	*Aspergillus welwitschiae* (99%)	16.67
		Aspergillus pulvericola (99%)	13.64
		Aspergillus ochraceus (100%)	1.52
		Aureobasidium pullulans (100%)	3.03
		Botrytis cinerea (100%)	1.52
		Cladosporium cladosporioides (100%)	42.42
		Cladosporium sphaerospermum (100%)	4.55
		Penicillium flavigenum (100%)	10.61
		Penicillium pancosmium (99%)	6.06
Cascade Hall (CCH)	190 (30)	*Arthrinium arundinis* (99%)	36.11
		Penicillium olsonii (99%)	5.56
		Penicillium sumatrense (98%)	13.89
		Penicillium glabrum (100%)	5.56
		Penicillium raistrickii (99%)	11.11
		Penicillium brevicompactum (100%)	25.00
		Scopulariopsis brevicaulis (100%)	2.78
Ghosts Hall (GH)	420 (40)	*Arthrinium arundinis* (99%)	4.00
		Aspergillus welwitschiae (99%)	1.33
		Aspergillus spelaeus (99%)	4.00
		Aspergillus ustus (100%)	18.67
		Cladosporium sphaerospermum (100%)	5.33
		Penicillium glabrum (100%)	20.00
		Penicillium dierckxii (100%)	9.33
		Penicillium flavigenum (100%)	1.33
		Aspergillus oryzae (100%)	1.33
		Talaromyces brunneus (100%)	34.67
Cataclysm Hall (CTH)	180 (60)	*Aspergillus spelaeus* (99%)	6.06
		Aspergillus pulvericola (99%)	15.15
		Cladosporium sphaerospermum (100%)	12.12
		Penicillium flavigenum (100%)	33.33
		Penicillium glabrum (100%)	30.30
		Talaromyces brunneus (100%)	3.03
Kitchen Hall (KH)	2170 (130)	*Aspergillus baeticus* (100%)	26.91
		Aspergillus spelaeus (99%)	30.18
		Penicillium flavigenum (100%)	42.91
Heracles Hall (HH)	1330 (40)	*Aspergillus spelaeus* (99%)	23.50
		Aspergillus ustus (100%)	21.50
		Penicillium flavigenum (100%)	55.00
Immensity Hall (IH)	140 (20)	*Aspergillus ustus* (100%)	55.56
		Aspergillus spelaeus (99%)	22.22
		Cladosporium sphaerospermum (100%)	3.70
		Penicillium flavigenum (100%)	18.52

Table 3. *Cont.*

Sampling	CFU m^{-3} (SD) *	Identification (% Similarity)	Abundance (%)
Lance Hall (LH)	30 (10)	*Aspergillus spelaeus* (99%) *Aureobasidium pullulans* (100%) *Cutaneotrichosporon mucoides* (100%) *Scopulariopsis brevicaulis* (100%)	50.00 16.67 16.67 16.67
Mountain Hall (MH)	280 (0)	*Aspergillus ustus* (100%) *Aspergillus spelaeus* (99%) *Aspergillus versicolor* (100%) *Cladosporium sphaerospermum* (100%) *Penicillium flavigenum* (100%) *Scopulariopsis brevicaulis* (100%)	25.00 32.69 5.77 1.92 32.69 1.92
Exterior air (EA)	230 (20)	*Alternaria tenuissima* (100%) *Aspergillus spelaeus* (99%) *Aspergillus versicolor* (100%) *Botrytis cinerea* (100%) *Cladosporium cladosporioides* (100%) *Cutaneotrichosporon mucoides* (100%) *Monocillium indicum* (99%) *Penicillium flavigenum* (100%)	2.33 13.95 20.93 2.33 30.23 4.65 11.63 13.95

* CFU: colony forming units; SD: standard deviation.

3.2.2. Winter Campaign, 12 January 2017

The higher ventilation period (winter) showed a great number of different species inside the cave (25), but less diversity outside (8) (Table 3). Seventeen out of 25 species attained abundance above 10% in any of the halls (Table 3). The sampling was dominated by species of *Penicillium* (9), *Aspergillus* (9), and *Cladosporium* (2). Single species of *Arthrinium, Aureobasidium, Botrytis, Cutaneotrichosporon, Scopulariopsis,* and *Talaromyces* were also identified. The major diversity of fungal species was found in the halls of the Touristic Galleries (7 to 10 species), middle in the New Galleries (4 to 6 species), and low in the High Galleries (3 and 4 species).

Abundances above 50% in any hall were observed for *P. flavigenum*, *A. ustus*, and *A. spelaeus*, and between 50 and 25% for *C. cladosporioides*, *Arthrinium arundinis*, *Penicillium brevicompactum*, *Talaromyces brunneus*, *Penicillium glabrum*, and *Aspergillus baeticus*.

Two fungi, *A. spelaeus* and *P. flavigenum* were isolated from eight halls, *A. ustus* from four, and *Aspergillus pulvericola* and *P. glabrum* from 3 out of 10 halls. *Aspergillus* was not found in Cascade Hall, nor *Penicillium* in Lance Hall.

Two halls, Kitchen and Heracles, showed huge spore concentrations (2170 and 1330 CFU m^{-3}, respectively). *P. flavigenum* and *A. spelaeus* dominated in both cases, in addition to *A. baeticus* in Kitchen and *A. ustus* in Heracles halls.

In the sampling outside the cave, 6 out of 8 fungal species were also retrieved inside, and only *Alternaria tenuissima* and *Monocillium indicum* were not present in the cave air. However, *A. tenuissima* was also recovered inside and outside the cave in summer.

4. Discussion

The microclimatological study in Nerja Cave shows two well defined ventilation patterns corresponding to summer (low ventilation) and winter (high ventilation). The type of ventilation regime is closely related with the concentration of airborne fungi in the cave. In fact, the higher number of spores was observed in summer due to air stagnation or low ventilation, whereas in winter, ventilation helped to remove and/or dilute spore concentrations. Similar phenomena were described in other caves [7,19].

Figure 3 shows the distribution of the most abundant fungi across the different galleries and seasons. Several patterns can be observed, namely the practical absence of *Basidiomycota*. This pattern is opposite to that found in caves from Northern Spain, where the input of wood-rotting *Basiomycota* spores related to *Quercus* forests predominates in the ventilation period (winter) [18]. In Nerja Cave's surroundings, forests and dead trees are not prevalent.

Another well-defined pattern is the higher diversity of *Ascomycota* in the air in winter with relatively medium spore abundances, whereas in the summer period, the abundances of a few *Ascomycota* (*Aspergillus versicolor, Parengyodontium album, Penicillium flavigenum*) were remarkable, but they were not retrieved outside the cave.

The presence of *Ascomycota* in the two seasons, irrespective of the high or low ventilation period, could be associated with the phototrophic biofilms. Dominguez–Moñino et al. [30] related the abundance of *Parengyodontium, Aspergillus,* and *Penicillium* in the air of Tesoro Cave (Rincon de la Victoria, Malaga) at some 40 km from Nerja, with the abundant biofilms of cyanobacteria and algae originated by the artificial lighting used for allowing the visits. Nerja Cave houses also abundant phototrophic biofilms near the entrance, just in the Touristic Galleries. A similar association between phototrophs and *Ascomycota* was reported by other authors [31,32].

The senescence and lysis of old phototrophic biofilms can provide nutrients for the fungi and lead to the abundance of airborne spores in the halls close to the cave entrance. Jurado et al. [24] reported complex trophic interactions in Nerja Cave between the components of the biofilms, including processes of predation of bacteria and amoebas, and the occurrence of nematodes, fungi, and other *Eukaryota*.

The higher diversity and relative abundance of *Ascomycota* spores in winter are consistent with the entry of air according to the ventilation mode of Nerja Cave during this period of the annual cycle. During the winter campaign, the difference ΔT is negative. The colder and denser external air enters Nerja Cave through the lower entrances (E1, E2 and E3 in Figure 1) and moves from the tourist to nontourist sector (UAF-mode). So, the strong ventilation during this period contributes to the entry of atmospheric air inside the cave, to its rapid exit through the Pintada Cave, and thus, to the rapid renovation of the air cave. This is exemplified by the abundant occurrence of *Cladosporium cladosporioides, Aspergillus spelaeus,* and *Penicillium flavigenum* in the halls of the Touristic Galleries and outside.

Parengyodontium album (syn. *Engyodontium album, Beauveria alba, Tritirachium album*) [33], identified only in summer, was limited to the air of the visited area (Touristic Galleries), which could correspond to the presence of insects in spring and summer, since the fungus is a parasite of arthropods, and also to the presence of phototrophic biofilms in these galleries. Novakova et al. [34] found this species only in the air of the cave, but not in the sediments.

Jurado et al. [35] and Leplat et al. [36] reviewed the entomopathogenic fungi related to cultural heritage sites and associated *Pa. album* with the presence of arthropods. This was also reported in Lascaux Cave [37]. Data on the role of fungi in caves suggested that the species related to arthropods contribute to their mortality and favor the dispersal of spores. Jurado et al. [35] recommended the control of arthropod populations as a method to reduce fungal contamination and colonization of substrates, sediments, and rocks [35].

The spores of *Pa. album* were also widely distributed in the air of Castañar de Ibor, Ardales, Gruta de las Maravillas, and Tesoro caves [30,38,39]. In addition, *Pa. album* is common in other subterranean environments [40–43] and wall paintings, and in this last case associated to arthropods [44–46]. A review on cave fauna reported the presence of many different classes of arthropods in Nerja, including endemisms [47]. The abundant presence of spores of *Pa. album* and other entomopathogenic fungi in the air in Ardales and Tesoro caves [30] emphasizes the likely role of arthropods as a vector of these fungi.

The genus *Aspergillus* is one of the most abundant in the air of Nerja Cave. *Aspergillus* showed greater diversity in winter, with the species *A. versicolor, A. spelaeus, A. baeticus, A. ustus, A. pulvericola, A. welwitschiae, A. oryzae,* and *A. ochraceus,* while in summer were only recovered *A. versicolor* and *A. spelaeus.*

The abundance of *Aspergillus* was shown in previous publications, although the species could not be identified due to the protocol used [26,27]. Novakova et al. [32] isolated 72 fungi from sediments and air in Nerja, among them 18 different species of *Aspergillus,* from which two were only present in the air and three in both sediments and air. This suggests that an important source of *Ascomycota* in the cave, the sediments, could not be disregarded. We isolated seven species of *Aspergillus* from the cave air, and the most abundant species present in the two seasonal samplings were *A. versicolor*, *A. spelaeus*, *A. ustus*, and *A. baeticus*. The first two showed a marked seasonality, so while *A. versicolor* was abundant in summer and scarce in winter, *A. spelaeus* showed the inverse trend, being highly represented in almost all the halls in winter and occasionally in summer (8.7% in Kitchen Hall).

According to Gunde–Cimerman et al. [48] *A. versicolor* is a xerotolerant species frequently detected even in extreme environments. *Aspergillus versicolor* was relatively common in the air of caves all over the world [10,35,49–54], while *Aspergillus ustus* presented a more restricted distribution [49,50,52,55,56].

Aspergillus versicolor, relatively abundant outside in winter, was only found in the deepest Mountain Hall. Thus, according to the ventilation mode of the winter period (UAF-mode), their relationship with the external atmosphere is discarded. On the contrary, *A. versicolor* is not detected outside the cave in summer, but its concentration is high inside the cave. The highest frequency of *A. versicolor* in summer, unlike other species of the genus *Aspergillus*, points to a relationship with the increasing of nutrient sources, i.e., phototrophic biofilms or bat droppings, and could be explained by the ventilation mode of Nerja Cave during this period of the year (DAF-mode). The colder and denser Nerja Cave air is removed through the lower entrances (E1, E2 and E3 in Figure 1) and external air is sucked in through the Pintada Cave and other nonaccessible entrances located in the Almijara range. This suction effect causes the air previously stored in the PC and the GC to reach Nerja Cave. Thus, the source of these fungi would not be the external atmosphere but the cave air previously stored in PC, GC, and other possible subterranean voids connected to Nerja Cave and located in higher altitude.

This also applies to *A. spelaeus*, found in 2011 with some frequency in the sediments of Nerja Cave. The study of the isolated strains established the bases for the description of these strains as a new species [57]. A subsequent visit in 2012 allowed the isolation of new strains [32,57]. As far as we know, no other records on this *Aspergillus* were published.

Another new species of *Aspergillus* isolated in Nerja was *A. baeticus*. This was described in 2012 [58], and the type strain was isolated from sediments from Gruta de las Maravillas (Aracena, Huelva) along with other strains from Tesoro Cave. In addition, this fungus was isolated from air, cave sediments, and a dead spider (*Agraecina cristiani*) in Movile Cave, as well as from a bat cadaver in Demänovska Peace Cave [59], and from cave sediments in Nerja Cave [32].

The new species *Aspergillus europaeus* was isolated in 2011 from soils near Nerja and Movile caves, although no isolates were obtained directly from the caves [57]. We isolated this fungus in the air of the Entrance Hall in winter, suggesting an origin from the soil outside the cave and their transport inside, according to the UAF-mode ventilation [28].

Less abundant *Aspergillus* species identified in the air of Nerja Cave were also recorded in different caves such as *A. ochraceus* [55,60–62]. No cave record was found for *A. pulvericola* and *A. welwitschiae*.

The genus *Penicillium* is widely distributed in caves, and in Nerja, it was recorded in the two seasonal samplings in almost all the halls; although generally with little abundance in summer, not so in winter. In summer, six species of *Penicillium* were recorded in all the halls except in the Lance, while in winter, a total of 10 species were recovered in all the halls but not in the Lance. This is the same trend recorded for *Aspergillus*; very abundant in winter, and scarce in summer.

The most abundant *Penicillium* species was *P. flavigenum*, both in summer and winter, although with marked differences in their total abundances and presence in the halls (Tables 2 and 3), and *P. glabrum*, which was only identified in winter. *P. flavigenum* and *P. glabrum* are relatively common in caves [8,49,50,55,63–66]. Madsen et al. [67] found that *P. glabrum* was among the three most frequent species of fungi present inside Danish houses.

The presence of spores of *Penicillium*, in autumn and winter, in the air of Tesoro Cave, was important [30], as it suggests that this genus is abundant in the air of the caves of Malaga province in periods when ventilation is greater.

Relatively less abundant species compared with *P. flavigenum* and *P. glabrum* were *P. sumatrense*, *P. steckii*, *P. pancosmium*, *P. citrosulfuratum*, *P. nodositatum* in summer, and *P. brevicompactum*, *P. rubens P. pancosmium*, *P. olsonii*, *P. sumatrense*, *P. raistrickii*, and *P. fellutanum*, in winter. Moreover, similar to *Aspergillus*, *Penicillium* showed higher diversity in winter.

Talaromyces brunneus was isolated in winter from Ghosts and Cataclysm halls. The basionym is *Penicillium brunneum*. The references to this fungus in the literature are scarce, only to the original isolation from miller rice from Thailand [68], the production of anthraquinones [69] and the cellulolytic activity of a strain isolated from a cave [70].

In Nerja Cave, the spores of the genus *Cladosporium* were detected in the two seasonal samplings, both outside and inside the cave. In summer, *C. cladosporioides* accounted for 43.1% of all spores outdoor, but only 5.3% in the Ghosts Hall; *C. lebrasiae* was only found in the Entrance Hall, with 6.5% abundance, denoting clearly the existence of a low ventilation period. However, in winter, the situation changed drastically, since the concentration of *C. cladosporioides* spores outside was 30.2%, but its abundance increased in the different halls, such as the Entrance Hall (33.3%) and in Nativity Hall (42.2%), as correspondent with a higher ventilation period. In addition, *C. sphaerospermum* was represented in Nativity (4.6%), Ghosts (5.3%), Cataclysm (12.1%), Immensity (3.7%), and Mountain (1.9%) halls.

Garcia–Anton et al. [17] (2014) proved that in Altamira Cave *Cladosporium* spores had a clear external origin, as the data obtained for Nerja Cave also suggest that the presence of these fungi in the cave was due to transport from outside through air currents. It also stands out for its abundance in the air of Nerja Cave when the ventilation in the cave is greater (winter) compared to its practical absence in summer, with less or little ventilation. Likewise abundant were the species of the genus *Cladosporium* in the seasonal samplings (autumn and winter) in Ardales Cave [30].

The genus *Cladosporium* and, specifically, the species *C. cladosporioides*, was previously identified in the air of different European and Chinese caves [5,9,17,37,50,62,64,66,71], while *C. sphaerospermum* was retrieved from Slovakian and Chinese caves [49,50,66]. *Cladosporium lebrasiae* was first described from contaminated foods and forms a sister clade with *Cladosporium dominicanum* [72,73], but no reports on caves or air were found.

Other remarkable fungal species were *Aureobasidium pullulans*, *Botrytis cinerea*, *Arthrinium arundinis*, *Alternaria tenuissima*, *Bipolaris cynodontis*, all plant pathogens, but also isolated from caves [5,32,50,61,62,74,75]. The origin inside the cave may be due to the vegetation around the cavity. In addition, other noticeable fungi were *Amphichorda felina*, *Cutaneotrichosporon mucoides*, and *Scopulariopsis brevicaulis*. *Amphichorda* is one of the most ubiquitous entomopathogenic fungal genera. The species *A. felina* (syn. *Beauveria felina*, *Isaria felina*) appeared only in the Kitchen Hall with 8.7% abundance, but it deserves to be noticed for their ecology, since is a fungus colonizing animal feces in Spanish caves [76] and isolated from other caves [77,78]. *Scopulariopsis brevicaulis* is often recorded in indoor environments [79] and isolated from bat skin [80] as well as *Cutaneotrichosporon mucoides* (syn. *Trichosporon mucoides*), the later common in caves populated by bats [81,82]. *Scopulariopsis brevicaulis* was previously isolated from the air and sediments of Nerja Cave [32].

The abundance of *Cladosporium cladosporioides* spores in the air outside the cave is notable. Species of the genus *Cladosporium* are very common outdoor, at any time of the year, and also inside buildings, along with *Alternaria*, *Aspergillus*, and *Penicillium*. Outdoor, according to Gómez de Ana et al. [83], (2006), *Cladosporium* and *Penicillium* are very common in the four seasons of the year, but *Penicillium* was more abundant in winter and *Aspergillus* in summer. In Nerja Cave, both *Penicillium* and *Aspergillus* showed outdoor higher abundances in winter than in summer, as well as a high diversity in the cave air. However, *A. versicolor* stood out for their abundance in the cave in summer. This showed the influence that the outside air can have on the dispersal of the spores of these fungi inside the cave in winter and the influence of the air previously stored in other caves connected to the ventilation of Nerja Cave in summer.

Of the 18 species identified in the two seasonal air samples outside the cave, only 8 were not found inside, Tremateia chromolaenae, Victoriomyces antarcticus, Microascus croci, Hyphodermella rosae, Purpureocillium lilacinum, Monocillium indicum, Aspergillus tubingensis, and Aspergillus melleus. Hyphodermella rosae is a white-rot corticioid fungus previously isolated from La Garma Cave air [18], Purpureocillium lilacinum is an entomopathogenic fungus often isolated from caves [50,52,62,64], Monocillium indicum and Victoriomyces antarcticus were found in soils [84,85], Tremateia chromolaenae was associated with plants [86], and Microascus croci with aquatic sediments [87]. The remaining two Aspergillus spp. were isolated from Brazilian caves [88,89].

It is of interest to notice that many of the fungi isolated were reported as entomopathogens and/or associated with arthropods, including *A. versicolor, A. ustus, A. ochraceus, A. olsonii, A. welwitschiae, A. tubingensis, B. cinerea, C. cladosporioides, C. sphaerospermum, Pa. album, P. flavigenum, P. chrysogenum, P. raistrickii, P. brevicompactum, P. steckii, Pu. lilacinum*, and *S. brevicaulis*, among others [90–93].

5. Conclusions

The studies in Nerja and other Andalusian caves were essential for the description of new species of the genus *Aspergillus* [57,58], and specifically for *A. spelaeus*, which type strain was isolated from Nerja [94]. This suggests that these caves are reservoirs of new species of fungi as well as bacteria [95,96], and their study contributes to increasing the knowledge of biodiversity on the planet.

Considering the results, the influence that the external air has on Nerja Cave during the winter is demonstrated. Cold air enters in winter, and this is confirmed by the abundant presence of *Aspergillus* and *Penicillium* spores inside the cave, without ruling out that the reservoirs of these fungi were already installed in the cave itself (sediments, phototrophic biofilms) and that the increased air flow carries away the spores. In summer, the abundance of fungal spores in Nerja Cave, which are not detected outside, as in the case of *A. versicolor*, suggests a different source for these microorganisms, linked to the nutrient sources and other subterranean voids and caves near Nerja Cave. Other seasonal events, such as the presence of entomopathogenic fungal spores in summer, were also reflected in the air of the cave.

The data reported in this work are in agreement with the findings of Docampo et al. [26,27]. These authors reported that spores of *Aspergillus* and *Penicillium* were, by far, the most abundant in the air of the cave and were retrieved throughout the year, followed by *Cladosporium*. The influence of the ventilation is clearly shown by the high abundances of *Cladosporium* spores outside the cave both in summer and winter. The high occurrence in the halls near the entrance in winter is related to ventilation, and the practical absence inside the cave in summer is due to low ventilation or stagnation period.

The conditions of access and visit to a cave have an impact due to the mobilization of air and dispersal of aerial microorganisms. However, this is minimized in the summer season, when there is a greater influx of visitors, as shown by the low concentration of spores found inside. However, the high abundance of *Aspergillus* and *Penicillium* spores in the Kitchen and Heracles halls in winter requires a more detailed study to understand their origin and nature and to be able to reduce the concentrations of spores, which are much larger than in other neighboring halls (Cataclysm and Immensity).

Author Contributions: Investigation, V.J., Y.D.R., C.L., T.M.-P., J.L.G.-P., writing—original draft preparation, C.S.-J., Y.D.R., C.L., writing—review and editing, C.S.-J. All authors read and agreed to the published version of the manuscript.

Funding: This research was funded by Nerja Cave Foundation. This work is part of an Interdisciplinary Research Project for the conservation of Nerja Cave, authorized by the Ministry of Culture of the Andalusian Government.

Institutional Review Board Statement: Not applicable.

Informed Consent Statement: Not applicable.

Data Availability Statement: Data, under accession numbers MZ374012-MZ374055 and MZ375763-MZ375863, were deposited into the GenBank database.

Acknowledgments: This paper is dedicated to the memory of Professor Wolfgang E. Krumbein (1937–2021) (https://de.wikipedia.org/wiki/Wolfgang_E._Krumbein), an outstanding scientist and dear friend and colleague. The authors acknowledge Nerja Cave Foundation for the support and facilities and CSIC Interdisciplinary Thematic Platform Open Heritage: Research and Society (PTI-PAIS) for the professional support.

Conflicts of Interest: The authors declare no conflict of interest.

References

1. Adetutu, E.M.; Thorpe, K.; Bourne, S.; Cao, X.; Shahsavari, E.; Kirby, G.; Ball, A.S. Phylogenetic diversity of fungal communities in areas accessible and not accessible to tourists in Naracoorte Caves. *Mycologia* **2011**, *103*, 959–968. [CrossRef] [PubMed]
2. Martin-Sanchez, P.M.; Nováková, A.; Bastian, F.; Alabouvette, C.; Saiz-Jimenez, C. Use of biocides for the control of fungal outbreaks in subterranean environments: The case of the Lascaux Cave in France. *Environ. Sci. Technol.* **2012**, *46*, 3762–3770. [CrossRef]
3. Di Piazza, S.; Isaia, M.; Vizzini, A.; Badino, G.; Voyron, S.; Zotti, M. First mycological assessment in hydrothermal caves of Monte Kronio (Sicily, southern Italy). *Webbia. J. Plant Taxon. Geogr.* **2017**, *72*, 277–285. [CrossRef]
4. Man, B.; Wang, H.; Yun, Y.; Xiang, X.; Wang, R.; Duan, Y.; Cheng, X. Diversity of fungal communities in Heshang Cave of Central China revealed by mycobiome-sequencing. *Front. Microbiol.* **2018**, *9*, 1400. [CrossRef] [PubMed]
5. Ogórek, R.; Kozak, B.; Visnovská, Z.; Tancinová, D. Phenotypic and genotypic diversity of airborne fungal spores in Demänovská Ice Cave (Low Tatras, Slovakia). *Aerobiologia* **2018**, *34*, 13–28. [CrossRef] [PubMed]
6. Pusz, W.; Król, M.; Zwijacz-Kozica, T. Airborne fungi as indicators of ecosystem disturbance: An example from selected Tatra Mountains caves (Poland). *Aerobiologia* **2018**, *34*, 111–118. [CrossRef] [PubMed]
7. Wang, W.; Ma, X.; Ma, Y.; Mao, L.; Wu, F.; Ma, X.; An, L.; Feng, H. Seasonal dynamics of airborne fungi in different caves of the Mogao Grottoes, Dunhuang. China. *Int. Biodeter. Biodegr.* **2010**, *64*, 461–466. [CrossRef]
8. Porca, E.; Jurado, V.; Martin-Sanchez, P.M.; Hermosin, B.; Bastian, F.; Alabouvette, C.; Saiz-Jimenez, C. Aerobiology: An ecological indicator for early detection and control of fungal outbreaks in caves. *Ecol. Indic.* **2011**, *11*, 1594–1598. [CrossRef]
9. Fernandez-Cortes, A.; Cuezva, S.; Sanchez-Moral, S.; Porca, E.; Jurado, V.; Saiz-Jimenez, C. Detection of human-induced environmental disturbances in a show cave. *Environ. Sci. Pollut. Res.* **2011**, *18*, 1037–1045. [CrossRef]
10. Martin-Sanchez, P.M.; Jurado, V.; Porca, E.; Bastian, F.; Lacanette, D.; Alabouvette, C.; Saiz-Jimenez, C. Airborne microorganisms in Lascaux Cave (France). *Int. J. Speleol.* **2014**, *43*, 295–303. [CrossRef]
11. Carrasco, F.; Vadillo, I.; Liñan, C.; Andreo, B.; Durán, J.J. Control of environmental parameters for management and conservation of Nerja Cave (Malaga, Spain). *Acta Carsologica* **2002**, *31*, 105–122. [CrossRef]
12. Liñán, C.; Vadillo, I.; Carrasco, F. Carbon dioxide concentration in air within the Nerja Cave (Malaga, Andalusia, Spain). *Int. J. Speleol.* **2008**, *37*, 99–106. [CrossRef]
13. Liñán, C.; Ojeda, L.; Benavente, J.; Del Rosal, Y.; Vadillo, I.; Carrasco, F. Coupling air temperature records and gravimetric data to interpret ventilation patterns in a Mediterranean karstic system (Nerja-Pintada caves, southern Spain). *Sci. Total Environ.* **2020**, *730*, 139147. [CrossRef] [PubMed]
14. Benavente, J.; Vadillo, I.; Liñan, C.; Carrasco, F.; Soler, A. Ventilation effects in a karstic show cave and in its vadose environment, Nerja, Southern Spain. *Carbonates Evapotites* **2011**, *26*, 11–17. [CrossRef]
15. Dueñas, C.; Fernández, M.C.; Cañete, S.; Pérez, M.; Gordo, E. Seasonal variations of radon and the radiation exposure levels in Nerja cave, Spain. *Radiat. Meas.* **2011**, *46*, 1181–1186. [CrossRef]
16. Del Rosal, Y.; Liñan, C.; Hernandez-Marine, M. The conservation of the Nerja Cave: Preserving anthropogenic impact in a tourist cave. In *The Conservation of Subterranean Cultural Heritage*; Saiz-Jimenez, C., Ed.; CRC Press/Balkema: Leiden, The Netherlands, 2014; pp. 193–206.
17. Ojeda, L.; Vadillo, I.; Etiope, G.; Benavente, J.; Liñán, C.; Del Rosal, Y.; Tapia, S.T.; Moríñigo, M.A.; Carrasco, F. Methane sources and sinks in karst systems: The Nerja cave and its vadose environment (Spain). *Geochim. Cosmochim. Acta* **2019**, *259*, 302–315. [CrossRef]
18. Garcia-Anton, E.; Cuezva, S.; Jurado, V.; Porca, E.; Miller, A.Z.; Fernandez-Cortes, A.; Saiz-Jimenez, C.; Sanchez-Moral, S. Combining stable isotope ($\delta^{13}C$) of trace gases and aerobiological data to monitor the entry and dispersion of microorganisms in caves. *Environ. Sci. Pollut. Res.* **2014**, *21*, 473–484. [CrossRef] [PubMed]
19. Sanchez-Moral, S.; Jurado, V.; Fernandez-Cortes, A.; Cuezva, S.; Martin-Pozas, T.; Gonzalez-Pimentel, J.L.; Ontañon, R.; Saiz-Jimenez, C. Environment-driven control of the dispersal of fungi in subterranean ecosystems: The case of La Garma Cave (northern Spain). *Int. Microbiol.* **2021**, in press.

20. Ruíz, J.; Marín-Girón, F.; Ojeda, F.; Marón-Olalla, F.; Berros, J.; Marín-Olalla, E. Estudio macroscópico "in situ" y microscópico-ecológico de pequeñas zonas de flora verde (algas verdes y verdeazuladas) del interior de la Cueva de Nerja. In *Investigación Biológica y Edafológica de la Cueva de Nerja (Málaga)*; Marín, F., Carrasco, F., Eds.; Trabajos sobre la Cueva de Nerja: Málaga, Spain, 1992; Volume 2, pp. 113–125.
21. Del Rosal, Y.; Hernández-Mariné, M.; Roldán, M. Phototrophic microorganisms in the tourist cave of Nerja. In *Science, Technology and Cultural Heritage*; Rogerio-Candelera, M.A., Ed.; CRC Press/Balkema: Leiden, The Netherlands, 2014; pp. 229–234.
22. Del Rosal Padial, Y.; Jurado Lobo, V.; Hernández Mariné, M.; Roldán Molina, M.; Sáiz Jiménez, C. Biofilms en cuevas turísticas: La Cueva de Nerja y la Cueva del Tesoro. In *El Karst y el Hombre: Las Cuevas como Patrimonio Mundial*; Andreo, B., Durán, J.J., Eds.; Asociación de Cuevas Turísticas Españolas: Madrid, Spain, 2016; pp. 103–114.
23. Jurado, V.; Del Rosal, Y.; Hernández-Mariné, M.; Galocha Zapata, I.; Domínguez-Moñino, I.; Rogerio Candelera, M.A.; Sáiz-Jiménez, C. Biología molecular de los biofilms fotosintéticos de la Cueva de Nerja. In *Análisis, Impacto y Evolución de los Biofilms Fotosintéticos en Espeleotemas. El caso de la Cueva de Nerja (Málaga, España)*; Del Rosal Padial, Y., Ed.; Fundación Pública de Servicios Cueva de Nerja: Nerja, Spain, 2017; pp. 229–270.
24. Jurado, V.; Del Rosal, Y.; Gonzalez-Pimentel, J.L.; Hermosin, B.; Saiz-Jimenez, C. Biological control of phototrophic biofilms in a show cave: The case of Nerja Cave. *Appl. Sci.* **2020**, *10*, 3448. [CrossRef]
25. Del Rosal, Y.; Martínez-Manzanares, E.; Marín, F.; Liñán, C.; Simón, M.D. Análisis de la calidad microbiológica aérea en el interior de la Cueva de Nerja. In *Cuevas Turísticas: Aportación al Desarrollo Sostenible*; Durán, J.J., Robledo, P.A., Vázquez, J., Eds.; Instituto Geológico y Minero de España: Madrid, Spain, 2007; pp. 181–190.
26. Docampo, S.; Trigo, M.M.; Recio, M.; Melgar, M.; Garcia-Sanchez, J.; Calderon-Ezquerro, M.C.; Cabezudo, B. High incidence of *Aspergillus* and *Penicillium* spores in the atmosphere of the cave of Nerja (Malaga, southern Spain). *Aerobiologia* **2010**, *26*, 89–98. [CrossRef]
27. Docampo, S.; Trigo, M.M.; Recio, M.; Melgar, M.; García-Sánchez, J.; Cabezudo, B. Fungal spores content of the atmosphere of the Cave of Nerja (southern Spain): Diversity and origin. *Sci. Total Environ.* **2011**, *409*, 835–843. [CrossRef]
28. Liñán, C.; Del Rosal, Y.; Carrasco, F.; Vadillo, I.; Benavente, J.; Ojeda, L. Highlighting the importance of transitional ventilation regimes in the management of Mediterranean show caves (Nerja-Pintada system, southern Spain). *Sci. Total Environ.* **2018**, *631–632*, 1268–1278. [CrossRef] [PubMed]
29. Docampo, S.; Recio, M.; Trigo, M.M.; Cabezudo, B. Risk of pollen allergy in Nerja (southern Spain): A pollen calendar. *Aerobiologia* **2007**, *23*, 189–199. [CrossRef]
30. Dominguez-Moñino, I.; Jurado, V.; Rogerio-Candelera, M.A.; Hermosin, B.; Saiz-Jimenez, C. Airborne fungi in show caves from Southern Spain. *Appl. Sci.* **2021**, *11*, 5027. [CrossRef]
31. Grbic, M.L.; Vukojevic, J.; Simic, G.S.; Krizmanic, J.; Stupar, M. Biofilm forming cyanobacteria, algae and fungi on two historic monuments in Belgrade, Serbia. *Arch. Biol. Sci.* **2010**, *62*, 625–631. [CrossRef]
32. Pfendler, S.; Borderie, F.; Bousta, F.; Alaoui-Sosse, L.; Alaoui-Sosse, B.; Aleya, L. Comparison of biocides, allelopathic substances and UV-C as treatments for biofilm proliferation on heritage monuments. *J. Cult. Herit.* **2018**, *33*, 117–124. [CrossRef]
33. De Hoog, G.S. The genera *Beauveria*, *Isaria*, *Tritirachium* and *Acrodontium* gen. nov. *Stud. Mycol.* **1972**, *1*, 1–41.
34. Novakova, A.; Hubka, V.; Saiz-Jimenez, C. Microscopic fungi isolated from cave air and sediments in the Nerja Cave-preliminary results. In *The Conservation of Subterranean Cultural Heritage*; Saiz-Jimenez, C., Ed.; CRC Press/Balkema: Leiden, The Netherlands, 2014; pp. 239–246.
35. Jurado, V.; Sanchez-Moral, S.; Saiz-Jimenez, C. Entomogenous fungi and the conservation of the cultural heritage: A review. *Int. Biodeter. Biodegr.* **2008**, *62*, 325–330. [CrossRef]
36. Leplat, J.; François, A.; Bousta, F. *Parengyodontium album*, a frequently reported fungal species in the cultural heritage environment. *Fungal Biol. Rev.* **2020**, *34*, 126–135. [CrossRef]
37. Bastian, F.; Alabouvette, C.; Saiz-Jimenez, C. The impact of arthropods on fungal community structure in Lascaux Cave. *J. Appl. Microbiol.* **2009**, *106*, 1456–1462. [CrossRef] [PubMed]
38. Jurado, V.; Porca, E.; Cuezva, S.; Fernandez-Cortes, A.; Sanchez-Moral, S.; Saiz-Jimenez, C. Fungal outbreak in a show cave. *Sci. Total Environ.* **2010**, *408*, 3632–3638. [CrossRef]
39. Porca Belío, E. Aerobiología: Mecanismos de Dispersión de los Microorganismos en Cuevas Turísticas. Ph.D. Thesis, Facultad de Biología, Universidad de Sevilla, Seville, Spain, 2011.
40. Saarela, M.; Alakomi, H.-L.; Suihko, M.-L.; Maunuksela, L.; Raaska, L.; Mattila-Sandholm, T. Heterotrophic microorganisms in air and biofilm samples from Roman catacombs, with special emphasis on actinobacteria and fungi. *Int. Biodeter. Biodegr.* **2004**, *54*, 27–37. [CrossRef]
41. Simonovicová, A.; Gódyová, M.; Sevc, J. Airborne and soil microfungi as contaminants of stone in a hypogean cemetery. *Int. Biodeter. Biodegr.* **2004**, *54*, 7–11. [CrossRef]
42. Nugari, M.P.; Pietrini, A.M.; Caneva, G.; Imperi, F.; Visca, P. Biodeterioration of mural paintings in a rocky habitat: The Crypt of the Original Sin (Matera, Italy). *Int. Biodeter. Biodegr.* **2009**, *63*, 705–711. [CrossRef]
43. Mang, S.M.; Scrano, L.; Camele, I. Preliminary studies on fungal contamination of two rupestrian churches from Matera (Southern Italy). *Sustainability* **2020**, *12*, 6988. [CrossRef]

44. Saiz-Jimenez, C.; Samson, R.A. Microorganisms and environmental pollution as deteriorating agents of the frescoes of "Santa María de la Rábida", Huelva, Spain. In *Sixth Triennial Meeting ICOM*; Paper 81/15/5; Committee for Conservation: Ottawa, ON, Canada, 1981; p. 14.
45. Gorbushina, A.A.; Petersen, K. Distribution of microorganisms on ancient wall paintings as related to associated faunal elements. *Int. Biodeter. Biodegr.* **2000**, *46*, 277–284. [CrossRef]
46. Berner, M.; Wanner, G.; Lubitz, W. A comparative study of the fungal flora present in medieval wall paintings in the chapel of the castle Herberstein and in the parish church of St. Georgen in Styria, Austria. *Int. Biodeter. Biodegr.* **2004**, *40*, 53–61. [CrossRef]
47. Barranco, P.; Tinaut, A.; Baena, M. Entomofauna cavernícola de Andalucía. In *El Karst de Andalucía. Geoespeleología, Bioespeleología y Presencia Humana*; Calaforra Chordi, J.M., Berrocal Pérez, J.A., Eds.; Consejería de Medio Ambiente, Junta de Andalucía: Sevilla, Spain, 2008; pp. 75–85.
48. Gunde-Cimerman, N.; Ramos, J.; Plemenitas, A. Halotolerant and halophilic fungi. *Mycol. Res.* **2009**, *113*, 1231–1241. [CrossRef]
49. Nováková, A. Microscopic fungi in caves of the National Park Slovak Karst. *Phytopedon* **2004**, *3*, 26–31.
50. Nováková, A. Microscopic fungi isolated from the Domica Cave system (Slovak Karst National Park, Slovakia). A review. *Int. J. Speleol.* **2009**, *38*, 71–82. [CrossRef]
51. Domínguez-Moñino, I.; Jurado, V.; Hermosín, B.; Sáiz-Jiménez, C. Aerobiología de cuevas andaluzas. In *Las Cuevas Turísticas como Activos Económicos: Conservación e Innovación*; Durán, J.J., Robledo, P.A., Eds.; Asociación de Cuevas Turísticas Españolas: Madrid, Spain, 2012; pp. 299–307.
52. Taylor, E.L.S.; Resende-Stoianoff, M.A.A.; Lopes Ferreira, R. Mycological study for a management plan of a neotropical show cave (Brazil). *Int. J. Speleol.* **2013**, *42*, 267–277. [CrossRef]
53. Mitova, M.M.; Iliev, M.; Nováková, A.; Gorbushina, A.A.; Groudeva, V.I.; Martin-Sanchez, P.M. Diversity and biocide susceptibility of fungal assemblages dwelling in the Art Gallery of Magura Cave, Bulgaria. *Int. J. Speleol.* **2017**, *46*, 67–80. [CrossRef]
54. Mammola, S.; Di Piazza, S.; Zotti, M.; Isaia, M. Human-induced alterations of the cave mycobiota in an Alpine show cave (Italy, SW-Alps). *Acta Carsologica* **2017**, *46*, 111–123. [CrossRef]
55. Marvanová, L.; Kalouskova, V.; Hanuláková, D.; Schánĕl, L. Microscopic fungi in the Zbrašov aragonite caves. *Czech Mycol.* **1992**, *46*, 243–250.
56. Novakova, A.; Jurado, V.; Saiz-Jimenez, C. Are fungi a real threat for the conservation of Altamira Cave? In *The Conservation of Subterranean Cultural Heritage*; Saiz-Jimenez, C., Ed.; CRC Press/Balkema: Leiden, The Netherlands, 2014; pp. 223–228.
57. Hubka, V.; Nováková, A.; Peterson, S.W.; Frisvad, J.C.; Sklenář, F.; Matsuzawa, T.; Kubátová, A.; Kolaŕík, M. A reappraisal of *Aspergillus* section *Nidulantes* with descriptions of two new sterigmatocystin-producing species. *Plant Syst. Evol.* **2016**, *302*, 1267–1299. [CrossRef]
58. Nováková, A.; Hubka, V.; Saiz-Jimenez, C.; Kolarik, M. *Aspergillus baeticus* sp. nov. and *Aspergillus thesauricus* sp. nov.: Two new species in section Usti originating from Spanish caves. *Int. J. Syst. Evol. Microbiol.* **2012**, *62*, 2778–2785. [CrossRef]
59. Nováková, A.; Kubátová, A.; Sklenář, F.; Hubka, V. Microscopic fungi on cadavers and skeletons from cave and mine environments. *Czech Mycol.* **2018**, *70*, 101–121. [CrossRef]
60. Vanderwolf, K.J.; Malloch, D.; McAlpine, D.F.; Forbes, G.J. A world review of fungi, yeasts, and slime molds in caves. *Int. J. Speleol.* **2013**, *42*, 77–96. [CrossRef]
61. Popović, S.; Simić, G.S.; Stupar, M.; Unković, N.; Predojević, D.; Jovanović, J.; Grbić, M.L. Cyanobacteria, algae and microfungi present in biofilm from Božana Cave (Serbia). *Int. J. Speleol.* **2015**, *44*, 141–149. [CrossRef]
62. Kozlova, E.V.; Mazina, S.E. Biodiversity of fungi in the photic and aphotic zones of Montenegro caves. *Aerobiologia* **2020**, *36*, 589–604. [CrossRef]
63. Pusz, W.; Ogórek, R.; Knapik, R.; Kozak, B.; Bujak, H. The occurrence of fungi in the recently discovered Jarkowicka Cave in the Karkonosze Mts. (Poland). *Geomicrobiol. J.* **2015**, *32*, 59–67. [CrossRef]
64. Nováková, A.; Hubka, V.; Valinova, S.; Kolarik, M.; Hillebrand-Voiculescu, A.M. Cultivable microscopic fungi from an underground chemosynthesis-based ecosystem: A preliminary study. *Folia Microbiol.* **2018**, *63*, 43–55. [CrossRef]
65. Tavares, D.G.; Barbosa, B.V.L.; Ferreira, R.L.; Duarte, W.F.; Cardoso, P.G. Antioxidant activity and phenolic compounds of the extract from pigmentproducing fungi isolated from Brazilian caves. *Biocatal. Agric. Biotechnol.* **2018**, *16*, 148–154. [CrossRef]
66. Zhang, Z.F.; Zhou, S.Y.; Eurwilaichitr, L.; Ingsriswang, S.; Raza, M.; Chen, Q.; Zhao, P.; Liu, F.; Cai, L. Culturable mycobiota from Karst caves in China II, with descriptions of 33 new species. *Fungal Divers.* **2021**, *106*, 29–136. [CrossRef]
67. Madsen, A.M.; Larsen, S.T.; Koponen, I.K.; Kling, K.I.; Barooni, A.; Karottki, D.G.; Tendal, K.; Wolkoffa, P. Generation and characterization of indoor fungal aerosols for inhalation studies. *Appl. Environ. Microbiol.* **2016**, *82*, 2479–2493. [CrossRef]
68. Samson, R.A.; Yilmaz, N.; Houbraken, J.; Spierenburg, H.; Seifert, K.A.; Peterson, S.W.; Varga, J.; Frisvad, J.C. Phylogeny and nomenclature of the genus *Talaromyces* and taxa accommodated in *Penicillium* subgenus *Biverticillium*. *Stud. Mycol.* **2011**, *70*, 159–183. [CrossRef] [PubMed]
69. Shibata, S.; Udagawa, S. Metabolic products of fungi. XIX. Isolation of rugulosin from *Penicillium brunneum* Udagawa. *Chem. Pharm. Bull.* **1963**, *11*, 402–403. [CrossRef]
70. Paula, C.C.P.; Montoya, Q.V.; Meirelles, L.A.; Sanchez, C.F.; Rodrigues, A.; Seleghim, M.H.R. High cellulolytic activities in filamentous fungi isolated from an extreme oligotrophic subterranean environment (Catão cave) in Brazil. *An. Acad. Bras. Cienc.* **2019**, *91*, e20180583. [CrossRef] [PubMed]

71. Jiang, J.-R.; Cai, L.; Liu, F. Oligotrophic fungi from a carbonate cave, with three new species of *Cephalotrichum*. *Mycology* **2017**, *8*, 164–177. [CrossRef]
72. Razafinarivo, J.; Jany, J.-L.; Crous, P.W.; Looten, R.; Gaydou, V.; Barbier, G.; Mounier, J.; Vasseur, V. *Cladosporium lebrasiae*, a new fungal species isolated from milk bread rolls in France. *Fungal Biol.* **2016**, *120*, 1017–1029. [CrossRef]
73. Zalar, P.; de Hoog, G.S.; Schroers, H.-J.; Crous, P.W.; Groenewald, J.Z.; Gunde-Cimerman, N. Phylogeny and ecology of the ubiquitous saprobe *Cladosporium sphaerospermum*, with descriptions of seven new species from hypersaline environments. *Stud. Mycol.* **2007**, *58*, 157–183. [CrossRef]
74. Manamgoda, D.S.; Rossman, A.Y.; Castlebury, L.A.; Crous, P.W.; Madrid, H.; Chukeatirote, E.; Hyde, K.D. The genus *Bipolaris*. *Stud. Mycol.* **2014**, *79*, 221–288. [CrossRef]
75. Wang, M.; Tan, X.-M.; Liu, F.; Cai, L. Eight new *Arthrinium* species from China. *MycoKeys* **2018**, *34*, 1–24. [CrossRef]
76. Saiz-Jimenez, C. Cave Conservation: A Microbiologist's Perspective. In *Cave Microbiomes: A Novel Resource for Drug Discovery*; Cheepham, N., Ed.; SpringerBrief in Microbiology 1: New York, NY, USA, 2013; pp. 69–84.
77. Zhang, Z.F.; Liu, F.; Zhou, X.; Liu, X.Z.; Liu, S.J.; Cai, L. Culturable mycobiota from Karst caves in China, with descriptions of 20 new species. *Persoonia* **2017**, *39*, 1–31. [CrossRef]
78. Belyagoubi, L.; Belyagoubi-Benhammou, N.; Jurado, V.; Dupont, J.; Lacoste, S.; Djebbah, F.; Ounadjela, F.Z.; Benaissa, S.; Habi, S.; Abdelouahi, D.E.; et al. Antimicrobial activities of culturable microorganisms (actinomycetes and fungi) isolated from Chaabe Cave, Algeria. *Int. J. Speleol.* **2018**, *47*, 189–199. [CrossRef]
79. Woudenberg, J.H.C.; Meijer, M.; Houbraken, J.; Samson, R.A. *Scopulariopsis* and scopulariopsis-like species from indoor environments. *Stud. Mycol.* **2017**, *88*, 1–35. [CrossRef]
80. Larcher, G.; Bouchara, J.P.; Pailley, P.; Montfort, D.; Beguin, H.; De Bièvе, C.; Chabasse, D. Fungal biota associated with bats in Western France. *J. Med. Mycol.* **2003**, *13*, 29–34.
81. Vaughan-Martini, A.; Angelini, P.; Zacchi, L. The influence of human and animal visitation on the yeast ecology of three Italian caverns. *Ann. Microbiol.* **2000**, *50*, 133–140.
82. Sugita, T.; Kikuchi, K.; Makimura, K.; Urata, K.; Someya, T.; Kamei, K.; Niimi, M.; Uehara, Y. *Trichosporon* species isolated from guano samples obtained from bat-inshabited caves in Japan. *Appl. Environ. Microbiol.* **2005**, *71*, 7626–7629. [CrossRef] [PubMed]
83. Gómez de Ana, S.; Torres-Rodríguez, J.M.; Ramírez, E.A.; Belmonte-Soler, J. Seasonal distribution of *Alternaria, Aspergillus, Cladosporium* and *Penicillium* species isolated in homes of fungal allergic patients. *J. Investig. Allergol. Clin. Immunol.* **2006**, *16*, 357–363.
84. Hurst, C.J. *Understanding Terrestrial Microbial Communities*; Springer: Cham, Switzerland, 2019.
85. Davolos, D.; Pietrangeli, B.; Persiani, A.M.; Maggi, O. *Victoriomyces antarcticus* gen. nov., sp. nov., a distinct evolutionary lineage of the *Cephalothecaceae* (*Ascomycota*) based on sequence-based phylogeny and morphology. *Int. J. Syst. Evol. Microbiol.* **2019**, *69*, 1099–1110. [CrossRef] [PubMed]
86. Mapook, A.; Hyde, K.D.; McKenzie, E.H.C.; Jones, E.B.G.; Bhat, D.J.; Jeewon, R.; Stadler, M.; Samarakoon, M.C.; Malaithong, M.; Tanunchai, B.; et al. Taxonomic and phylogenetic contributions to fungi associated with the invasive weed *Chromolaena odorata* (Siam weed). *Fungal Divers.* **2020**, *101*, 1–175. [CrossRef]
87. Sandoval-Denis, M.; Gené, J.; Sutton, D.A.; Cano-Lira, J.F.; de Hoog, G.S.; Decock, C.A.; Wiederhold, N.P.; Guarro, J. Redefining *Microascus, Scopulariopsis* and allied genera. *Persoonia* **2016**, *36*, 1–36. [CrossRef]
88. Melo, A.G.; Souza, P.N.C.; Maia, N.C.; Thomas, A.B.; Silva, L.B.R.; Batista, L.R.; Ferreira, R.L.; Cardoso, P.G. Screening and identification of tannase-producing fungi isolated from Brazilian caves. *African J. Microbiol. Res.* **2013**, *7*, 483–487.
89. Cunha, A.O.B.; Bezerra, J.D.P.; Oliveira, T.G.L.; Barbier, E.; Bernard, E.; Machado, A.R.; Souza-Motta, C.M. Living in the dark: Bat caves as hotspots of fungal diversity. *PLoS ONE* **2020**, *15*, e0243494. [CrossRef] [PubMed]
90. Trovão, J.; Mesquita, N.; Paiva, D.S.; Carvalho, H.P.; Avelar, L.; Portugal, A. Can arthropods act as vectors of fungal dispersion in heritage collections? A case study on the archive of the University of Coimbra, Portugal. *Int. Biodeter. Biodegr.* **2013**, *79*, 49–55. [CrossRef]
91. Jaber, S.; Mercier, A.; Knio, K.; Brum, S.; Kambris, Z. Isolation of fungi from dead arthropods and identification of a new mosquito natural pathogen. *Parasite Vector* **2016**, *9*, 491. [CrossRef] [PubMed]
92. Yun, T.-S.; Park, S.Y.; Yu, J.; Hwang, Y.; Hong, K.-J. Isolation and identification of fungal species from the insect pest *Tribolium castaneum* in rice processing complexes in Korea. *Plant Pathol. J.* **2018**, *34*, 356–366. [CrossRef]
93. Jenoh, E.M.; de Villiers, E.P.; de Villiers, S.M.; Okoth, S.; Jefwa, J.; Kioko, E.; Kaimenyi, D.; Hendrickx, M.; Dahdouh-Guebas, F.; Koedam, N. Infestation mechanisms of two woodborer species in the mangrove *Sonneratia alba* J. Smith in Kenya and co-occurring endophytic fungi. *PLoS ONE* **2019**, *14*, e0221285. [CrossRef]
94. Hubka, V.; Nováková, A.; Kolarík, M.; Jurjević, A.; Peterson, S.W. Revision of *Aspergillus* section *Flavipedes*: Seven new species and proposal of section *Jani* sect. nov. *Mycologia* **2015**, *107*, 169–208. [CrossRef]
95. Dominguez-Moñino, I.; Jurado, V.; Hermosin, B.; Saiz-Jimenez, C. *Paracoccus cavernae* sp. nov., isolated from a show cave. *Int. J. Syst. Evol. Microbiol.* **2016**, *66*, 1–6. [CrossRef] [PubMed]
96. Dominguez-Moñino, I.; Jurado, V.; Gonzalez-Pimentel, J.L.; Miller, A.Z.; Hermosin, B.; Saiz-Jimenez, C. *Bacillus onubensis* sp. nov. isolated from the air of two Andalusian caves. *Syst. Appl. Microbiol.* **2018**, *41*, 167–172. [CrossRef] [PubMed]

applied sciences

MDPI

Article

FTIR Spectroscopy for Identification and Intra-Species Characterization of *Serpula lacrymans*

Rony Barboux [1,2], Faisl Bousta [2] and Patrick Di Martino [1,*]

1 Laboratoire ERRMECe, CY Cergy Paris University, 95000 Neuville sur Oise, France; rony.barboux@u-cergy.fr
2 Laboratoire de Recherche des Monuments Historiques, Ministère de la Culture et de la Communication, Centre de Recherche sur la Conservation (CRC-USR 3224), Muséum National d'Histoire Naturelle, CNRS, Sorbonne Universités, 29 Rue de Paris, 77420 Champs-sur-Marne, France; faisl.bousta@culture.gouv.fr
* Correspondence: patrick.di-martino@cyu.fr

Featured Application: The method described in this study consists of the preparation and analysis of mycelium samples for the identification of *Serpula lacrymans* by FTIR.

Abstract: The dry rot fungus *Serpula lacrymans* is the most destructive fungal agent of wood building materials in Europe, Russia, North America, and Japan. The identification of this wood-deteriorating agent and the discrimination of different fungal isolates is very important for the control of buildings in general and for the preservation of cultural heritage in particular. The objective of the study was to develop a Fourier transform infrared (FTIR) method coupled with a partial least square discriminant analysis (PLS-DA) for the sample preparation and identification of *S. lacrymans*. Five distinct *S. lacrymans* strains were analysed and compared to two strains of unrelated fungal species. Different methods of mycelial growth, sample preparation, and FTIR spectral data normalisation were compared. FTIR analysis of a harvested mycelium grown on the surface of a polyether sulfone microfiltration membrane deposited on a malt extract agar medium, followed by vector normalization and PLS-DA statistical analysis, resulted in 100% correct attribution at phylum, species, and strain level, regardless of the type of standardization used. This study confirms the applicability of FTIR spectroscopy for the identification of *S. lacrymans* and the discrimination of different strains belonging to this species.

Keywords: *Serpula lacrymans*; biodegradation; wood; cultural heritage; FTIR; identification

Citation: Barboux, R.; Bousta, F.; Di Martino, P. FTIR Spectroscopy for Identification and Intra-Species Characterization of *Serpula lacrymans*. *Appl. Sci.* **2021**, *11*, 8463. https://doi.org/10.3390/app11188463

Academic Editors: Filomena De Leo and Daniela Isola

Received: 31 July 2021
Accepted: 9 September 2021
Published: 12 September 2021

1. Introduction

Serpula lacrymans is a Basidiomycota of the order Boletales, which is the main agent of brown rot in buildings and historical monuments in Europe, Russia, North America, and Japan [1–5]. *S. lacrymans* is rarely detected alone in a building; other wood-degrading fungi accompany it which makes its detection difficult [6,7]. The observation of brown rot and the analysis of the mycelium and fruiting bodies can direct the identification of wood-damaging fungi to a presence of *S. lacrymans* [8]. When the fruiting bodies or microstructures of the mycelial cords (rhizomorph) characteristic of *S. lacrymans* cannot be found, a sample must be taken. Two identification approaches are then possible: a classical agar culture technique or a molecular biological approach. The cultivation of the fungus on a nutrient agar medium allows for the specification of its macro and microscopic morphological characters. The main advantage of this cultural approach is low cost, but it is time-consuming and tedious, requires strong expertise, and may not be very discriminating for the related species [3,9]. As the cultivation of *S. lacrymans* from samples is often unsuccessful, a molecular biological approach is of interest. Polymerase chain reaction (PCR) amplification and the sequencing analysis of molecular markers in the internal transcribed spacer (ITS) region can identify fungi with varying degrees of discrimination depending on the taxonomical group considered [10]. The universal primer pair, ITS1 and

ITS4, is used generically for fungal identification; the primer pair, ITS1-F and ITS4-B, can be used to specifically highlight Basidiomycota, and the *S. lacrymans*-specific SL primer can be combined with ITS1 to detect this species [11–13]. The advantage of this method is that it allows the identification of a fungus regardless of its developmental stage [14]. However, some published studies using species-specific primers to detect *S. lacrymans* from environmental samples recorded some failures [6,7,15].

Fourier transform infrared spectroscopy (FTIR) is an alternative technique for the characterization and identification of different types of microorganisms, particularly fungi [16–20]. FTIR is based on the acquisition of a specific spectral signature of each microorganism after cultivation under given conditions. This spectral signature corresponds mainly to the composition of proteins, lipids, nucleic acids and carbohydrates in the biomass. The vibrational spectra of different fungal species have many signals in common and contain a large number of variables requiring the coupling to multivariate statistical analysis methods for their comparison [20]. The reproducibility of FTIR for fungal identification requires the standardization of each step of the analytical protocol: culture conditions, sample preparation and spectral acquisition parameters [21].

Several protocols using FTIR spectroscopy have been described to identify moulds [19,20,22–25]. Only a few studies have used FTIR to identify wood-decay fungi, but none to identity *S. lacrymans* [26,27]. In this study, we used different fungal growth conditions and different mycelial harvesting methods to select the best protocol for the identification of *S. lacrymans* by Attenuated total reflectance (ATR)-FTIR spectroscopy. The goal of this study was to standardize, optimize, and show that FTIR spectroscopy is a suitable method for the identification and discrimination of different strains of *S. lacrymans*.

2. Materials and Methods

2.1. Fungal Strains and Growth Conditions

Seven fungal strains (5 *S. lacrymans*, 1 *Rhodonia placenta* and 1 *Trichoderma harzianum*) obtained from the Culture Collection of Université de Bretagne Occidentale (UBOCC, Plouzané, France) and the culture collection of the Laboratoire de Recherche des Monuments Historiques (LRMH, Champs-sur-Marne, France), were used (Table 1). *S. lacrymans* and *R. placenta* were brown-rot fungi. *T. harzianum* was phylogenetically distant from *S. lacrymans* and served as an easily distinguished negative control of this species. Precultures of fungi were performed at 23 °C on Malt Extract Agar (MEA) (Malt extract 20.0 g, Agar 20.0 g per litre distilled water) for *R. placenta* and *T. harzianum*, and on Malt Extract Agar with Peptone (MEAP) (Malt extract 20.0 g, Peptone 1.0 g, Glucose 20.0 g, Agar 20.0 g per litre distilled water) for *S. lacrymans*.

Table 1. Fungal strains.

Species	Phylum	Strain Number [1]
Serpula lacrymans	Basidiomycota	UBOCC-A-110074
Serpula lacrymans	Basidiomycota	UBOCC-A-110108
Serpula lacrymans	Basidiomycota	UBOCC-A-111008
Serpula lacrymans	Basidiomycota	LRMH-RX-01
Serpula lacrymans	Basidiomycota	LRMH-SF-Ec3-003
Rhodonia placenta	Basidiomycota	LRMH-Op-001
Trichoderma harzianum	Ascomycota	LRMH-LV-Ec1-001

[1] UBOCC, Université de Bretagne Occidentale Culture Collection. LRMH, Laboratoire de Recherche des Monuments Historiques.

One mL of sterile saline solution (NaCl 0.9%) was added to a fungal pre-culture on the surface of a Petri dish. The mycelium was scraped off and the liquid containing the suspended mycelium fragments was collected and used to inoculate broth or agar media. The 100 µL of mycelium suspension was placed in a tube containing 8 mL of malt extract peptone broth (Malt extract 20.0 g, Peptone 1.0 g, Glucose 20.0 g per litre distilled water) or plated onto a polyether sulfone (PES) membrane (Merck, France) placed on the surface of a new MEA plate. Inoculated broth and agar media were incubated for 12 days at 23 °C. Broth cultures were then centrifuged for 1.5 min at 13,000× *g*. After the supernatant removal, the mycelium was washed with 10 mL of saline and centrifuged as above. The obtained pellet was dried for 48 h at 50 °C. Mycelia cultured on filtration membranes were prepared in two ways. Either the membrane carrying the mycelium was dried directly or the mycelium was harvested by scraping the surface of the membrane before drying. The mycelium collection was performed to exclude possible membrane interference on infrared signals. Drying was carried out for 48 h at 50 °C. For each strain and each culture condition, at least five independent cultures were performed.

2.2. FTIR Analysis

The FTIR analysis was carried out using an ALPHA II spectrometer equipped with an ATR Platinum crystal diamond module (Bruker Optics, Marne la Vallée, France). The analysis was performed between 4000 and 400 cm^{-1} with 64 accumulations per sample and a spectral resolution of 2 cm^{-1}. The infrared data were cleaned and processed with OPUS (version 6.5, Marne la Vallée, France) software as shown below. The spectrum was first trimmed by removing the values on either side of the interval between 3700 and 600 cm^{-1}, and then normalised. Normalisation of the spectra was conducted according to two models: a minimum–maximum (min–max) normalisation and a vector normalisation. For the min–max normalisation of the spectra, the data were smoothed (25 smoothing points), a straight line was generated between 2800 and 1800 cm^{-1} and the baseline was corrected before normalisation. All values of the spectra were set between 0 and 2, while keeping the distances between them. For the vector normalisation of the spectra, the data were second-derivative smoothed (25 smoothing points), a straight line was generated between 2800 and 1800 cm^{-1}, and then a vector normalisation was performed. The second derivative allowed for better resolution of the near and small absorption peaks. This was completed by calculating the average absorbance values of each spectrum in the selected spectral range. Each absorbance value in the spectrum was subtracted from this average value. The corrected value was then divided by the root of the sum of the squares of all the absorbances in the spectral range.

2.3. Statistical Analysis

The FTIR data were subjected to statistical analysis using partial least square discriminant analysis (PLS-DA) developed in XLSTAT (version 2018.1, Paris, France) software. This supervised linear analysis method using the algorithm of multivariate PLS regression permitted the classification of samples in the spectral ranges 3700–2800 cm^{-1} and 1800–600 cm^{-1}, corresponding to the major biochemical functions present in the mycelial samples. Each class was coded by a combination of 0 and 1, depending on whether the sample belonged to a given class or not. The explained data corresponding to the defined variables of the samples were correlated with a matrix of class properties corresponding to the different classes of the samples. A confusion matrix was generated to summarise information about the reclassifications after analysis by the algorithm. The percentage of correct attribution corresponding to the ratio of the number of well-classified observations to the total number of observations was calculated.

3. Results

3.1. Fourier Transformed Infrared (FTIR) Spectroscopy

The mycelial biomass obtained after growth and collection under different conditions was analysed by ATR-FTIR. The IR spectra obtained after normalisation of the data are presented in Figure 1, Figures S1 and S2. The two methods of data normalisation allow the visualisation of the signals corresponding to the chemical groups of the main biological macromolecules of the biomass in the spectral ranges 3600–2800 cm^{-1} and 1800–600 cm^{-1} (Figure 1 and Figure S1). Two regions dominate the fungal spectra: the signals between 1200 and 900 cm^{-1} corresponding to saccharides and nucleic acids, and the signals between

1650 and 1550 cm^{-1} corresponding to proteins. Signals corresponding to fatty acids were also observed between 3100 and 2800 cm^{-1}.

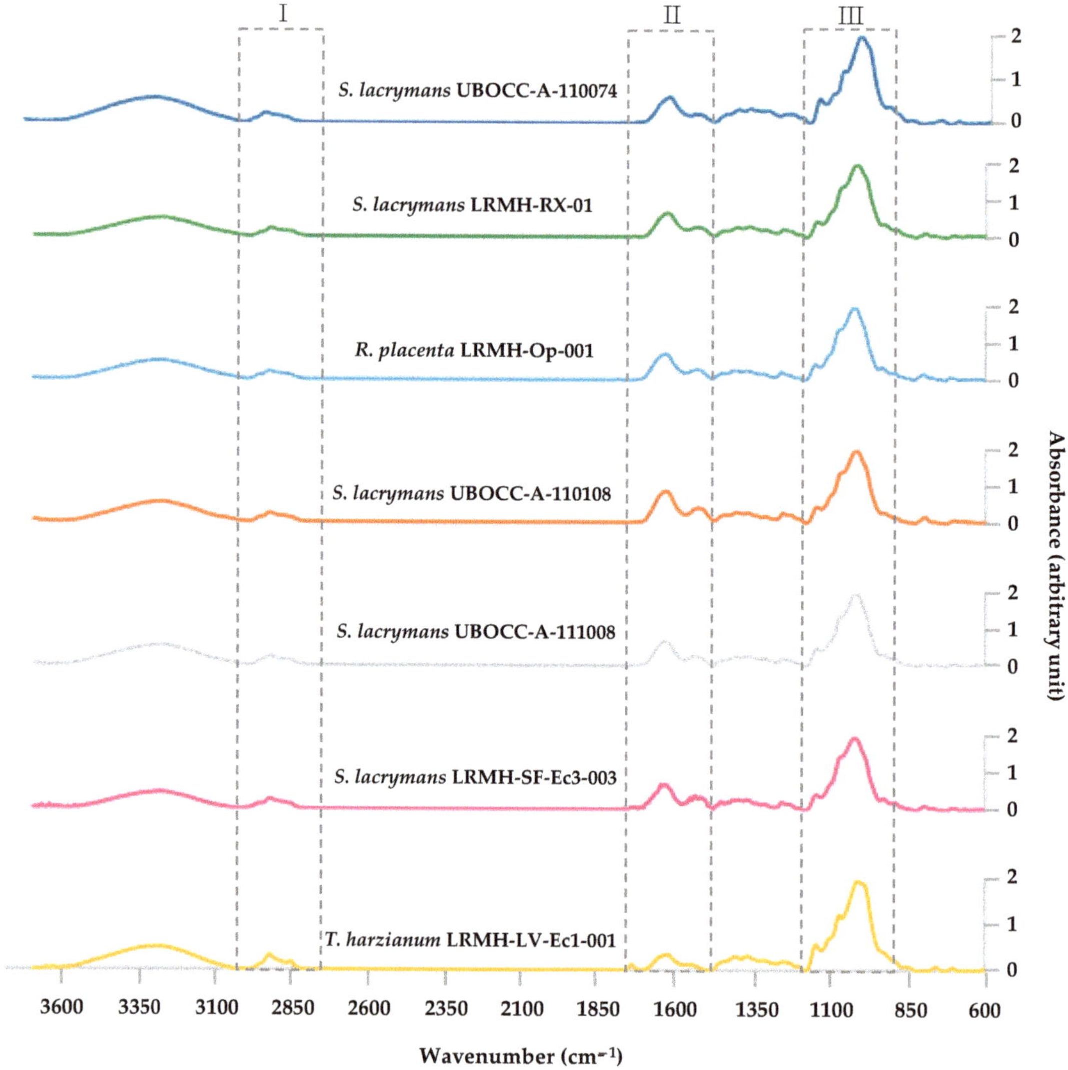

Figure 1. Fourier transform infrared (FTIR) spectra of fungi obtained after growth in liquid medium and min–max normalisation. Characteristic spectral ranges corresponding to major macromolecules are indicated by dotted boxes (I, fatty acids; II, proteins; III, polysaccharides and nucleic acids). Each spectrum is the average of at least five spectra obtained after repeating independent experiments under identical conditions.

3.2. *Partial Least Square Discriminant Analysis*

After the acquisition of the FTIR spectra of the different mycelium samples, the data were subjected to min–max and vector normalisation in parallel to determine the most efficient technique. The normalisation data were then processed by the same statistical analysis, PLS-DA.

The PLS-DA method carried out on the FTIR spectra of the different fungi after growth on the PES membrane and the harvesting of the mycelium allows the classification of the spectra with a percentage of correct attribution of 100% at the level of the phylum, species, and strain for the two methods of data normalisation (Figures 2 and 3).

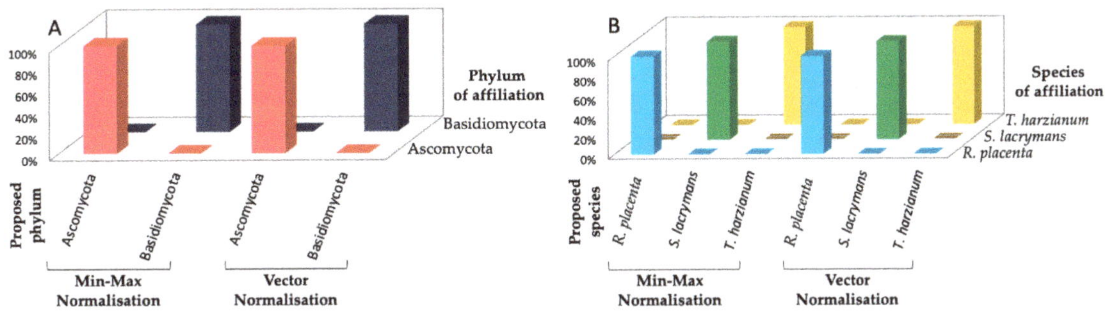

Figure 2. Confusion matrix obtained for the discrimination of the phylum (**A**) and the species (**B**) after growth on the polyether sulfone (PES) membrane and scraping the surface of the membrane to harvest the mycelium.

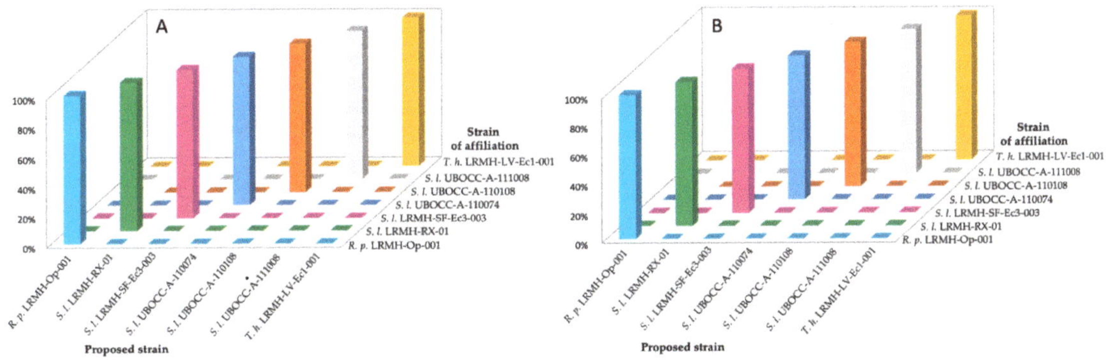

Figure 3. Confusion matrix obtained for the discrimination of the strains after growth on the PES membrane and scraping the surface of the membrane to harvest the mycelium. Min–max normalization (**A**), Vector normalization (**B**). *R. p.*, *Rhodonia placenta*. *S. l.*, *Serpula lacrymans*. *T. h.*, *Trichoderma harzianum*.

The PLS-DA method performed on the FTIR spectra of mycelia after fungal culture on the PES membrane and direct analysis on the membrane (Figure 4) classifies the spectra with a 100% correct attribution percentage at phylum as well as species level for both data normalisation methods. In the case of min–max normalisation, the percentage of correct attribution is 94.44% at the strain level within the species *S. lacrymans*. In this case, 16.7% of the spectra of *S. lacrymans* UBOCC-A-110074 are attributed to strain *S. lacrymans* UBOCC-A-110108 and 20% of the spectra of *S. lacrymans* UBOCC-A-110108 are attributed to strain *S. lacrymans* UBOCC-A-110074. 83.3% of the spectra of *S. lacrymans* UBOCC-A-110074 and 80% of the spectra of *S. lacrymans* UBOCC-A-110108 are correctly attributed to each of these two strains, respectively. The percentage of correct attribution is 100% for all strains when vector normalisation is performed, indicating a higher efficiency than min–max normalisation for this sample type (Figure 4).

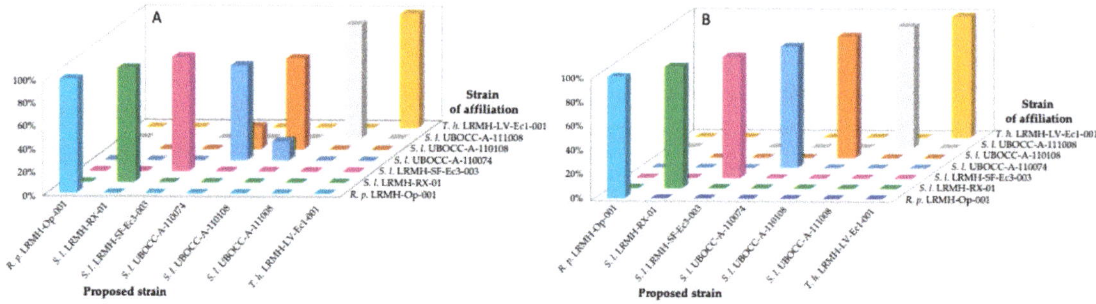

Figure 4. Confusion matrix obtained for the discrimination of the strains after growth on the PES membrane, and direct analysis on the membrane. Min–max normalization (**A**), Vector normalization (**B**). *R. p., Rhodonia placenta. S. l., Serpula lacrymans. T. h., Trichoderma harzianum.*

The method is less efficient for identifying fungi when cultured in liquid medium with a variable result according to the type of normalisation (Figure 5). A percentage of the correct attribution at the level of the strain of 80.95% and 99.05% is obtained for min–max and vector normalisation, respectively. After min–max normalisation, 6.7% of the spectra of *S. lacrymans* LRMH-RX-01 were attributed to *R. placenta* LRMH-Op-001, which belonged to another species, and the attribution of the spectra of the *S. lacrymans* strains UBOCC-A-110074, UBOCC-A-110108 and UBOCC-A-111008 were sometimes incorrect, with some spectra being associated with the wrong strain. After vector normalisation, the percentage of correct attribution was 99.05% (Figure 5). In this case, 6.7% of the spectra of *S. lacrymans* UBOCC-A-110074 were wrongly attributed to *S. lacrymans* LRMH-SF-EC3-003.

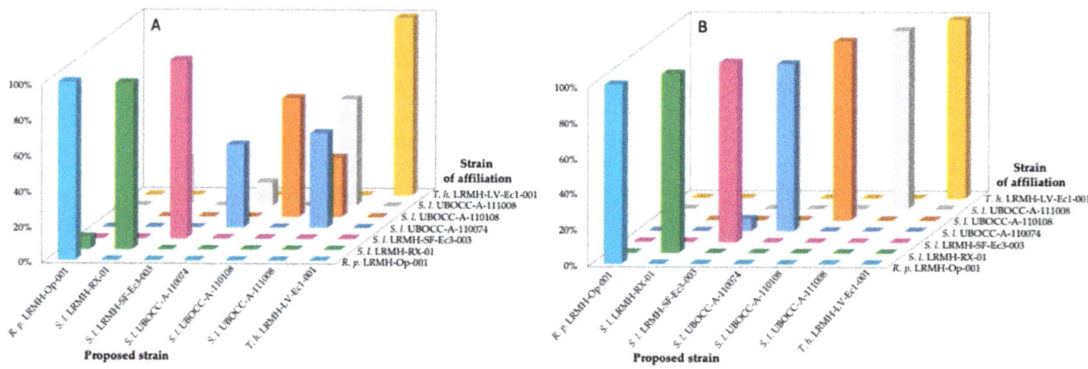

Figure 5. Confusion matrix obtained for the discrimination of the strains after growth in liquid medium. Min–max normalization (**A**), or Vector normalization (**B**). *R. p., Rhodonia placenta. S. l., Serpula lacrymans. T. h., Trichoderma harzianum.*

4. Discussion

The treatment of fungal growth in a building begins with identifying and eliminating the source of the moisture to stop the growth [1]. Then, it is necessary to determine the extent of the damage by inspecting all areas of the building where the fungus may have spread [7]. This also helps to characterise the type of wood decay (brown rot, white rot, and/or soft rot) and guide the identification of the fungi involved. This identification is essential as the course of action will depend on whether or not *S. lacrymans* is present. If *S. lacrymans* is not involved, the damage is minor and the fungal growth is limited to the areas where the wood is degraded. Therefore, treatment can be limited to allowing the building to dry out by ventilating it. However, if the source of the moisture cannot be removed immediately or the building cannot be allowed to dry out, fungicidal treatments are applied and the contaminated woodwork and plasterwork is destroyed. In the case of *S. lacrymans*, uncontaminated woodwork adjacent to areas where the fungus is present is also destroyed. Furthermore, in France, the occupant or owner of the building affected by *S. lacrymans* has a legal obligation to declare the presence of the fungus in the building to the city hall [28]. In the case of a historic building which may have valuable woodwork and walls decorated with sculptures or paintings, care should be taken and biocidal treatments with the least possible impact on the materials should be used. In this context, the identification of *S. lacrymans* in the early stages is essential.

The identification of fungi involved in wood biodegradation through culture and molecular methods could be tedious, time-consuming and/or expensive. FTIR spectroscopy has shown its suitability for the identification of fungi of medical or food interest, but few data exist for wood-rot fungi [19,22,27,29]. This identification method is based on the global analysis of the mycelium or spores of a fungus to determine its spectral profile and to compare it to reference spectra [19,20]. The IR spectrum obtained for a fungus corresponds to all the biomolecules present in the biomass at the time of analysis. The differences in the composition and quantity of these biomolecules between species, genera or strains make it possible to differentiate fungi by FTIR spectroscopy. These parameters vary for the same individual depending on the stage and conditions of growth [30].

In this study, the application of FTIR spectroscopy to the identification and discrimination of *S. lacrymans* strains was evaluated. Reference FTIR spectra were made from different strains of *S. lacrymans*, *R. placenta* and *T. harzianum* after growth under standardised conditions. The analysis was carried out using a wide spectral range extending from 3700 to 2800 and from 1800 to 600 cm^{-1}, which corresponded to the signals obtained for the main biological macromolecules carbohydrates, proteins, lipids and nucleic acids [16,19,31]. The analysis of FTIR spectra was completed from fungal mycelium and not from fungal spores in order to obtain more complex spectra and a better discrimination [19,20,22,23]. Furthermore, *R. placenta* and *S. lacrymans* did not sporulate under the laboratory conditions used.

Several culture protocols have been described for FTIR application. A mycelium can be analysed after liquid culture and harvested by centrifugation, filtration, or freeze-drying, or obtained after growth on the surface of an agar medium or a membrane deposited on an agar medium and harvested from its surface [18,19,23,26,32]. The advantage of liquid culture is that it prevents the sporulation of many fungi [18,33]. In the present study, three conditions for obtaining mycelium were compared: growth in broth and harvesting by centrifugation, growth on a microfiltration membrane deposited on the surface of an agar medium, and direct analysis of the mycelium on the membrane or collection of the mycelium from the surface of the membrane by scraping before analysis. The culture time was set at 12 days to obtain the sufficient biomass for slow-growing fungi such as *S. lacrymans*. At longer culture times when there is limited or no growth in the study area, the production of biomolecules tends to decrease in the mycelium leading to an increase in the reproducibility of FTIR analyses [18]. After 2, 3, 4 or 5 days' of growth of moulds on malt extract agar, the highest reproducibility of FTIR spectra was obtained for the longest cultivation time [18].

After the spectral data acquisition from mycelium samples, a pre-processing of the data is necessary before their statistical analysis. In different studies using FTIR spectroscopy for fungal identification, a vector normalisation of the spectral data is performed [19,20,22,26,34]. Vector normalisation brings the spectra back to the same intensity without altering the spectral profile by performing a second derivative which increases the spectral resolution. In the present study, the results obtained after vector normalisation were compared to the results obtained after min–max normalisation. In the case of min–max normalisation, the baseline of the data was corrected and the data for each spectrum were multiplied so that for each spectrum the minimum was 0 and the maximum was 2. After the normalisation of the spectral data by each of the two methods, the identification of the fungi was carried out using PLS-DA analysis of the spectral data. PLS-DA is a supervised statistical analysis method unlike other statistical methods such as Principal Component Analysis or Hierarchical Ascendant Classification (cluster analysis) [23,35–37]. In the case of a supervised statistical analysis method, each spectrum is assigned a class so that quantitative spectral data are supplemented by qualitative data. From the annotated reference IR spectra, the algorithm creates a prediction function of the class to which the analysed spectrum belongs.

The results obtained showed that the choice of culture conditions and the type of data normalisation were essential for the identification of *S. lacrymans* by FTIR spectroscopy. The min–max normalisation was less resolving than the vector normalisation which was widely used for FTIR fungal identification [20,22,37]. Regarding the mode of mycelium culture, although liquid culture may be in some cases the simplest to implement [18,24], it was not relevant for the identification of *S. lacrymans* by FTIR. Indeed, broth growth was the culture condition that led to the lowest resolution for discriminating different *S. lacrymans* strains, with even an error at the species level, as a spectrum of the *S. lacrymans* strain (LRMH-RX-01) was assigned to *R. placenta* (LRMH-Op-001). In the other mycelium growth conditions tested (on the surface of a microfiltration membrane), only strain affiliation errors within the *S. lacrymans* species were observed. When the analysis was performed on the IR data obtained after harvesting the mycelium from the membrane, the assignment of IR spectra was consistently correct at the phylum, species and strain level of *S. lacrymans*. This 100% correct assignment percentage must be put into perspective because the number of references we used was small, as the number of strains analysed was limited. Kümmerle et al. studied 722 unknown yeast isolates with a library of reference spectra based on 332 strains belonging to 18 genera and 74 species and obtained a correct identification rate of 97.5% [38]. In a study analysing 25 mould strains with a reference spectra library based on 106 strains belonging to 14 genera and 32 species, the correct identification rates of 98.97% and 98.77% were obtained at the genus and species level, respectively [19]. In another study of 105 strains, with a reference spectra library based on 288 strains belonging to 26 genera and 68 species, the correct identification rates of 99.17% and 92.3% were obtained at genus and species level, respectively [20].

In the conclusion of the present study, the most efficient fungal sample preparation and analysis protocol to identify an isolate suspected to be *S. lacrymans* is a culture on the surface of a filtration membrane deposited on a malt extract agar medium; a collection of the mycelium by scraping, followed by drying of the mycelium by heat treatment; FTIR analysis; the processing of the spectral data by vector normalisation, and PLS-DA statistical analysis. The constitution of a library of reference spectra based on a larger number of strains of *S. lacrymans* and other fungi and the analysis of field isolates of wood-decay fungi (particularly brown-rot fungi such as *Gloeophyllum sepiarium, Fibroporia (Antrodia) vaillantii, Coniophora puteana, Serpula himantoides*) is necessary to confirm the effectiveness of this protocol. The next step in this study is to adapt the use of FTIR spectroscopy to the identification of *S. lacrymans* by direct in situ analysis without fungal culture. This will involve performing FTIR spectra of fungi after growth on different native substrates such as wood, and materials found in buildings.

Supplementary Materials: The following are available online at https://www.mdpi.com/article/10.3390/app11188463/s1, Figure S1: FTIR spectra of fungi obtained after growth under different conditions, from different sample preparations, and two normalisations of the spectral data. Figure S2: FTIR spectra of fungi obtained after growth in different conditions illustrating the reproducibility of the method and the heterogeneity of the mycelial matter.

Author Contributions: Conceptualization, P.D.M. and F.B.; methodology, P.D.M.; validation, P.D.M. and F.B.; investigation, R.B.; data curation, R.B.; writing-original draft preparation, P.D.M.; writing-review and editing, R.B. and F.B.; supervision, P.D.M.; project administration, P.D.M.; funding acquisition, P.D.M. and F.B. All authors have read and agreed to the published version of the manuscript.

Funding: This study was part of the project "Biodétérioration du bois dans les bâtiments historiques: biodiversité microbienne et évaluation in vitro de traitements alternatifs" funded by "La Fondation des Sciences du Patrimoine", LabEx PATRIMA (AAP_2016_01), Agence nationale de la recherche ANR-10-LABX-0094-01.

Conflicts of Interest: The authors declare no conflict of interest.

References

1. Krzyzanowski, N.; Oduyemi, K.; Jack, N.; Ross, N.; Palfreyman, J. The management and control of dry rot: A survey of practitioners' views and experiences. *J. Environ. Manag.* **1999**, *57*, 143–154. [CrossRef]
2. Singh, J. Dry rot and other wood-destroying fungi: Their occurrence, biology, pathology and control. *Indoor Built Environ.* **1999**, *8*, 3–20. [CrossRef]
3. Schmidt, O. Indoor wood-decay basidiomycetes: Damage, causal fungi, physiology, identification and characterization, prevention and control. *Mycol. Prog.* **2007**, *6*, 261–279. [CrossRef]
4. Watkinson, S.; Eastwood, D. Serpula lacrymans, Wood and Buildings. In *Advances in Clinical Chemistry*; Elsevier BV: Amsterdam, The Netherlands, 2012; Volume 78.
5. Reinprecht, L. *Biological Degradation of Wood, in Wood Deterioration, Protection and Maintenance*; John Wiley & Sons, Ltd.: Oxford, UK, 2016; pp. 62–125.
6. Maurice, S.; Le Floch, G.; Le Bras-Quéré, M.; Barbier, G. Improved molecular methods to characterise Serpula lacrymans and other Basidiomycetes involved in wood decay. *J. Microbiol. Methods* **2011**, *84*, 208–215. [CrossRef]
7. Slimen, A.; Barboux, R.; Mihajlovski, A.; Moularat, S.; Leplat, J.; Bousta, F.; Di Martino, P. High diversity of fungi associated with altered wood materials in the hunting lodge of "La Muette", Saint-Germain-en-Laye, *France. Mycol. Prog.* **2020**, *19*, 139–146. [CrossRef]
8. Huckfeldt, T.; Schmidt, O. Identification key for European strand-forming house-rot fungi. *Mycologist* **2006**, *20*, 42–56. [CrossRef]
9. Gabriel, J.; Švec, K. Occurrence of indoor wood decay basidiomycetes in Europe. *Fungal Biol. Rev.* **2017**, *31*, 212–217. [CrossRef]
10. Schoch, C.L.; Robbertse, B.; Robert, V.; Vu, D.; Cardinali, G.; Irinyi, L.; Meyer, W.; Nilsson, R.H.; Hughes, K.; Miller, A.N.; et al. Finding needles in haystacks: Linking scientific names, reference specimens and molecular data for Fungi. *Database* **2014**, *2014*, bau061. [CrossRef] [PubMed]
11. Gardes, M.; Bruns, T.D. ITS primers with enhanced specificity for basidiomycetes—Application to the identification of mycorrhizae and rusts. *Mol. Ecol.* **1993**, *2*, 113–118. [CrossRef]
12. Schmidt, O.; Moreth, U. Species-specific PCR primers in the rDNA-ITS region as a diagnostic tool for Serpula lacrymans. *Mycol. Res.* **2000**, *104*, 69–72. [CrossRef]
13. Bellemain, E.; Carlsen, T.; Brochmann, C.; Coissac, E.; Taberlet, P.; Kauserud, H. ITS as an environmental DNA barcode for fungi: An in silico approach reveals potential PCR biases. *BMC Microbiol.* **2010**, *10*, 189. [CrossRef] [PubMed]
14. Manter, D.K.; Vivanco, J.M. Use of the ITS primers, ITS1F and ITS4, to characterize fungal abundance and diversity in mixed-template samples by qPCR and length heterogeneity analysis. *J. Microbiol. Methods* **2007**, *71*, 7–14. [CrossRef] [PubMed]
15. Horisawa, S.; Sakuma, Y.; Doi, S. Qualitative and quantitative PCR methods using species-specific primer for detection and identification of wood rot fungi. *J. Wood Sci.* **2009**, *55*, 133–138. [CrossRef]
16. Naumann, D.; Helm, D.; Labischinski, H. Microbiological characterizations by FT-IR spectroscopy. *Nat. Cell Biol.* **1991**, *351*, 81–82. [CrossRef] [PubMed]
17. Santos, C.; Fraga, M.E.; Kozakiewicz, Z.; Lima, N. Fourier transform infrared as a powerful technique for the identification and characterization of filamentous fungi and yeasts. *Res. Microbiol.* **2010**, *161*, 168–175. [CrossRef]
18. Shapaval, V.; Møretrø, T.; Suso, H.P.; Asli, A.W.; Schmitt, J.; Lillehaug, D.; Martens, H.; Böcker, U.; Kohler, A. A high-throughput microcultivation protocol for FTIR spectroscopic characterization and identification of fungi. *J. Biophotonics* **2010**, *3*, 512–521. [CrossRef]
19. Lecellier, A.; Mounier, J.; Gaydou, V.; Castrec, L.; Barbier, G.; Ablain, W.; Manfait, M.; Toubas, D.; Sockalingum, G.D. Differentiation and identification of filamentous fungi by high-throughput FTIR spectroscopic analysis of mycelia. *Int. J. Food Microbiol.* **2014**, *168–169*, 32–41. [CrossRef]
20. Lecellier, A.; Gaydou, V.; Mounier, J.; Hermet, A.; Castrec, L.; Barbier, G.; Ablain, W.; Manfait, M.; Toubas, D.; Sockalingum, G. Implementation of an FTIR spectral library of 486 filamentous fungi strains for rapid identification of molds. *Food Microbiol.* **2015**, *45*, 126–134. [CrossRef]
21. Essendoubi, M.; Toubas, D.; Lepouse, C.; Leon, A.; Bourgeade, F.; Pinon, J.M.; Manfait, M.; Sockalingum, G.D. Epidemio-logical investigation and typing of Candida glabrata clinical isolates by FTIR spectroscopy. *J. Microbiol. Methods.* **2007**, *71*, 325–331. [CrossRef]
22. Fischer, G.; Braun, S.; Thissen, R.; Dott, W. FT-IR spectroscopy as a tool for rapid identification and intra-species character-ization of airborne filamentous fungi. *J. Microbiol. Methods* **2006**, *64*, 63–77. [CrossRef]
23. Garon, D.; El Kaddoumi, A.; Carayon, A.; Amiel, C. FT-IR Spectroscopy for Rapid Differentiation of Aspergillus flavus, Aspergillus fumigatus, Aspergillus parasiticus and Characterization of Aflatoxigenic Isolates Collected from Agricultural Environments. *Mycopathologia* **2010**, *170*, 131–142. [CrossRef]

24. Shapaval, V.; Schmitt, J.; Møretrø, T.; Suso, H.P.; Skaar, I.; Åsli, A.W.; Lillehaug, D.; Kohler, A. Characterization of food spoilage fungi by FTIR spectroscopy. *J. Appl. Microbiol.* **2013**, *114*, 788–796. [CrossRef] [PubMed]
25. Kogkaki, E.A.; Sofoulis, M.; Natskoulis, P.; Tarantilis, P.A.; Pappas, C.S.; Panagou, E.Z. Differentiation and identification of grape-associated black aspergilli using Fourier transform infrared (FT-IR) spectroscopic analysis of mycelia. *Int. J. Food Microbiol.* **2017**, *259*, 22–28. [CrossRef] [PubMed]
26. Naumann, A. A novel procedure for strain classification of fungal mycelium by cluster and artificial neural network analysis of Fourier transform infrared (FTIR) spectra. *Analyst* **2009**, *134*, 1215–1223. [CrossRef]
27. Girometta, C.; Dondi, D.; Baiguera, R.M.; Bracco, F.; Branciforti, D.S.; Buratti, S.; Lazzaroni, S.; Savino, E. Characterization of mycelia from wood-decay species by TGA and IR spectroscopy. *Cellulose* **2020**, *27*, 6133–6148. [CrossRef]
28. LOI n° 2014-366 du 24 mars 2014 pour l'accès au Logement et un Urbanisme Rénové. Available online: https://www.cnle.gouv.fr/loi-no-2014-366-du-24-mars-2014-1240.html (accessed on 24 June 2021).
29. Bastert, J.; Korting, H.C.; Traenkle, P.; Schmalreck, A.F. Identification of dermatophytes by Fourier transform infrared spectroscopy (FT-IR). *Mycoses* **1999**, *42*, 525–528. [CrossRef]
30. Cantu, D.; Carl Greve, L.; Labavitch, J.M.; Powell, A.L.T. Characterization of the cell wall of the ubiquitous plant patho-gen Botrytis cinerea. *Mycol. Res.* **2009**, *113*, 1396–1403. [CrossRef]
31. Szeghalmi, A.; Kaminskyj, S.; Gough, K.M. A synchrotron FTIR microspectroscopy investigation of fungal hyphae grown under optimal and stressed conditions. *Anal. Bioanal. Chem.* **2006**, *387*, 1779–1789. [CrossRef]
32. Mularczyk-Oliwa, M.; Bombalska, A.; Kaliszewski, M.; Włodarski, M.; Kwaśny, M.; Kopczyński, K.; Szpakowska, M.; Trafny, E.A. Rapid discrimination of several fungus species with FTIR spectroscopy and statistical analysis. *Biuletyn W. A. T.* **2013**, *62*, 71–80.
33. Dahlberg, K.R.; Etten, J.L.V. Physiology and Biochemistry of Fungal Sporulation. *Annu. Rev. Phytopathol.* **1982**, *20*, 281–301. [CrossRef]
34. Salman, A.; Tsror, L.; Pomerantz, A.; Moreh, R.; Mordechai, S.; Huleihel, M. FTIR spectroscopy for detection and identifi-cation of fungal phytopathogenes. *Spectroscopy* **2010**, *24*, 261–267. [CrossRef]
35. Cao, Z.; Wang, Z.; Shang, Z.; Zhao, J. Classification and identification of Rhodobryum roseum Limpr. and its adulterants based on fourier-transform infrared spectroscopy (FTIR) and chemometrics. *PLoS ONE* **2017**, *12*, e0172359. [CrossRef] [PubMed]
36. Mariey, L.; Signolle, J.; Amiel, C.; Travert, J. Discrimination, classification, identification of microorganisms using FTIR spectroscopy and chemometrics. *Vib. Spectrosc.* **2001**, *26*, 151–159. [CrossRef]
37. Pena, R.; Lang, C.; Naumann, A.; Polle, A. Ectomycorrhizal identification in environmental samples of tree roots by Fouri-er-transform infrared (FTIR) spectroscopy. *Front. Plant Sci.* **2014**, *5*, 229. [CrossRef]
38. Kümmerle, M.; Scherer, S.; Seiler, H. Rapid and Reliable Identification of Food-Borne Yeasts by Fourier-Transform Infrared Spectroscopy. *Appl. Environ. Microbiol.* **1998**, *64*, 2207–2214. [CrossRef] [PubMed]

applied sciences

MDPI

Review

Electromagnetic Field as Agent Moving Bioactive Cations. A New Antimicrobial System in Architecture Technology

Andrzej Chlebicki [1,*], Wojciech Spisak [2], Marek W. Lorenc [3], Lucyna Śliwa [1] and Konrad Wołowski [1]

1 W. Szafer Institute of Botany, Polish Academy of Sciences, Lubicz 46, 31-512 Kraków, Poland; l.sliwa@botany.pl (L.Ś.); k.wolowski@botany.pl (K.W.)
2 Research & Development Centre, Alcor Ltd., Kępska 12, 45-130 Opole, Poland; spisak@alcor.pl
3 Department of Landscape Architecture, Wroclaw University of Environmental and Life Sciences, Grunwaldzka 55, 50-357 Wrocław, Poland; marek.lorenc@upwr.edu.pl
* Correspondence: a.chlebicki@botany.pl

Featured Application: Paint with Zn and Cu (patent Galvi PL229012) can be used inside damp lodgings (butchers, winery, bathrooms, cellars, storehouses, museums etc.) as well as elevations of buildings. Using paint with Zn, Bi, Cu (three electrode systems) for covering the lower side of the stretchers is suggested. This system is the source of bioactive "ions cocktail" and it works only at a very fine level on the lower surface of the stretcher. According to a pessimistic prediction, building material in architecture could be largely destroyed at the end of XXI century. Our last investigations can help in this matter. Two-phase media with SNA (synthetic low-nutrient agar) medium + sterile grandiorite, melabasanite, sandstone, marl and sodium glass (as control) were used. This enables the investigation of the processes of colonization, course, and development of selected microorganisms in the emerging rock microhabitats (patent pending). For all analyzed rocks, their color may also be important. A dark or even black rock (i.e., melabasanite) exposed to solar radiation heats up much more than a light or even white one. In this particular case, granodiorite is light gray, sandstone is beige, and marl is pale beige.

Abstract: There is a new described antifungal system (GALVI) involving the moving of bioactive ions of Zn, Cu and Bi for the protection of cultural heritage objects such as buildings, sculptures and stretchers. There were two kinds of galvanic cells that were used: the first composed of a two-electrodes system, Zn, Cu, and second one composed of a three-electrodes system, Zn, Bi and Cu. Moreover, two-phase media are proposed with various kinds of rocks used in architectonical objects. Microorganisms inhabit the boundaries of two liquid and solid phases. This enables the investigation of the process of rock colonization. Possible applications of the suggested GALVI system are mentioned.

Keywords: galvanic cells; GALVI technology; Zn; Bi; Cu; antimicrobial technology; two-phase media

Citation: Chlebicki, A.; Spisak, W.; Lorenc, M.W.; Śliwa, L.; Wołowski, K. Electromagnetic Field as Agent Moving Bioactive Cations. A New Antimicrobial System in Architecture Technology. *Appl. Sci.* **2021**, *11*, 8320. https://doi.org/10.3390/app11188320

Academic Editors: Filomena De Leo and Daniela Isola

Received: 24 July 2021
Accepted: 27 August 2021
Published: 8 September 2021

1. Introduction

Cultural heritage contains not only buildings but also mines, caves, graves, old canals, bridges, ships, boats, historical glass, ceramics, sculptures, and artistic painting. Historical architectural objects are mostly made of stone and wood and are rarely made out of glass, metals and ceramics. We will try to describe methods of building protection against fungi. Various microorganisms are involved in the biodeterioration of the architecture objects. These are algae, lichens, fungi, and bacteria. Damp buildings are mostly colonized by hydrophilic fungi [1].

In architectural objects, various kinds of rocks such as granite, basalt, sandstones, and calcareous stones were used. Among the various rock types, calcareous stones are the most widely used, while other kinds are less common in architecture technology [2]. Embry & Klovan [3] and more recently Flügel [4] presented a classification of carbonate rocks (limestones, dolomites, marls, etc.). Siegesmund and Török [2] put these rocks in chemically and biologically precipitated rocks. Gadd [5] pointed out that microcolonial fungi can significantly change a rock's surficial appearance, such as discoloration, staining, biodeterioration and the formation of new biogenic minerals. Calcareous stones are often investigated by mycologists [6–15]. Some publications are devoted to other stones: basalt [16], granite [17,18], sandstone [19,20] and minerals such as löllingite $FeAs_2$ and arsenopyrite FeAsS [21,22]. Brandl [23] cited 82 articles that were devoted to bioleaching of metals from mineral resources (albite, amphibolite, andesite, apophylite, argentite, armtone, augite, basalt, bauxite, biotite, calamine, calcium, carbonaceous ore, chalcocite, chrysocolla, chrysolite, copper molybdenum ore, cobalt, cuprite, diorite, dunite, dysprosium, ferromanganese sea nodules, galena, garnierite, genthite, geothite, glauconite, gold bearing ore, gypsum, grandiorite, granite, halloysite, harmtone, hematites, hemetitic lateritic ore, heulandite, illite, kaolonite, labradorite, lateritic nickel ore, leucite, limescale, limonite, lydite, malachite, manganese, manganese ore, microcline, montmorillonmite, muscovite, natrolite, nepheline, neodymium, olivine, orthoclase, pearceite, pegmatite, peridoptite, phlogopite, phosphates of aluminium, polybasite, piritic ore, proutite, pyromorphite, quartz sand, rhodochrosite, rhyolite, samarium, sandstone, saponite, scheelite, serpentine, silicate nickel ore, spodumene, stilbite, vermiculite, wollastonite, yttrium, zinc oxide, zinc). Among stone colonizing fungi we can find mostly anamorphic states of ascomycete and basidiomycete fungi, and Zygomycetes rarely occurs. Coincidentally, teleomorphs of Basidiomycetes occur [24]. There are different names for rock-inhabiting fungi (lithobionts), such as microcolonial fungi = MCF [5,11], rock-eating fungi = REF [25], rock-inhabiting fungi = RIF [14], and rock-building fungi = RBF [26]. Most of them belong to black yeasts and their relatives [27,28]. Black yeasts are mostly extremotolerant anamorphs that lost their teleomorphic stadia. These RIF microorganisms are present both in hot deserts and cold Arctic and Antarctic places. Stone deterioration in the presence of microorganisms is approximately ten thousand times faster than without their presence [29]. Cyanobacteria creating biofilms are often reported from diverse architectural buildings (Table 1).

Table 1. The most frequent strains of Cyanobacteria identified in diverse architectural buildings.

Name of Species	Citation	Kind of Substratum	Climate
Chroococcus sp. *Chlorococcus* sp.	[30,31]	historic church in Porto Alegre, RS, Brazil Europe, UK. Historical building Windermere	Tropical Moderate
Fischerella sp., *Gloeocapsa*,	[32]	cyanobacteria from external stone walls, Mexico	Equatorial and subequatorial zone
Leptolyngbya gracillima	[33]	cooling tower of the power plant at Bełchatów, Poland	Moderata
Scytonema myochrous	[33]	cooling tower of the power plant at Bełchatów, Poland;	Moderate
Scytonema sp.	[34]	limestone walls at other low-pollution sites, the Great Jaguar Pyramid at Tikal, Guatemala	Equatorial and subequatorial zone
Gloeothece rupestris	[33]	wall of the cooling tower of the power plant et Bełchatow	Moderate
Nostoc sp., *Gloeothece rupestris* (Lyngbye)	[33]	wall of the cooling tower of the power plant at Bełchatów	Moderate
Trentepohlia sp.	[32]	stone altar in the architectural zone of Becan, Campeche state, Mexico	Tropical
Gloecapsa sp. *Gloeothece*, *Aphanocapsa*, and *Chroococcus*, and filamentous species like *Scytonema* and *Tolypothrix*.	[35]	on urban building surfaces	Tropics

Fungi are extremely corrosive endoliths [27,29,36,37]. They can produce acids (Figure 1) that can dissolve minerals. Fungi can cause the biodeterioration of stone elevations, surfaces, various kinds of monuments, buildings, sculptures, and other objects made of stone as well as wooden elements, artistic paintings, and even concrete walls [38]. There are a huge number of publications concerning the fungal deterioration of rock in architecture objects. Nielsen [39] noted 746 such publications in his own reference list. There are only some important articles [7,13,19,20,22,40–55]. Recently Adams et al. [1] noted 111 species of fungi in damp buildings. Table 2 presents only the most common (according to Nielsen, [39] and phenomenal fungi noted in buildings, sculptures, tombs, and mines.

Table 2. The most frequent strains of fungi identified in diverse architectural buildings, with published references, focusing on the diverse climate of Earth. The most common fungal species are highlighted in italics.

Name of Fungi	Citations	Kind of Substratum	Climate
Alternaria	very common	gypsum, tombstone, damp buildings	widely distributed
Aspergillus	very common	walls	widely distributed
Aureobasidium	common, [15,56]	marble, damp buildings	widely distributed
Baudoinia	[50,57]	buildings, cars	temperate
Beauveria	[58]	biofilm in Roman catacombs	mediterranean
Botryomyces	[8]	stone monuments	mediterranean
Chaetomium	[1,59]	damp buildings	temperate
Cladosporium	very common	damp buildings, tombstone	widely distributed
Constantinomyces	[15]	tombstone	temperate
Epicoccum	[1]	damp buildings	widely distributed
Exophiala	[15,60]	gold mine, tombstones	temperate
Geomyces	[21]	gold mine	temperate
Hortea	[29]	marble, monuments Delos island	mediterranean
Knufia	[15,56]	marble, monuments, tombstone	temperate, mediterranean
Lecanicillium	[58]	biofilm in Roman catacombs	mediterranean
Neocatenulostroma	[15]	tombstone	temperate
Neodevresia	[15]	tombstone	temperate
Penicillium	very common	damp buildings	widely distributed
Phoma	very common	damp buildings	widely distributed
Rhizopus stolonifer	Ref. [1], Chlebicki personal inf.	damp wall after flood, packing houses	widely distributed
Rhodotorula	[15,39]	paints, tombstone	widely distributed
Sarcinomyces	[8]	marble	mediterrenean
Stachybotrys	very common	gypsum	temperate
Torrubiella	[58]	biofilm in Roman catacombs	mediterranean
Trichoderma	very common	damp buildings, tombstone	widely distributed
Trimmatostroma	[29]	rock	temperate
Zasmidium	[61,62]	vine cellar wall	temperate

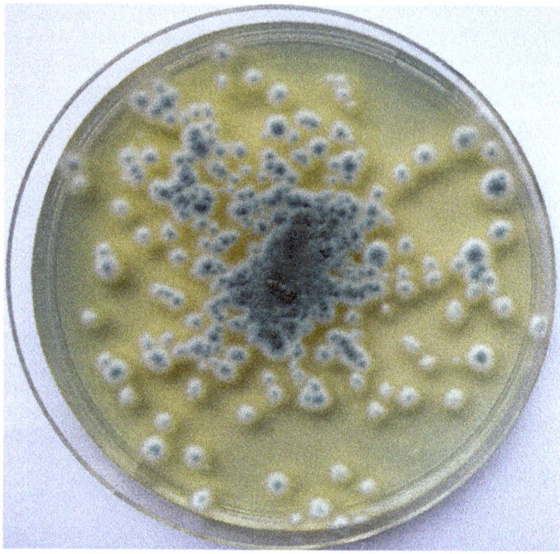

Figure 1. Production of acid by *Talaromyces diversus* isolated from Golubinka anchialaine cave, strain cultivated on YMA—(Yast Malt Agar) + 20 drops of dibromothymolsulphonephthalein and 10 drops of 10% KOH. Such media were green at the beginning (pH = 7). Paler places indicate an acid production, photo A. Chlebicki.

2. Biofilms

Natural biofilms are composed of many various species, crustose epilitic lichens, algae, cyanobacteria. Fungi are also present in biofilms as propagules [21], but they are not forms of biofilms. However, *Aureobasidium pullulans* sensu lato forms biofilm with extracellular polymers (EPS) composed of pullulan [63]. Biofilms are composed from bacterial cells embedded in an organic polymer matrix [64], known as extracellular polymers EPS [65]. The chemical composition of EPS is extremely diverse [65]. Moreover, natural biofilms can be composed of some layers with algae, lichens, bacteria and fungi. The bacteriological investigation of rock biofilm in Gertruda adit [66,67] showed that the Gram and DAPI staining methods showed all the bacterial shapes known in literature, whereas 16S rRNA gene amplification from total DNA showed more than one hundred OTUs, dominated by *α-Proteobacteria* [68]. Only seven species of fungi were isolated from this biofilm [10]. Among Antarctic sponge-associated bacteria, the *Roseovarius* genus plays a pivotal role in biofilm production [69], enabling adhesion and ecological competition with other microorganisms [70]. Artificial biofilms can be composed of some species of bacteria, algae and black yeasts [71]. The vertical distribution of these microorganisms was different: algae and black yeasts were more abundant in the outer layer, while in the inner layer only few algae were observed. However, it appeared that the inner layer harbored a higher amount of microorganisms than the outer one. Miao et al. [72] observed more complex bacterial networks on artificial substrates than on natural substrates. However, the keystone species on natural substrates were more abundant. Among Antarctic sponge-associated bacteria, the *Roseovarius* genus plays a pivotal role in biofilm production, enabling adhesion and ecological competition with other microorganisms.

3. GALVI Technology

The rapid development of drug resistance in pathogenic fungi forced us to develop a new antifungal technology [73]. Snethlage [74] described antimicrobial drugs that we can use at the present time in architectural technology, as well as the stone conservation of historic and artistic objects. Among these drugs, conservative materials such as water repellents, biocides, and consolidants are mentioned. Then, chemicals such as water glass, ethyl silicates, fluorosilicates, epoxy resins, acrylic resins, and silicon organic compounds are presented. Moreover, Castanier et al. [44] described a biomineralization process with a suitable bacterial suspension culture that creates a coating scale, the "biocalcin", whereas Koo et al. [75] noted a wide variety of biofilm-targeting strategies, among them a surface charge, roughness and topography, hydrophobicity, as well as using inorganic metallic nanoparticles of Ag and Cu and organic nanoparticles (inhibitors such as liposomes and aptamers), and biofilm removal by using mechanical, energy- and light disruption (ultrasound, acoustic, electric waves, laser shockwaves). This can protect stone objects from deterioration. The next methods, new antifungal discovery—system RNA interference or the exogenous application of synthetic RNA, are in preparation [73]. Recently, De Leo et al. [71] suggested the application of surface active ionic liquids (SAILs), based on cholinium cations and dodecylbenzenesulfonate, as anions in the protection of cultural heritage.

The toxicity of metal ions (mostly silver and copper) is well documented, and results have been presented in many papers [46,76–78]. Zinc has been shown to inhibit the growth and respiration of fungi [79,80], as well as the germination of fungal spores [81]. The action of silver is proportional to the bioactive released silver ion (Ag^+). Silver interacts with bacterial or fungal cell membranes [46]. We proposed using a new antimicrobial technology—an electromagnetic field created by galvanic cells that influences the moving of metal ions [82,83]. This method was named GALVI system. We used two kinds of galvanic cells: the first composed of a two-electrodes system, Zn, Cu [82], and the second one composed of a three-electrodes system, Zn, Bi and Cu [83] (Figure 2). The three-electrodes system has a stronger antibacterial activity than the first one (Figure 2). Unfortunately, this system is much more expensive because the use of expensive rheological additives securing the fall of bismuth is necessary.

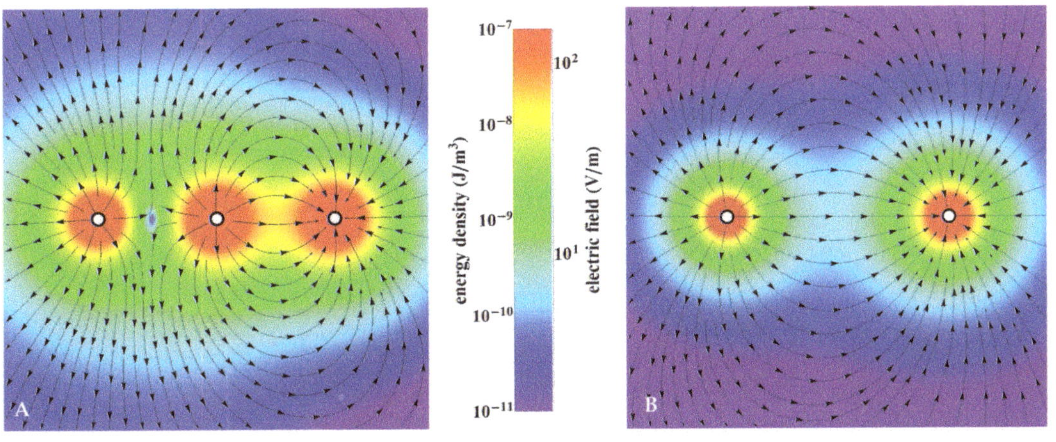

Figure 2. The geometry of electric fields: (**A**)—three-electrode systems (Zn, Bi, Cu), (**B**)—two-electrode systems (Cu, Zn), visible colors show the energy density. A finite element modeling (FEM) package was used to create models of the electric field: see Spisak et al., 2016, prepared by Alcor R & D Center, unpublished report, project No. POIR.01.01.01-00-0004/15.

The inhibition zones produced by the first system were compared with Cassini ovals [82]. However, the last investigation showed an additional agent that also inhibited the growth of fungi (Figure 3). Using a pH indicator (dibromothymolsulphonaphtalein), we can observe changes in the pH distribution and the influence of pH on fungal growth. H^+ distributed in half of the Petri plate inhibited fungal growth similarly to ions of Zn^{2+}.

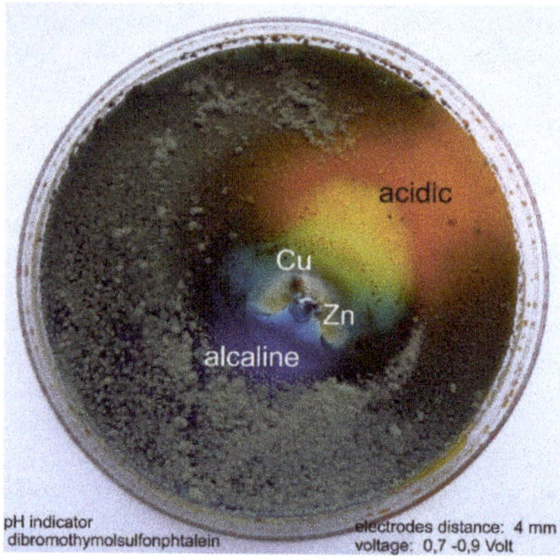

Figure 3. Influence of pH on fungal growth (*Penicillium spinulosum*) as an effect of galvanic action provoked by two electrodes of GALVI systems, photo A. Chlebicki.

Inhibition zones are larger when this system works in an environment with a higher salinity (Figure 4). Moreover, in such an environment, Cu electrodes work more effectively than those made of Zn (Figure 4). Babich & Stotzky [84] noted that the anionic composition of the environment influenced the toxicity of heavy metals. In particular, high-concentration NaCl can provoke the formation of complex Zn-Cl species ($ZnCl^+$, $ZnCl_2$, $ZnCl_3^-$, $ZnCl_4^{2-}$) with increased toxicity.

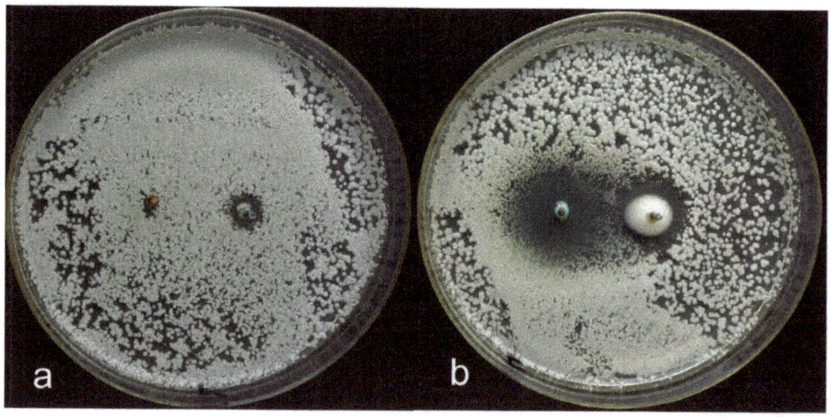

Figure 4. Influence of NaCl on the size of the inhibition zone for *Aureobasidium pullulans*: (**a**)—control, (**b**)—3% NaCl, distance between electrodes of Zn and Cu—5 mm, photo A. Chlebicki.

Some metals produce ions that inhibit or kill bacteria and other microorganisms. This is known as the oligodynamic effect, discovered by von Nageli in 1893 [85]. The galvanic system (which is used by us) enables one to move metal ions over a much longer distance and creates a large, inhibited zone. Nine different metals for the experiments were used. Of them, Zn and Cu created the biggest inhibition zone (Figure 5). The classical galvanic system of Cu Zn is known very well, but the optimal effect of the electromagnetic field depends on the distance between electrodes. Thirty-two distances between these electrodes were investigated. Very common indoor fungi such as *Aureboasidium pullulans*, *Alternaria atra*, *Cladosporium cladosporioides*, and *Rhodotorula mucilaginosa* were used in the experiments as indicators of the size of the inhibition zone. It appears that a distance of 4.2 mm creates the biggest inhibition zone [82].

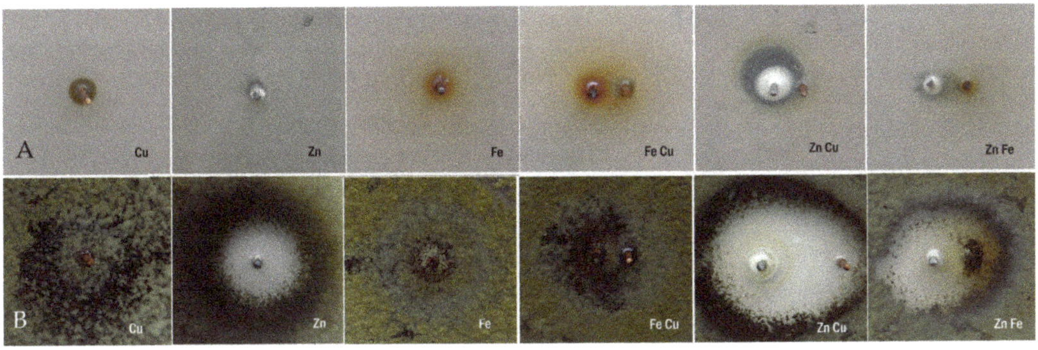

Figure 5. Inhibition zones created by single metals and galvanic cells. (**A**)—*Aureobasidium pullulans*, (**B**)—*Alternaria atra*, photo A. Chlebicki.

In the case of the three-electrode systems, their configurations (Figure 6) have fundamental implications for obtaining an optimal inhibition zone [83].

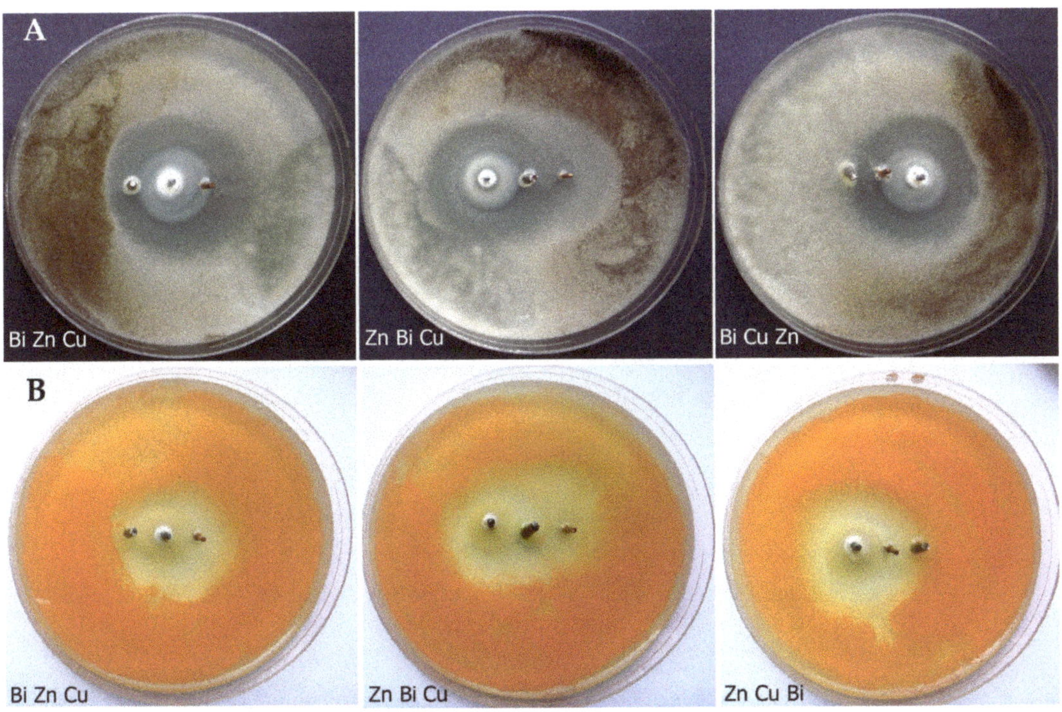

Figure 6. Influence of electrode configurations on the inhibition zones. (**A**)—*Aspergillus tubingensis*, (**B**)—*Rhodotorula mucilaginosa*, photo A. Chlebicki.

4. Two-Phase Media

The use of GALVI technology in natural building stones requires knowledge of the stone composition and its deterioration by microorganisms. Earlier microbiologists isolated microorganisms and cultivated them on artificial one-phase media afterwards [13,25,37,86]. Mirocha & DeVay [87] observed the autotrophical growth of *Cephalosporium* sp. and *Fusarium* sp. on a medium devoid of any known organic carbon source. Recently started experiments with media would allow us to observe the development of fungi in a two-phase environment. These two-phase media were used with SNA (synthetic low-nutrient agar) medium + sterile grandiorite[1], melabasanite[2], sandstone[3], marl[4], and sodium glass (as control). Microorganisms inhabit the boundaries of two liquid and solid phases. This enables the investigation of the processes of colonization, course, and development of selected microorganisms in the emerging rock microhabitats (Figure 7). Two-phase media with granodiorite and sandstone were much more colonized by fungi than media with melabasanite and marl. Both marl and melabasanite have a lower porosity and higher content of Ca; moreover, melabasanite is dark, has a low water absorption, high pH (45% SiO_2), and contains toxic Fe (see Figure 5). In addition, melabasanite can be a good habitat for special microorganisms that, in addition to Ca, still require Mg, Fe, and/or Al [88–90].

The content of elements in the mentioned rocks is very rich, and possible interactions between cations and anions during changing pH are difficult to interpret. For example, a ferric hydroxide can be created at pH > 4.5, whereas poor coagulation occurs in the pH range between 7 and 8.5 [91]. The influence of environs on cations' toxicity was mentioned earlier by Babich & Stotzky [85]. Metal transformations were thoroughly described by Gadd [92].

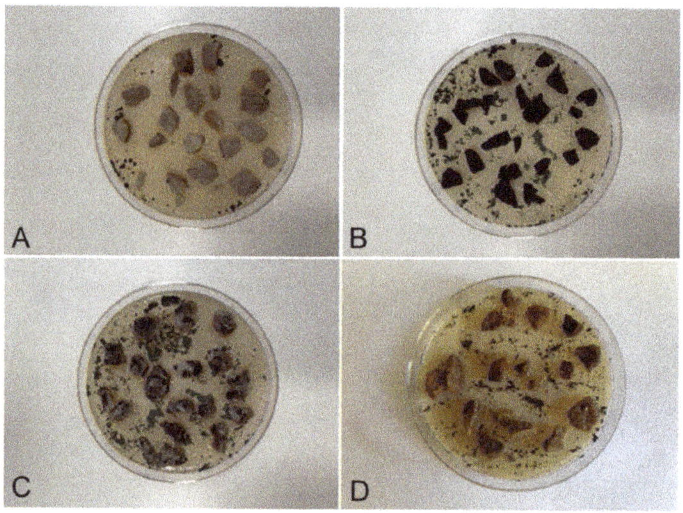

Figure 7. Fungi growing on SNA medium after two months: (**A**)—marl, (**B**)—melabasanite, (**C**)—grandiorite, and (**D**)—sandstone, photo A. Chlebicki.

1. Granodiorite is a light gray igneous acidic rock, the main components of which are quartz (SiO_2), potassium feldspar ($KAlSi_3O_8$), sodium plagioclase ($Na[AlSi_3O_8]$), and biotite ($K(Mg,Fe_2)_3(OH,F)_2AlSi_3O_{10}$). The average analysis of the chemical composition shows the dominance of silica (71–74%), followed by large amounts of aluminium (12–16%) and smaller but significant amounts of sodium (3–4%) and potassium (2–3%).

2. Melabasanite (with a transition to ankaratrite) is a dark volcanic rock belonging to ultrabasic rocks (<45% SiO_2). The main minerals are pyroxenes (Ca silicates, Mg and Fe silicates), olivines $(Mg,Fe)_2[SiO_4]$, amphiboles (hydrated Ca-silicates, Mg, Fe, and Na), and calcium plagioclases $Ca[Al_2Si_2O_8]$. In a variety closer to ankaratrite (a type of nephfinite), one of the main minerals is nepheline (sodium and potassium aluminosilicate ($KNa_3(AlSiO_4)_4$). The average chemical composition analysis shows, apart from silica (40–42%), large amounts of aluminium (13–14%), iron (approx. 12%), magnesium (approx. 12%), calcium (10–12%), and significant amounts of sodium (3–4%).

3. Sandstone is a coarse-grained beige clastic rock in which the spaces between quartz grains (SiO_2) are filled with silica-iron-clay binder (SiO_2/Fe_2O_3/hydrated aluminosilicates). Its technical parameters are variable, depending on the content of Fe_2O_3. Therefore, the main components of this sandstone are silica, aluminum, and iron.

4. Marl is a pale beige sedimentary rock with variable proportions of two main components: calcite ($CaCO_3$) or dolomite ($CaMg(CO_3)_2$) in the amount of 50–75% and clay minerals (hydrated aluminosilicates) in the amount of 25–50%. Marl may also contain 0–50% SiO_2. Its main components are therefore calcium, aluminium ± silica.

This was elaborated on the basis of the following books: [93–95].

Each of the analyzed types of rock, due to its chemical composition and color, is likely to be suitable for microorganisms requiring appropriate elements. Although marl and melabasanite abound in Ca, both may have the same calcium-demanding microorganism. For all the analyzed rocks, their color may also be important. A dark or even black rock, exposed to solar radiation, heats up much more than a light or even white one. For the samples analyzed, low-porous and low-absorbing melabasanite, which is rich in minerals Al, Fe, Mg, Ca and Na, will heat up more than slightly less porous and less absorbing red sandstone containing Si, Al, and Fe minerals. The lower temperature will reach a bright, low-porous, and very absorbing marl, built mostly of $CaCo_3$, while the relatively lowest one will reach also bright, but much less porous and less absorbing granodiorite, which in addition to SiO_2 contains minerals rich in Al, Na and K.

5. Possible Application of Galvanic Systems

Galvanic macrocells. Our first experiments were constructed with electrodes in the form of long bars of 2–3 mm-diameter and 5–10 cm length. This GALVI system can be used in wooden buildings against colonization by *Serpula lacrymans*, *Gloeophyllum trabeum*, and *Schizophyllum commune*.

Galvanic microcells. This next GALVI system was formed with electrodes that were in the form of metal grains of Zn and Cu deposited randomly on the surface. The zinc grains (anode) had diameters under 90 μm and copper grains (cathode) with a diameter under 63 μm [84]. Galvanic microcells were used in the paints production of GALVI technology in many variants:

1. Paint with Zn and Cu that can be used inside damp lodgings (butchers, winery, bathrooms, cellars, storehouses, museums etc.).

2. Paints with Zn and Cu that can be used outside in modern building elevations. This is a watertight system for use on facades that are at risk of fouling, with harmful microorganisms in objects with a higher exposure to biocorrosion. Grains of metals will be dipped in the matrix (Figures 8–10), but their tops will be free and stand up. This is necessary for the function of the electromagnetic field when the paint surface will be in damp conditions.

3. We also suggest using paint with Zn, Bi, and Cu (three-electrode systems) on the lower side of the stretchers. This system is the source of bioactive "ions cocktail" [78], and it works only at a very fine level on the lower surface of the stretcher. Metal ions cannot come through the canvas and damage the upper part of the stretcher.

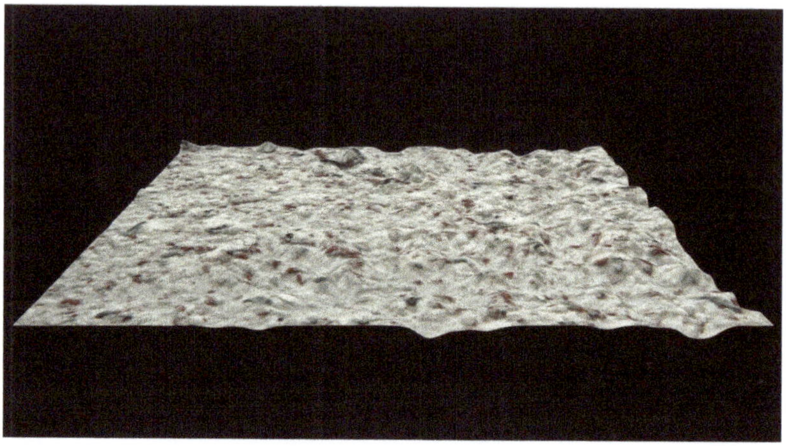

Figure 8. Paint surface with metal grains, 3D front view, Alcor R&D Center unpublished report, project No. POIR.01.01.01.00-636/20.

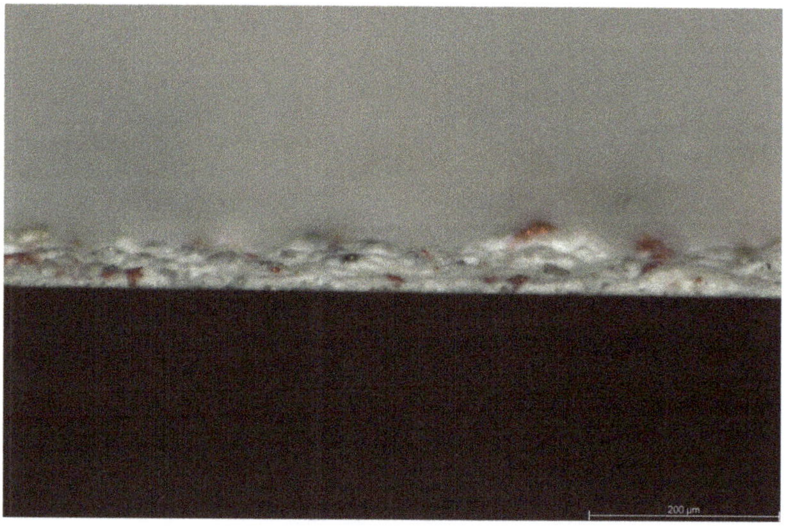

Figure 9. Paint surface, cross section, visible metal grains on undulate surface, scale bar = 200 µm, Alcor R&D Center unpublished report, project No. POIR.01.01.01.00-636/20.

Metal grains should be added to the GALVI systems by the users directly before the application. Then, the paint must be mixed with a manual or mechanical stirrer to ensure an even distribution of electrodes in the coating. All GALVI systems do not work in very dry conditions, but in such conditions harmful biodeterioration fortunately does not occur. When a place with GALVI coating is damp, then the action of the electromagnetic field starts.

We are planning future experiments, forming a GALVI system with a transparent base composed of ethyl silicates, fluorosilicates, epoxy resins, or acrylic resins. It can be used in decorative stone walls or elevations, as well as in artistic concrete. It is possible that our methods could be used in antimicrobial technology in architecture and art. Microorganisms have never adapted to the high electromagnetic field's strength, which contrasts with traditional biocides.

Figure 10. Paint surface, top view, scale bar = 200 µm, Alcor R&D Center unpublished report, project No. POIR.01.01.01.00-636/20.

Author Contributions: A.C.: carry on experiments, collected data from experiments of Galvi and two-faze media, produced photographs and wrote the paper, participation in preparing the patent application, corresponding author. W.S.: invention and developing of GALVI technology, prepared materials for two-faze media experiments, analysed data, elaborated of two patents and application, produced photographs 8, 9, 10. M.W.L.: petrographic study and chemical composition of the analyzed rocks. Meaning of their color and porosity for microorganisms that require appropriate elements and humidity L.Ś.: ensuring of working condition in laboratories, financial support of publication. Participation in the elaboration of a patent application K.W.: preparation of the experiment, research and analysis of results on the use of two-faze media for culture of algae and other microorganisms; participation in preparing the patent application. All authors have read and agreed to the published version of the manuscript.

Funding: Financial support under Project No. POIR.01.01.01-00-0004/15 and POIR.01.01.01.00-636/20 was provided by the Ministry of Science and Higher Education (Poland). The studies were also financially supported by the W. Szafer Institute of Botany, Polish Academy of Sciences through statutory funds.

Institutional Review Board Statement: Not applicable.

Informed Consent Statement: This study not involving investigation of humans.

Data Availability Statement: Both articles in Scientific Reports, see reference list Spisak et al., 2016 and 2020, patents: PL229012, PL43076-A1, Web of Sciences: Derwent Primary Accession Number 2021-594776.

Conflicts of Interest: The authors declare no conflict of interest.

References

1. Adams, R.I.; Sylvain, I.; Spilak, M.P.; Taylor, J.W.; Waring, M.S.; Mendel, M.J. Fungal signature of moisture damage in buildings: Identyfication by target and untarget approaches with mycobiome data. *Appl. Environ. Microbiol.* **2020**, *86*, e01047-20. [CrossRef] [PubMed]
2. Siegesmund, S.; Török, Á. Building Stones. In *Stone in Architecture*; Siegesmund, S., Snethlage, R., Eds.; Springer: Berlin/Heidelberg, Germany, 2014; pp. 11–95. [CrossRef]
3. Embry, A.F.; Klovan, J.E. Absolute water depths limits of late Devonian paleoecological zones. *Geol. Rundschau.* **1972**, *61*, 672–686. [CrossRef]
4. Flügel, E. *Microfacies of Carbonate Rocks, Analysis, Interpretation and Application*; Springer: New York, NY, USA, 2004; p. 976.

5. Gadd, G.M. Fungi, rocks, and minerals. *Elements* **2017**, *13*, 171–176. [CrossRef]
6. Sterflinger, K. Geomicrobiological Investigations on the Alteration of Marble Monuments by Dematiaceous Fungi (Sanctuary of Delos, Cyclades, Greece). Ph.D. Thesis, Universitety of Oldenburg, Oldenburg, Germany, 1995; p. 138.
7. Wollenzien, U.; Hoog, G.S.; Krumbein, W.E.; Urzi, C. On the isolation of microbial fungi occurring on and in marble and other calcareous rocks. *Sci. Total Environm.* **1995**, *167*, 287–294. [CrossRef]
8. Wollenzien, U.; Hoog, G.S.; Krumbein, W.E.; Ulithof, J.M.J. *Sarcinomyces petricola*, a new microcolonial fungus from marble in the Mediterranean basin. *Anton Leeuw. Int. J. G.* **1996**, *71*, 281–288. [CrossRef]
9. Diakamaku, E.; Gorbushina, A.A.; Krumbein, W.E.; Panina, L.; Soukharjevski, S. Black fungi in marble and limestones—An aestetical, chemical, physical problem for the conservation of monuments. *Sci. Total Environ.* **1995**, *167*, 295–304. [CrossRef]
10. Gorbushina, A.A.; Panina, L.K.; Vlasov, D.Y.; Krumbein, W.E. Fungi deteriorating marble in Chersonessus. *Mikol. Fitopatol.* **1996**, *30*, 23–27.
11. Staley, J.T.; Palmer, F.E.; Adams, J.B. Microcolonial common inhabitants on desert rocks? *Science* **1982**, *215*, 1093–1095. [CrossRef]
12. Sterflinger, K.; De Baere, R.; De Hoog, G.S.; De Wachter, R.; Krumbein, W.E.; Haase, G. Coniosporium perforans and Coniosporium apollinis, two new rock-inhabiting fungi isolated from marble in the Sanctuary of Delos (Cyclades, Greece). *Antonie Leeuwenhoek* **1997**, *72*, 349–363. [CrossRef]
13. Sterflinger, K.; Krumbein, W.E. Dematiaceous fungi as a major agent of biopitting for Mediterranean marbles and limestnes. *Geomicrobiol. J.* **1997**, *14*, 219–230. [CrossRef]
14. Ruibal, C.; Gueidan, C.; Selbman, L.; Gorbushina, A.A.; Crous, P.; Groenewald, J.Z.; Muggia, L.; Grube, M.; Isola, D.; Schoch, C.L.; et al. Phylogeny of rock-inhabiting fungi related to Dothideomycetes. *Stud. Mycol.* **2009**, *64*, 123–133. [CrossRef]
15. Owczarek-Kościelniak, M.; Krzewicka, B.; Piątek, J.; Kapusta, P. Is there a link between the biological colonization of the gravestone and its decoloration? *Int. Biodeterior. Biodegrad.* **2020**, *148*, 104879. [CrossRef]
16. Mehta, A.P.; Torma, A.E.; Murr, L.E. Effect of environmental parameters on the efficiency of biodegradation of basalt rock by fungi. *Biotechnol. Bioenergy* **1979**, *21*, 875–885. [CrossRef]
17. Turian, G. Maxi-toxitolerance d'une Moissure-Dematiée algicorticole du genre *Coniosporium*. *Ber. Schweiz. Bot. Ges.* **1975**, *85*, 204–209.
18. Blazquez, F.; Garcia-Vallez, M.; Krumbein, W.E.; Sterflinger, K.; Vendrell-Saz, M. Microstromatolitic deposits on granitic monuments: Development and decay. *Eur. J. Mineral.* **1997**, *9*, 889–901. [CrossRef]
19. Braams, J. Ecological Studies of the Fungal Microflora Inhabiting Historical Sandstone Monuments. Ph.D. Thesis, University of Oldenburg, Oldenburng, Germany, 1992.
20. Urzi, C.; Realini, M. Colour changes of Noto's calcareous sandstone as related with its colonization by microorganisms. *Int. Biodeterior. Biodegrad.* **1998**, *42*, 45–54. [CrossRef]
21. Chlebicki, A.; Zielenkiewicz, U.; Wilczek, A. Fungi are not involved in biofilm formation on rock wall in subterranean arsenic mine in Poland. *Nova Hedwig.* **2014**, *99*, 255–269. [CrossRef]
22. Chlebicki, A.; Godzik, B.; Lorenc, M.W.; Skłodowska, A. Fungi and arsenic-tolerant Bacteria in the hypogean environment of an ancient gold mine in Lower Silesia SW Poland. *Pol. Bot. Stud.* **2005**, *19*, 81–95.
23. Brandl, H. Heterotrophic Leaching. In *Fungi in Bioremediation*; Gadd, G.M., Ed.; Cambridge University Press: Cambridge, UK, 2001; pp. 383–423.
24. Chlebicki, A.; Lorenc, M.W. The troglophile fungus, *Physisporinus vitreus* on a mine wall at Złoty Stok, Poland. *Pol. Bot. J.* **2006**, *22*, 149–154.
25. Jongmans, A.G.; van Breemen, N.; Lundström, U.; van Hees, P.A.W.; Finlay, R.D.; Srinivasan, M.; Unestam, T.; Giesler, R.; Melkerud, P.A.; Olsson, M. Rock-eating fungi. *Nature* **1997**, *389*, 682–683. [CrossRef]
26. Fomina, M.; Burford, E.P.; Hillier, S.; Kiernas, M.; Gadd, G.M. Rock-building fungi. *Geomicrobiol. J.* **2010**, *27*, 624–629. [CrossRef]
27. Sterflinger, K.; De Hoog, G.S.; Haase, G. Phylogeny and ecology of meristematic ascomycetes. *Stud. Mycol.* **1999**, *43*, 5–22.
28. Gueidan, C.; Ruibal, C.; de Hoog, G.S.; Schneider, H. Rock-inhabiting fungi originated during periods of dry climate in the late Devonian and middle Triassic. *Fungal Biol.* **2011**, *115*, 987–996. [CrossRef] [PubMed]
29. Steiger, M.; Charola, E.; Sterflinger, K. Weathering and deterioration. In *Stone in Architecture*; Siegesmund, S., Snethlge, R., Eds.; Springer: Berlin/Heidelberg, Germany, 2014; pp. 225–316.
30. Crispin, C.A.; Gaylarde, C.C. Cyanobacteria and biodeterioration of cultural heritage: A review. *Microb. Ecol.* **2005**, *49*, 1–9. [CrossRef] [PubMed]
31. Canter Lund, H.; Lund, J.W.G. *Freshwater Algae*; Biopress Ltd.: Bristol, UK, 1998; p. 360.
32. Gaylarde, C.C. Influence of Environment on Microbial Colonization of Historic Stone Buildings with Emphasis on Cyanobacteria. *Heritage* **2020**, *3*, 1469–1482. [CrossRef]
33. Hindak, F.; Wołowski, K.; Hindakova, A. The epilithon of a cooling tower of the power plant at Bełchatów, Poland. *Oceanol. Hydrobiol. Stud.* **2011**, *40*, 38–43. [CrossRef]
34. García de Miguel, J.M.; Sánchez-Castillo, L.; Ortega-Calvo, J.J.; Gil, J.A.; Saiz-Jimenez, C. Deterioration of building materials from the Great Jaguar Pyramid at Tikal, Guatemala. *Build. Environ.* **1995**, *30*, 591–598. [CrossRef]
35. Tripathi, S.N.; Tiwari, B.S.; Talpasayi, E.R.S. Growth of cyanobacteria (blue-green algae) on urban buildings. *Energy Build.* **1990**, *15*, 499–505. [CrossRef]

36. Sterflinger, K. Fungi as geological agents. *Geomicrobiol. J.* **2000**, *17*, 97–124. [CrossRef]
37. Ciferri, O.; Tiano, P.; Mastromei, G. *Of Microbes and Art: The Role of Microbial Communities in the Degradation and Protection of Cultural Heritage*; Kluwer Academic: Moscow, Russia, 2000; p. 250. ISBN 978-1-4613-6904-2. [CrossRef]
38. Sand, W.; Bock, E. Biodeterioration of mineral materials by microorganisms—biogenic sulfuric and nitric acid corrosion of concrete and natural stone. *Geomicrobiol. J.* **1991**, *9*, 129–138. [CrossRef]
39. Nielsen, K.F. Mould Growth on Building Materials. Secondary Metabolites, Mycotoxins and Biomarkers. Ph.D. Thesis, Danish Buking and Urban Research, Hørsholm, Denmark, 2002.
40. Warscheid, T.; Braams, J. Biodeterioration of stone: A review. *Int. Biodeterior. Biodegrad.* **2000**, *46*, 343–368. [CrossRef]
41. Dorniedent, T.; Gorbushina, A.A.; Krumbein, W.E. Anderungen der physikalischen Eigenschaften von Marmor durch *Pilzbewuchs. Int. Bauinstandsetz.* **1997**, *3*, 441–454.
42. Sterflinger, K.; Gorbushina, A.A. Morphological and molecular characterisation of a rock inhabiting and rock decaying dematiaceous fungus isolated from antique monuments of Delos (Cyclades, Greece) and Chersonesus (Crimea, Ukraine). *Syst. Appl. Microbiol.* **1997**, *20*, 329–335. [CrossRef]
43. Castanier, S.; Le Metayer-Levrel, G.; Orial, G.; Loubiere, J.-F.; Perthuisot, J.-P. Bacterial carbonatogenesis and applications to preservation and restoration of historic property. In *Of Microbes and Art. The Role of Microbial Communities in the Degradation and Protection of Cultural Heritage*; Ciferri, O., Tiano, P., Mastromei, G., Eds.; Springer: Boston, MA, USA, 2000; pp. 203–218.
44. Eppard, M.; Krumbein, W.E.; Koch, C.; Rhiel, E.; Stanley, J.T.; Stackebrandt, E. Morphological, physiological and molecular characterization of actinomycetes isolated from dry soil, rocks, and monument surfaces. *Arch. Microbiol.* **1996**, *166*, 12–22. [CrossRef]
45. Mottershead, D.; Gorbushina, A.A.; Lucas, G.; Wright, J. The influence of marine salt, aspect and microbes in the weathering of sandstone in two historic structures. *Build. Environ.* **2003**, *38*, 1193–1204. [CrossRef]
46. Lansdown, A.B.G. Silver in health care: Antimicrobial effects and safety in use. *Curr. Probl. Dermatol.* **2006**, *33*, 17–34.
47. Lorenc, M.W. Deterioracja obiektów kamiennych i metody jej zapobiegania. *Biul. Inf. Konserw. Dzieł Szt.* **2003**, *14*, 44–48.
48. Macedo, M.F.; Miller, A.Z.; Dionísio, A.; Saiz-Jimenez, C. Biodiversity of cyano-bacteria and green algae on monuments in the Mediterranean Basin: An overview. *Microbiology* **2009**, *155*, 3476–3490. [CrossRef] [PubMed]
49. Scott, J.A.; Untereiner, W.A.; Ewaze, O.J.; Doyle, D. *Baudoinia*, a new genus to accomodate *Torula compniacensis*. *Mycologia* **2007**, *99*, 592–601. [CrossRef] [PubMed]
50. Isola, D.; Zucconi, L.; Onofri, S.; Caneva, G.; De Hoog, G.S.; Selbmann, L. Extremotolerant rock inhabiting black fungi from Italian monumental sites. *Fungal Divers.* **2016**, *76*, 75–96. [CrossRef]
51. Urzi, C.; De Leo, F.; De Hoog, S.; Sterflinger, K. Recent advances in the molecular biology and ecophysiology of meristematic stone-inhabiing fungi. In *Of Microbes and Art*; Ciferri, O., Tiano, P., Mastromei, G., Eds.; Springer: Boston, MA, USA, 2000; pp. 3–19.
52. Krumbein, W.E. Microbial interactions with mineral materials. In *Biodeterioration*; Houghton, D.R., Smith, R.N., Eggins, H.O., Eds.; Springer Science & Business Media: Dordrecht, The Netherlands, 2012; Volume 7, pp. 78–100.
53. Krumbein, W.E.; Urzi, C.; Gehrmann, C. On the biocorrosion and biodeterioration of antique and medieval glass. *Geomicrobiol. J.* **1991**, *9*, 139–160. [CrossRef]
54. Fassatiova, O. Micromycetes inhabiting the mines of Přibram (Czechoslovakia). *Česka Mykol.* **1970**, *24*, 162–165.
55. Seves, A.M.; Sora, S.; Ciferri, O. The microbial colonization of oil paintings. A laboratory investigation. *Int. Biodeter. Biodegrad.* **1996**, *37*, 215–224. [CrossRef]
56. De Hoog, G.S. Ecology and evolution of black yeasts and their relatives. *Stud. Mycol.* **1999**, *43*, 1–208.
57. Ewaze, J.O.; Summerbell, R.C.; Scott, J.A. Physiological studies of the warehouse staining fungus, *Baudoinia compniacensis*. *Mycol. Res.* **2007**, *111*, 1422–1430. [CrossRef] [PubMed]
58. Saarela, M.; Alakomi, H.L.; Suihko, M.L.; Maunuksel, L.; Raaska, L.; Mattilla-Sandholm, T. Heterotrofic microorganisms in air and biofilm samples from Roman catacombs, with special emphasis on actinobacteria and fungi. *Int. Biodeterior. Biodegrad.* **2004**, *54*, 27–37. [CrossRef]
59. McMullin, D.; Sumarah, M.; Miller, D. Chaetoglobosins and azaphilones produced by Canadian strains of *Chaetomium globosum* isolated from the indoor environment. *Mycotoxin Res.* **2013**, *29*, 47–54. [CrossRef] [PubMed]
60. Seyedmousavi, S.; Badali, H.; Chlebicki, A.; Zhao, J.; Prenafeta-Boldú, F.X.; de Hoog, G.S. *Exophiala sideris*, a novel black yeast isolated from environments polluted with toxic alkyl benzenes and arsenic. *Fungal Biol.* **2011**, *115*, 1030–1037. [CrossRef] [PubMed]
61. Tribe, H.T.; Thines, E.; Weber, R.W.S. Moulds that should be better known: The wine cellar mould. *Mycologist* **2006**, *20*, 171–175. [CrossRef]
62. Chebicki, A.; Majewska, M. *Zasmidium cellare* in Poland. *Acta Mycol.* **2010**, *45*, 139–142.
63. Sinhg, R.S.; Saini, G.K.; Kennedy, J.F. Pullulan: Microbial sources, production and applications. *Carbohydr. Polym.* **2008**, *73*, 515–531.
64. Characklis, W.G.; Marshall, K.C. *Biofilms*; John Wiley & Sons Inc.: Hoboken, NJ, USA, 1990; p. 816.
65. Konhauser, K. *Introduction to Geomicrobiology*; Blackwell Publishing: Hong Kong, China, 2007.
66. Drewniak, Ł.; Skłodowska, A. Isolation and characterization of psychrotolerant arsenite-oxidizing bacterium from gold mine in Złoty Stok. *Adv. Mat. Res.* **2007**, *20*, 576.

67. Drewniak, Ł.; Styczek, A.; Skłodowska, A. Arsenic hypertolerant bacteria isolated from gold mine rocks biofilms. *Adv. Mat. Res.* **2007**, *20*, 575.
68. Tomczyk-Żak, K.; Kaczanowski, S.; Drewniak, Ł.; Dmoch, Ł.; Skłodowska, A.; Zielenkiewicz, U. Bacteria diversity and arsenic mobilization in rock biofilm from anancient gold and arsenic mine. *Sci. Total Environ.* **2013**, *461–462*, 330–340. [CrossRef] [PubMed]
69. Savoca, S.; Lo Giudice, A.; Papale, M.; Mangano, S.; Caruso, C.; Spanò, N.; Michaud, L.; Rizzo, C. Antarctic sponges from the Terra Nova Bay (Ross Sea) host a diversified bacterial community. *Sci. Rep.* **2019**, *9*, 1–15. [CrossRef] [PubMed]
70. Luo, H.; Moran, M.A. Evolutionary ecology of the marine *Roseobacter* clade. *Microbiol. Mol. Biol. Rev.* **2014**, *78*, 573–587. [CrossRef] [PubMed]
71. De Leo, F.; Marchetta, A.; Capillo, G.; Germanà, A.; Primerano, P.; Schiavo, S.L.; Urzì, C. Surface active ionic liquids based coatings as subaerial anti-biofilms for stone built cultural heritage. *Coatings* **2021**, *11*, 1–22.
72. Miao, L.; Wang, C.; Adyel, T.M.; Zhao, J.; Yan, N.; Wu, J.; Hou, J. Periphytic biofilm formation on natural and artificial substrates: Comparison of microbial compositions, interactions, and functions. *Front. Microbiol.* **2021**, *12*, 684903. [CrossRef] [PubMed]
73. Fisher, M.C.; Hawkins, N.J.; Sanglard, D.; Gurr, S.J. Worldwide emergence of resistance to antifungal drugs challenges human health and food security. *Science* **2018**, *360*, 739–742. [CrossRef]
74. Snethlage, R. Stone conservation. In *Stone in Architecture*; Siegesmund, S., Snethlage, R., Eds.; Springer: Berlin/Heidelberg, Germany, 2014; Volume 7, pp. 415–550.
75. Koo, H.; Allan, R.N.; Howlin, R.P.; Stoodley, P.; Hall-Stoodley, L. Targeting microbial biofilms: Current and prospective therapeutic strategies. *Nat. Rev. Microbiol.* **2017**, *15*, 740–755. [CrossRef]
76. Gadd, G.M. Effect of copper on *Aureobasidium pullulns* in solid medium: Adptation not necessary for tolerant behaviour. *Trans. Br. Mycol. Soc.* **1984**, *82*, 546–549. [CrossRef]
77. Gadd, G.M. Interactions of fungi with toxic metals. *New Phytol.* **1993**, *124*, 25–60. [CrossRef]
78. Gadd, G.M.; Griffiths, A.J. Microorganisms and heavy metal toxicity. *Microb. Ecol.* **1978**, *4*, 303–317. [CrossRef] [PubMed]
79. Nickerson, W.J. Inhibition of fungus respiration: A metabolic bioassay method. *Science* **1946**, *103*, 484–486. [CrossRef] [PubMed]
80. Smith, W.H.; Staskawicz, B.J.; Harkov, R.S. Trace-metal pollutants and urban-tree leaf pathogens. *Trans. Br. Mycol. Soc.* **1978**, *70*, 29–33. [CrossRef]
81. Somers, E. The fungi toxicity of metal ions. *Ann. Appl. Biol.* **1961**, *49*, 246–253. [CrossRef]
82. Spisak, W.; Chlebicki, A.; Kaszczyszyn, M. Galvanic microcells as control agent of indoor microorganisms. *Sci. Rep.* **2016**, *6*, 35847. [CrossRef]
83. Spisak, W.; Chlebicki, A.; Kaszczyszyn, M.; Szar, M.; Kozak, J.; Olma, A. Three-electrode galvanic microcells as a new antimicrobial tool. *Sci. Rep.* **2020**, *10*, 7341. [CrossRef]
84. Babich, H.; Stotzky, G. Toxicity of Zinc to Fungi, Bacteria, and Coliphages: Influence of Chloride Ions. *Appl. Environ. Microbiol.* **1978**, *36*, 906–914. [CrossRef]
85. Kraemer, H. The oligodynamic action of copper foil on certain intestinal organisms. *Proc. Am. Phil. Soc.* **1905**, *44*, 51–65.
86. Ametrano, C.G.; Selbmann, L.; Muggia, L. A standardized approach for co-culturing dothidealean rock-inhabiting fungi and lichen photobionts in vitro. *Symbiosis* **2017**, *73*, 1–10. [CrossRef]
87. Mirocha, C.J.; DeVay, J.E. Growth of fungi on an inorganic medium. *Can. J. Microbiol.* **1971**, *17*, 1373–1378. [CrossRef] [PubMed]
88. Neilands, J.B. Microbial iron compounds. *Ann. Rev. Biochem.* **1981**, *50*, 715–731. [CrossRef] [PubMed]
89. Sayer, J.A.; Kierans, M.; Gadd, G.M. Solublization of some naturally occurring metal-bearing minerals, limescale and lead phosphate by *Aspergillus niger*. *FEMS Microbiol. Lett.* **1997**, *154*, 29–35. [CrossRef] [PubMed]
90. Gharieb, M.M.; Gadd, G.M. Influence of nitrogen source on the solublization of natural gypsum ($CaSO_4.2H_2O$) and the formation of calcium oxalate by different oxalic and nitric acid-producing fungi. *Mycol. Res.* **1999**, *103*, 473–481. [CrossRef]
91. Scholz, M. Coagulation and flocculation. In *Wetland Systems to Control Urban Runoff*; Scholz, M., Ed.; Elsevier Science & Technology: Edinburgh, UK, 2006; pp. 43–50.
92. Gadd, G.M. Metal Transformation. In *Fungi in Bioremediation*; Gadd, G.M., Ed.; Cambridge University Press: Cambridge, UK, 2001; pp. 359–382.
93. Dziedzic, K.; Kozłowski, S.; Majerowicz, A.; Sawicki, L. (Eds.) *Surowce mineralne Dolnego Śląska, Zakład Imienia Ossolińskich Wydawnictwo*; Polskiej Akademii Nauk: Wrocław, Poland, 1979; p. 510.
94. Kozłowski, S. *Surowce Skalne Polski*; Wydawnictwa Geologiczne: Warszawa, Poland, 1975; p. 344.
95. Ryka, W.; Maliszewska, A. *Słownik Petrograficzny*; Wydawnictwa Geologiczne: Warszawa, Poland, 1982; p. 403.

applied sciences

MDPI

Review

Fungi Affecting Wall Paintings of Historical Value: A Worldwide Meta-Analysis of Their Detected Diversity

Laura Zucconi [1,†], Fabiana Canini [1,†], Daniela Isola [1,2,*] and Giulia Caneva [2]

1 Department of Ecological and Biological Sciences (DEB), University of Tuscia, 01100 Viterbo, Italy; zucconi@unitus.it (L.Z.); canini.fabiana@unitus.it (F.C.)
2 Department of Sciences, Roma Tre University, 00146 Rome, Italy; giulia.caneva@uniroma3.it
* Correspondence: daniela.isola@uniroma3.it or isola@unitus.it
† These authors contributed equally to this work.

Abstract: Wall paintings have been a cultural expression of human creativity throughout history. Their degradation or destruction represents a loss to the world's cultural heritage, and fungi have been identified as a major contributor to their decay. We provide a critical review of fungi isolated from worldwide wall paintings between 1961–2021. One-hundred three scientific papers were reviewed focusing on fungal diversity, isolation protocols, and spatial distribution of data. The study sites were grouped into five environmental categories on the basis of the expected major microclimatic conditions (temperature, relative humidity, ventilation), and the possible relationship with the species found was investigated. The highest number of records were localized in Europe, with 38 sites on a total of 74, 20 of which were from Italy. A total of 378 fungal entries were obtained, consisting of 1209 records, belonging to 260 different species and 173 genera. The accuracy level in taxa determination was highly variable among different papers analyzed. Data showed a dominance of Ascomycota, mainly of orders Eurotiales and Hypocreales probably due to their wide distribution and easily air dispersed spores and due to the possible pitfalls linked to the isolation methods, favoring rapidly growing taxa. Statistical analyses revealed that fungal communities were not strictly linked to environmental categories with different ventilation, temperature, and humidity. Such findings may be due to the wide geographical area, the wide heterogeneity of the data, and/or the absence of standardized sampling and analyses protocols. They could also be the result of the dominance of some prevailing factors in the various sites that mask the influence one of each other.

Keywords: frescoes deterioration; fungal diversity; fungal ecology; hypogean conservation; mural paintings biodeterioration; subterranean cultural heritage deterioration; wall paintings conservation

Citation: Zucconi, L.; Canini, F.; Isola, D.; Caneva, G. Fungi Affecting Wall Paintings of Historical Value: A Worldwide Meta-Analysis of Their Detected Diversity. *Appl. Sci.* **2022**, *12*, 2988. https://doi.org/10.3390/app12062988

Academic Editor: Cesareo Saiz-Jimenez

Received: 4 February 2022
Accepted: 10 March 2022
Published: 15 March 2022

1. Introduction

Wall paintings are among the most representative elements of figurative artworks and have been developed by human creativity since prehistoric times [1]. Their technique of execution requires a layered structure consisting of a support, a ground, and a paint layer, which changed over time across different cultures (secco, such as tempera, or frescoes) [2]. In secco technique, which is the earliest, the preparation layers are applied, but the colors remain on the surface, whereas in the frescoes the colors are applied before the mortar dries, allowing their in-depth penetration [2]. The employed colors usually have a mineral origin, but some pigments can also be derived from plants. Organic compounds can later be added during restoration or because of other human activities (e.g., firing candles in the churches) [3].

The observed deterioration phenomena of mural paintings depend largely on the materials used and the environmental conditions [4]. Indeed, mural paintings are subject to a variety of biodeterioration phenomena, which varies depending on the humidity, lighting, temperature, ventilation, and nutrients, which also select the occurring biological agents [1,5]. Furthermore, many environmental factors may synergistically or antagonistically contribute to the deteriorating actions of microorganisms [6]. Organisms belonging to all domains (bacteria, algae, fungi, animals and sometimes also lichens, mosses, ferns, and higher plants) have been isolated from the surfaces of mural paintings [4,7]. Indeed, given the indoor conditions of most mural paintings, photoautotrophs are highly limited, while fungi and bacteria are more frequent [8]. Bacteria with reduced nutritional needs have been often suggested to be the first colonizers. With their death and lyses, they release organic matter that promotes the growth of secondary colonizers, such as fungi [9–11]. Fungi, instead, can produce a large assortment of enzymes and have the remarkable ability to grow and thrive in a wide variety of environmental conditions, including low water activity [12]. Fungi have been rightly recognized as the most common cause of biodeterioration of painted surfaces and other artworks, causing both physical and chemical deterioration phenomena, with aesthetic and structural consequences [1,13,14].

Generally, damage is due to the mycelial growth on the substrate, hyphal penetration, and fruiting bodies production onto and into the substrate, all of which increases the volume and number of cracks, causing the rupture of the pigment layer and leading to surface fragments detachments [1,15]. Fungal colonization generally starts on the surface and then moves in-depth, up to decreasing painted layer cohesion and cause exfoliations and loss of the paint [9,16]. A study carried out by Dornieden and colleagues demonstrated that some fungi, as the so called microcolonial black fungi, are among the most dangerous for cultural heritages and can influence the resistance to shear and torsion stress of mortar and marble, contributing to the separation of different layers of material in mural paintings [17]. Aesthetic damages are also frequent, due to pigment discolorations, mycelial pigmentation, and/or the release of organic pigments of different colors, depending on the species involved. Moreover, secondary compounds such as extracellular enzymes and/or organic acids are generally released in the substrate from fungal hyphae, and this may cause chemical alterations of the mineral constituents of the surfaces as well as the original pigments [9,16]. The secretion of organic acids (e.g., oxalic, citric, succinic, formic, malic, acetic, fumaric, glyoxylic, gluconic, and tartaric acids) also plays a significant role in chemical attack, causing acidification of the substrate [18,19]. They can cause dissolution of cations and chelation of metal ions from mortar and mineral pigments, leading to the formation of stable metal complexes whose crystallization causes an increase of internal pressure resulting in cracking, peeling, and the eventual loss of mural fragments [20].

Awareness of the considerable role played by microorganisms in the preservation of art objects and historical buildings dates back to the 1950s [21]. Ionita and colleagues provided one of the first detailed descriptions of the mycoflora involved in the deterioration of mural paintings of monasteries in Moldavia, noting that it was favored by the various nutritional sources present in the materials used for the realization of the frescoes and by local environmental parameters [21]. This was perhaps one of the first statements of the importance of interdisciplinary studies to prevent and control deterioration processes and define restoration and preservation strategies. Two interesting mini-reviews were later published by Garg and Ciferri teams [1,16]. Many papers have been published after that, showing a growing awareness of the degradative role of fungi as well as the importance of mycological analyses as an integral part of the state-of-the-art system of wall painting safeguards [22].

Despite the fact that the fungal role in the deterioration of frescoes has been documented by a huge number of papers, a global inventory of fungal diversity and their optimal settlement conditions is not yet available. These paintings are mainly present in confined and semi-confined environments, both hypogean and non-hypogean. A fungal alteration pattern dependent on the environmental conditions of these different sites was expected. Those present in hypogean environments are often subjected to a constant extreme humidity, promoting fungal spores germination and mycelial growth. The amount and type of available nutrients also affects the fungal growth rate and the type of fungal taxa. Nutrients may arrive from the external environment as airborne particles, and the more confined are the environments, the lower are the air spores dispersion phenomena. With this contribution, we aimed to describe the diversity of fungal colonizers involved in the deterioration of wall paintings, as well as their distribution under different environmental conditions. Additionally, we aimed to determine if a correlation among the different species recorded and the different types of environments-hypogean, non-hypogean, confined, non-confined, and open-exists and to speculate on their preferential habitat and their possible origin. A dataset of all the fungal taxa occurring on wall paintings based on bibliographic references was created for these purposes.

2. Materials and Methods

2.1. The Bibliographic Search

An extensive search was made among peer reviewed literature, proceedings to conferences, and books. The literature was identified using international databases, such as Scopus (https://www.scopus.com, 29 December 2021), Science Direct (https://www.sciencedirect.com, 29 December 2021), Web of Science (http://www.webofknowledge.com, 29 December 2021) and Google Scholar (https://scholar.google.com, 29 December 2021), that were consulted by using keywords such as 'wall paintings', 'mural paintings', 'frescoes', 'fungi', 'biodeterioration', 'microbial deterioration', and 'biodeteriogenic agents'. The thematic databases of ICCROM (International Centre for the Study of the Preservation and Restoration of Cultural Property) and the Italian ISCR (Istituto Superiore per la Conservazione e il Restauro) were also consulted, being important reference institutions in the field. Such sources were fundamental in the search of literature related to congresses and reports, that are not found by the most common scientific reference tools. The search covered more than 50 years, dating back to the first papers published in the 1960s (Figure 1), even if mostly of the papers containing useful taxonomic information were published after the 1980s.

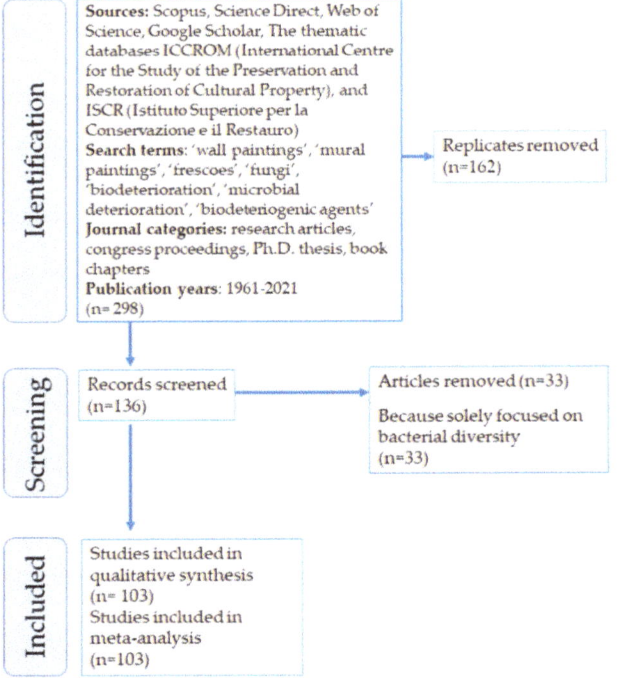

Figure 1. PRISMA flow diagram of the search process.

2.2. The Database

A list of all the entries retrieved, corresponding to taxa identified at both species and genus level, was compiled. Few were the entries referring to taxa above the rank of genus and they are listed at the end of the database. Current names of the taxa were reported according to the Index Fungorum (www.indexfungorum.org, 20 January 2022). The synonyms under which the different species were eventually reported in the analyzed papers were also indicated. Those entries recorded on paintings that have been the subject of multiple studies over the years, such as Takamatsuzuka and Kitora tumuli in Japan [23–25], were reported once accompanied by all bibliographic references.

2.2.1. The Geographic Localization of the Study Sites

The geographical locations and coordinates of the studied monuments were retrieved using Google Maps. The graphical representation on a map of sites distribution was performed using the 3D Map function of the Excel package. Some papers dealt with an unidentified number of monuments, as in the case of 'Monasteries in Romania' [21], 'several churches' in Northern Portugal [26], or 'Ajanta caves' in India [27–32]. In these cases, one or more sites were counted, depending on the details provided by the authors within the studied area and on their geographical distribution. Instead, some papers dealt with different monuments from the same area (e.g., different buildings in the historical site of Herculanum, Italy), that have been considered as a single site for the purpose of this study. This is why the total number of monuments is higher than the total number of geographical sites assessed in the statistical analyses.

2.2.2. Isolation and Identification Methods Used

The following information has been recorded: the type of culture media used, the growth temperature and incubation time, and the methods used for isolates identification.

2.2.3. The Types of Environments

The environments that housed the wall paintings object of the studies considered were grouped into five categories, based on whether they were hypogean or not, as well as the expected ventilation, confinement, and moisture conditions. The following are the definitions of the categories:

1. **C-HE:** Confined Hypogean Environment (isolated, without air circulation, with generally high humidity levels and relatively stable and low temperatures) as tumuli, close tombs, or prehistoric caves. They are often characterized by not negligible organic matter inputs from dripping waters, animals, and their fecal pellets and may be influenced by the presence of visitors, promoting the introduction and movement of airborne particulate.
2. **NC-HE**: Non-Confined Hypogean Environment (hypogean environments with reduced air circulation), such as underground crypts, catacombs, rupestrian churches, or Roman houses. These sub-aerial environments offer semi- or non-confined situations that are partially isolated from the external environment, with humidity and temperature comparatively more stable than the outdoor conditions but influenced by external day-night cycles and seasonal fluctuations.
3. **C-NHE**: Confined Non-Hypogean Environment: the sub-aerial environment of churches, monasteries, temples, refectories, and castle chapels represents a confined condition in which the microbial community is enclosed in a mesocosm. They are partially isolated from the external environment and have relatively more stable humidity and temperature than outdoor conditions, but they are influenced by external day-night cycles, seasonal variations in temperature, and relative humidity values. They are more prone to microbial attacks since they include more visited sites.
4. **SC-NHE**: Semi Confined Non-Hypogean Environment (open towards the outdoor environment but protected by wide changes in environmental parameters), such as ancient archaeological buildings and private homes. They are open to the outdoors

but sheltered from rain and ventilation; they all experience seasonal and daily relative humidity and temperature fluctuations.

5. **O-SPE**: Open and Semi-Protected or Protected Environment; they include buildings' walls or collapsed caves, which are among the most exposed sites to biodeterioration risks.

2.3. Statistical Analyses

A data dissimilarity matrix was inferred using the Jaccard's dissimilarity index [33], and a hierarchical cluster analysis was performed on this matrix using the UPGMA method. Two dendrograms relating dissimilarities between sites and entries were plotted. Entries identified at the genus level were considered as diverse elements. The Silhouette index was used to resolve the optimal number of clusters [34]. A contingency table between the obtained clusters and the environmental categories to which they belonged was also constructed to assess the relationship between their data. An indicator species analysis of the individual clusters was performed, which identifies associations between entries or combinations of entries and clusters, using the Indval index [35,36].

All analyses were performed with the R Software with the packages ade4, vegan, gclus, cluster, vegclust, and indicspecies.

3. Results

3.1. The Fungal Data Set

A total of 103 papers dealing with the fungal deterioration of wall paintings were collected, regarding 107 different monuments grouped in 74 sites. A total of 378 fungal entries were obtained, consisting of 1209 records belonging to 173 genera and 260 species (Table 1).

Table 1. List of the fungal entries retrieved from the different papers grouped by genera, in association with the corresponding references and the environmental categories where they have been registered.

Genus	Fungal Name	References	Environment
Acremoniella	*Acremoniella atra*	[21]	C-NHE
Acremonium	*Acremonium camptosporum*	[37]	NC-HE
	Acremonium charticola	[38–41]	NC-HE, C-NHE
	Acremonium masseei	[23,25]	C-HE
	Acremonium murorum (syn. *Gliomastix murorum*)	[23–25,28,32]	C-HE, NC-HE
	Acremonium rutilum (syn. *A. roseum*)	[21]	C-NHE
	Acremonium cf. *rutilum*	[39]	C-NHE
	Acremonium sp.	[23,38–47]	C-HE, NC-HE, C-NHE
Acrodontium	*Acrodontium crateriforme*	[48]	C-NHE
Acrophialophora	*Acrophialophora fusispora* (syn. *A. nainiana*)	[28,32]	NC-HE
Acrostalagmus	*Acrostalagmus luteoalbus* (syn. *Verticillium lateritium*)	[49]	C-NHE
Acrothecium	*Acrothecium* sp.	[50]	O-SPE
Actinomucor	*Actinomucor elegans*	[51]	C-HE
Akanthomyces	*Akanthomyces lecanii* (syn. *Verticillium lecanii*)	[39,43,52]	NC-HE, C-NHE
Allophoma	*Allophoma labilis* (syn. *Phoma labilis*)	[53]	SC-NHE

Table 1. *Cont.*

Genus	Fungal Name	References	Environment
Alternaria	*Alternaria alternata* (syn. *A. tenuis* and *Ulocladium alternariae*)	[21,26,28,29,31,32,40,46,49,51,53–59]	All environments
	Alternaria angustiovoidea	[60]	C-NHE
	Alternaria chartarum (syn. *Ulocladium chartarum*)	[21,61]	C-NHE
	Alternaria dianthi	[31]	C-NHE
	Alternaria longipes	[31]	C-NHE
	Alternaria longissima	[28,31,32]	NC-HE, C-NHE
	Alternaria oudemansii (syn. *Ulocladium oudemansii*)	[62]	NC-HE, C-NHE
	Alternaria tenuissima	[21,22,24,31,46,57,63]	NC-HE, C-NHE, O-SPE
	Alternaria sp. (syn. *Ulocladium* sp.)	[10,11,21,23,26,38,41,46,64,65]	C-HE, NC-HE, C-NHE, O-SPE
Amphinema	*Amphinema* sp.	[66]	C-NHE
Amyloporia	*Amyloporia sinuosa* (syn. *Antrodia sinuosa*)	[11]	C-NHE
Antrodia	*Antrodia* sp.	[66]	C-NHE
Apiotrichum	*Apiotrichum* sp. (syn. *Hyalodendron* sp.)	[43]	C-NHE
Arachnomyces	*Arachnomyces* sp.	[45]	NC-HE
Armillaria	*Armillaria* sp.	[66]	C-NHE
Arthrinium	*Arthrinium arundinis*	[40]	C-NHE
	Arthrinium phaeospermum (syn. *Papularia sphaerosperma*)	[28,32]	NC-HE
	Arthrinium sp.	[46,65]	C-NHE; O-SPE
Arthrobotrys	*Arthrobotrys* sp.	[23]	C-HE
Ascochyta	*Ascochyta medicaginicola* (syn. *Phoma medicaginis*)	[20,56]	O-SPE
	Ascochyta sp.	[63]	C-NHE
Ascotricha	*Ascotricha guamensis*	[32]	NC-HE
Aspergillus	*Aspergillus aeneus*	[67]	SC-NHE
	Aspergillus amstelodami (syn. *Eurotium amstelodami*)	[46,55,63]	C-NHE
	Aspergillus aureolatus	[20,56]	O-SPE
	Aspergillus auricomus	[22]	O-SPE
	Aspergillus candidus	[28,32,39,68]	C-HE, NC-HE, C-NHE
	Aspergillus clavatus	[37]	C-HE
	Aspergillus creber	[20,56]	O-SPE
	Aspergillus echinulatus	[21]	C-NHE
	Aspergillus europaeus	[20,56]	O-SPE
	Aspergillus fischeri (syn. *Neosartorya fischeri*)	[62]	NC-HE; C-NHE
	Aspergillus flavipes	[20,56]	O-SPE

Table 1. *Cont.*

Genus	Fungal Name	References	Environment
	Aspergillus flavus (syn. *A. oryzae*)	[10,17,20,28,30–32,56,57,68–75]	C-HE, NC-HE, C-NHE, O-SPE
	Aspergillus fumigatus	[30,31,38,46,61,76–79]	C-HE, NC-HE, C-NHE
	Aspergillus glaucus group	[79]	C-NHE
	Aspergillus ivoriensis	[67]	SC-NHE
	Aspergillus japonicus	[78]	C-HE
	Aspergillus melleus	[67]	SC-NHE
	Aspergillus multicolor	[67]	SC-NHE
	Aspergillus nidulans (syn. *Emericella nidulans*)	[28,30–32,46,51,61,70,71,78]	C-HE, NC-HE, C-NHE
	Aspergillus niger	[10,17,19,20,28,30–32,49,50,53–56,72,74,75,77,80–82]	All environments
	Aspergillus niger group	[46]	C-NHE
	Aspergillus ochraceus	[38,49,67]	C-HE, C-NHE, SC-NHE
	Aspergillus ostianus	[20,56,67]	SC-NHE, O-SPE
	Aspergillus pallidofulvus	[20,56]	O-SPE
	Aspergillus parasiticus	[20,56]	O-SPE
	Aspergillus penicilloides	[83]	C-HE
	Aspergillus petrakii	[67]	SC-NHE
	Aspergillus proliferans	[28,32]	NC-HE
	Aspergillus protuberus	[67]	SC-NHE
	Aspergillus puniceus	[67]	SC-NHE
	Aspergillus repens	[21,41]	C-NHE
	Aspergillus restrictus	[63,84]	C-HE, C-NHE
	Aspergillus sclerotiorum	[76]	C-NHE
	Aspergillus spectabilis (syn. *Emericella spectabilis*)	[67]	SC-NHE
	Aspergillus stellatus (syn. *Emericella variecolor*)	[67]	SC-NHE
	Aspergillus sydowii	[28,32,39,51,52,62,69]	C-HE, NC-HE, C-NHE
	Aspergillus terreus	[29–32,78]	C-HE, NC-HE, C-NHE
	Aspergillus unguis	[76]	C-NHE
	Aspergillus ustus	[46,67]	C-NHE, SC-NHE
	Aspergillus versicolor	[17,28,31,32,39,41,43,46,49,51,55,61,67–71,78,82,84–86]	C-HE, NC-HE, C-NHE, SC-NHE
	Aspergillus wentii	[28,31,32]	NC-HE, C-NHE
	Aspergillus sp.	[11,18,19,22–24,27,44–46,50,64,66,69,75,77,82,87–95]	C-HE, NC-HE, C-NHE, O-SPE
Aureobasidium	*Aureobasidium pullulans*	[31,41,43,45,49,52,57]	NC-HE, C-NHE
Beauveria	*Beauveria bassiana*	[48]	C-NHE
	Beauveria sp.	[45,52]	NC-HE, C-NHE
Bispora	*Bispora* sp.	[65]	O-SPE
Bjerkandera	*Bjerkandera adusta*	[77]	C-NHE

Table 1. *Cont.*

Genus	Fungal Name	References	Environment
Blastobotrys	*Blastobotrys aristatus*	[39]	C-NHE
Blastomyces	*Blastomyces* sp.	[84]	C-HE
Botryotrichum	*Botryotrichum atrogriseum*	[17,55,59]	NC-HE, C-NHE
	Botryotricum domesticum	[60]	C-NHE
	Botryotrichum murorum (syn. *Chaetomium murorum*)	[20,21,46,55,56]	C-NHE, O-SPE
Botrytis	*Botrytis cinerea*	[40,41,43,46,69]	C-NHE
Brunneochlamydosporium	*Brunneochlamydosporium nepalense* (syn. *Acremonium nepalense*)	[38,85]	C-HE
Burgoa	*Burgoa* sp.	[23]	C-HE
Candida	*Candida takamatsuzukensis*	[23,96]	C-HE
	Candida tumulicola	[23,96]	C-HE
	Candida sp.	[23,24,65]	C-HE, NC-HE
Capronia	*Capronia coronata*	[85]	C-HE
Cephalotrichum	*Cephalotrichum verrucisporum* (syn. *Doratomyces verrucisporus*)	[23,24]	C-HE
	Cephalotrichum sp. (syn. *Doratomyces* sp.)	[23,24]	C-HE
Cephalosporium	*Cephalosporium* sp.	[88,90]	C-NHE
Chaetomium	*Chaetomium ancistrocladum*	[20,56]	O-SPE
	Chaetomium elatum	[40]	C-NHE
	Chaetomium globosum	[21,22,26,31,32,38,41,46,51,55]	C-HE, NC-HE, C-NHE, O-SPE
	Chaetomium piluliferum (syn. *Botryotrichum piluliferum*)	[21,55]	C-NHE
	Chaetomium sp.	[9,22,27,31,39,40,43,46,65,70–72]	NC-HE, C-NHE, O-SPE
Chondrostereum	*Chondrostereum* sp.	[66]	C-NHE
Chrysosporium	*Chrysosporium pseudomerdarium*	[85]	C-HE
	Chrysosporium sp.	[47,52,62,97]	C-HE, NC-HE, C-NHE, O-SPE
Circinella	*Circinella muscae* (syn. *Circinella sydowii*)	[55]	C-NHE
Cladophialophora	*Cladophialophora tumulicola*	[24,98]	C-HE
Cladosporium	*Cladosporium cladosporioides*	[22,28–32,38,41,46,49,51,57,59,68,69, 85,99,100]	C-HE, NC-HE, C-NHE, O-SPE
	Cladosporium cucumerinum	[46,49,51]	C-HE, C-NHE
	Cladosporium herbarum	[21,28,31,32,39,46,51,55,58,63,82]	C-HE, NC-HE, C-NHE
	Cladosporium macrocarpum	[60]	NC-HE
	Cladosporium sphaerospermum	[5,7,9,22,28,31,32,39–43,46,48,51,52,62,63,69]	C-HE, NC-HE, C-NHE, O-SPE
	Cladosporium uredinicola	[20,56]	O-SPE
	Cladosporium xylophilum	[60]	C-NHE
	Cladosporium sp.	[15,18,19,22,23,27,42–45,52,58,62,64–66,71,75,77,80,82,86,87,90,91,94,101–103]	All environments

Table 1. *Cont.*

Genus	Fungal Name	References	Environment
Clonostachys	*Clonostachys rosea* (syn. *Gliocladium roseum*)	[58]	C-HE
Cochliobolus	*Cochliobolus geniculatus* (syn. *Curvulata geniculata*)	[32,76]	NC-HE, C-NHE
Collariella	*Collariella bostrychodes* (syn. *Chaetomium bostrychodes*)	[28,32]	NC-HE
Coltricia	*Coltricia* sp.	[66]	C-NHE
Coprinellus	*Coprinellus aokii* (syn. *Coprinus aokii*)	[67]	SC-NHE
Coprinopsis	*Coprinopsis atramentaria*	[38]	C-HE
	Coprinopsis cothurnata (syn. *Coprinus cothurnatus*)	[63]	C-NHE
Cordyceps	*Cordyceps farinosa* (syn. *Isaria farinosa*)	[68]	C-HE
Coriolopsis	*Coriolopsis* sp.	[66]	C-NHE
Cunninghamella	*Cunninghamella echinulata*	[9,28,32,40,55]	NC-HE, C-NHE
	Cunninghamella elegans	[38]	C-HE
	Cunninghamella sp.	[23,24]	C-HE
Curvularia	*Curvularia australiensis* (syn. *Drechslera australiensis*)	[28–32]	NC-HE, C-NHE
	Curvularia hawaiiensis (syn. *Drechslera hawaiiensis*)	[28–32]	NC-HE, C-NHE
	Curvularia lunata	[28,30–32,76,83]	NC-HE, C-NHE
	Curvularia pallescens	[29–32]	NC-HE, C-NHE
	Curvularia spicifera (syn. *Drechslera spicifera*)	[46]	C-NHE
	Curvularia sp.	[75]	C-NHE
Cutaneotrichosporon	*Cutaneotrichosporon mucoides* (syn. *Trichosporon mucoides*)	[63]	C-NHE
Cylindrocarpon	*Cylindrocarpon* sp.	[23,24]	C-HE
Cyphellophora	*Cyphellophora olivacea*	[42]	C-HE
	Cyphellophora sp.	[42]	C-HE
Cyphellostereum	*Cyphellostereum* sp.	[66]	C-NHE
Cystoderma	*Cystoderma* sp.	[66]	C-NHE
Devriesia	*Devriesia* sp.	[45]	NC-HE
Dichotomophilus	*Dichotomophilus indicus* (syn. *Chaetomium indicum*)	[55]	C-NHE
Didymella	*Didymella glomerata* (syn. *Phoma glomerata*)	[21,40]	C-NHE
Dipodascus	*Dipodascus geotrichum*	[55]	C-NHE
	Dipodascus sp. (syn. *Geotrichum* sp.)	[47,75]	NC-HE, C-NHE
Discostroma	*Discostroma corticola* (syn. *Seimatosporium lichenicola*)	[20,56]	O-SPE
Drechslera	*Drechslera* sp.	[65]	O-SPE

Table 1. *Cont.*

Genus	Fungal Name	References	Environment
Emericella	*Emericella ruber*	[31]	C-NHE
	Emericella sp.	[49,75]	C-HE, C-NHE
Engyodontium	*Engyodontium* sp.	[45,69]	C-HE, NC-HE, C-NHE
Epicoccum	*Epicoccum nigrum* (syn. *Epicoccum purpurascens*)	[20,28,31,32,46,49,56,58]	C-HE, NC-HE, C-NHE, O-SPE
	Epicoccum sp.	[27,65,87,90]	C-HE, NC-HE, C-NHE, O-SPE
Eurotium	*Eurotium halophilicum*	[104]	C-NHE
	Eurotium herbariorum	[68]	C-HE
	Eurotium sp.	[47,65,89,92]	C-HE, NC-HE, C-NHE
Exophiala	*Exophiala angulospora*	[42,98]	C-HE
	Exophiala moniliae	[85]	C-HE
	Exophiala sp.	[23,42]	C-HE
Fomitopsis	*Fomitopsis vinosa*	[63]	C-NHE
Fusarium	*Fusarium chlamydosporum*	[51]	C-HE
	Fusarium culmorum	[31]	C-NHE
	Fusarium equiseti	[53]	SC-NHE
	Fusarium fujikuroi (syn. *F. moniliforme*)	[29,30,32,74]	C-HE, NC-HE
	Fusarium oxysporum	[23,28,31,32,38,46,49,58,67,73,105]	C-HE, NC-HE, C-NHE, SC-NHE
	Fusarium proliferatum	[77]	C-NHE
	Fusarium sporotrichioides	[38]	C-HE
	Fusarium sp.	[10,23,24,27,31,39,43,47,62,64,82,91,94,106]	C-HE, NC-HE, C-NHE
Fuscoporia	*Fuscoporia* sp.	[66]	C-NHE
Fusidium	*Fusidium viride*	[48]	C-NHE
Ganoderma	*Ganoderma* sp.	[66]	C-NHE
Gliomastix	*Gliomastix tumulicola* (syn. *Acremonium tumulicola*)	[23,25]	C-HE
	Gliomastix sp.	[58,97]	C-HE
Gloiothele	*Gloiothele* sp.	[66]	C-NHE
Helminthosporium	*Helminthosporium* sp.	[65,75,82]	C-NHE, O-SPE
Humicola	*Humicola fuscoatra*	[49]	C-NHE
	Humicola udagawae	[38]	C-HE
	Humicola sp.	[31]	C-NHE
Hyphodontia	*Hyphodontia alutaria*	[66]	C-NHE
	Hyphodontia sp.	[66]	C-NHE
Hyphodontiella	*Hyphodontiella* sp.	[66]	C-NHE
Hypholoma	*Hypholoma* sp.	[66]	C-NHE
Idriella	*Idriella* sp.	[58]	C-HE
Kendrickiella	*Kendrickiella phycomyces*	[23,24,107]	C-HE
Kernia	*Kernia geniculotricha*	[67]	SC-NHE
	Kernia hippocrepida	[67]	SC-NHE

Table 1. *Cont.*

Genus	Fungal Name	References	Environment
Lactarius	*Lactarius* sp.	[66]	C-NHE
Lecanicillium	*Lecanicillium psalliotae*	[38,59]	C-HE, NC-HE
	Lecanicillium sp.	[68]	C-HE
Leptobacillium	*Leptobacillium muralicola*	[37]	NC-HE
Leptosphaeria	*Leptosphaeria* sp.	[65]	O-SPE
Leptosphaerulina	*Leptosphaerulina* sp.	[64]	NC-HE
Macrophomina	*Macrophomina phaseolina*	[28,31,32]	NC-HE
Malbranchea	*Malbranchea* sp.	[46]	C-NHE
Mammaria	*Mammaria echinobotryoides*	[83]	C-HE
Memnoniella	*Memnoniella* sp.	[31]	C-NHE
Metapochonia	*Metapochonia bulbillosa* (syn. *Verticillium bulbillosum*)	[47]	C-HE, NC-HE
	Metapochonia suchlasporia (syn. *Verticillium suchlasporium*)	[52]	NC-HE, C-NHE
Meyerozyma	*Meyerozyma guilliermondii*	[60]	NC-HE
Microascus	*Microascus brevicaulis* (syn. *Scopulariopsis brevicaulis*)	[21,52,68]	C-HE, NC-HE, C-NHE
	Microascus chartarum (syn. *Scopulariopsis chartarum*)	[62]	NC-HE, C-NHE
	Microascus cirrosus	[67]	SC-NHE
	Microascus sp.	[68]	C-HE
Microdochium	*Microdochium lycopodinum*	[38]	C-HE
Monilinia	*Monilinia* sp. (syn. *Monilia* sp.)	[82]	C-NHE
Monocillium	*Monocillium*-like	[23]	C-HE
Monodictys	*Monodictys castaneae* (syn. *Stemphylium macrosporoideum*)	[21]	C-NHE
	Monodictys sp.	[31,58]	C-HE, C-NHE
Mortierell	*Mortierella alpina*	[47]	NC-HE
	Mortierella ambigua	[38]	C-HE
	Mortierella parvispora (syn. *M. gracilis*)	[31]	C-NHE
	Mortierella sp.	[47,58]	C-HE, NC-HE
Mucor	*Mucor plumbeus* (syn. *M. spinosus*)	[21]	C-NHE
	Mucor racemosus (syn. *M. globosus*)	[28,32,47,60]	NC-HE
	Mucor silvaticus	[28,32]	NC-HE
	Mucor sp.	[23,46,58,64,82,93]	C-HE, NC-HE, C-NHE
Myxotrichum	*Myxotrichum stipitatum*	[46]	C-NHE
	Myxotrichum sp.	[46]	C-NHE
Nectria	*Nectria* sp.	[15]	C-NHE

Table 1. *Cont.*

Genus	Fungal Name	References	Environment
Neocosmospora	*Neocosmospora solani* (syn. *Fusarium solani*)	[23,28,31,76,97,105]	C-HE, NC-HE, C-NHE
Neodevriesia	*Neodevriesia modesta* (syn. *Devriesia modesta*)	[99,108]	O-SPE
	Neodevriesia simplex (syn. *Devriesia simplex*)	[99,108]	O-SPE
	Neodevriesia sp.	[45]	NC-HE
Neosartorya	*Neosartorya* sp.	[62]	NC-HE, C-NHE
Neosetophoma	*Neosetophoma cerealis* (syn. *Coniothyrium cerealis*)	[69]	C-NHE
Neurospora	*Neurospora intermedia*	[76]	C-NHE
	Neurospora sp.	[82]	C-NHE
Nigrospora	*Nigrospora oryzae* (syn. *N. sphaerica*)	[28,32,82]	C-HE, NC-HE, C-NHE
	Nigrospora sp.	[28,82]	NC-HE, C-NHE
Oidiodendron	*Oidiodendron cereale*	[69]	C-NHE
	Oidiodendron tenuissimum	[49]	C-NHE
Ophiostoma	*Ophiostoma* sp.	[23]	C-HE
Paecilomyces	*Paecilomyces variotii*	[28–32,61]	NC-HE, C-NHE
	Paecilomyces sp.	[27,46,68,75,82,97]	C-HE, NC-HE, C-NHE
Parengyodontium	*Parengyodontium album* (syn. *Beauveria alba*, *Tritirachium album*, and *Engyodontium album*)	[9,21,39–43,48,52,59,68,94,103,109–111]	C-HE, NC-HE, C-NHE
Penicillium	*Penicillium aethiopicum*	[67]	SC-NHE
	Penicillium albicans	[93]	C-NHE
	Penicillium aurantiogriseum (syn. *P. verrucosum* var. *cyclopium*)	[9,41,43,51]	C-HE, C-NHE
	Penicillium brevicompactum	[39,40,46,68,69,94]	C-HE, NC-HE, C-NHE
	Penicillium camemberti	[93]	C-NHE
	Penicillium canescens (syn. *P. raciborski*)	[40,93]	C-NHE
	Penicillium carneum	[67]	SC-NHE
	Penicillium chrysogenum (syn. *P. notatum*)	[9,21,40,41,46,51,54,60,67,69,81,87,93, 112,113]	C-HE, NC-HE, C-NHE, SC-NHE
	Penicillium citreonigrum (syn. *P. citreoviride*)	[22,58]	C-HE, O-SPE
	Penicillium citrinum	[31,32,40,58,76,93]	C-HE, NC-HE, C-NHE
	Penicillium commune	[51,53,68,93]	C-HE, C-NHE, SC-NHE
	Penicillium concentricum	[67]	SC-NHE
	Penicillium coprobium	[67]	SC-NHE
	Penicillium corylophilum	[44,91]	C-NHE
	Penicillium daleae	[91]	C-NHE
	Penicillium decumbens	[9,40,93]	C-NHE

Table 1. *Cont.*

Genus	Fungal Name	References	Environment
	Penicillium dierckxii (syn. *P. fellutanum*)	[46]	C-NHE
	Penicillium digitatum	[66]	C-NHE
	Penicillium dipodomyicola	[67]	SC-NHE
	Penicillium expansum	[46,49]	C-NHE
	Penicillium fuscoglaucum	[60]	C-NHE
	Penicillium glabrum (syn. *P. frequentans*)	[9,39,46,49,66,93,94]	NC-HE, C-NHE
	Penicillium granulatum	[31]	C-NHE
	Penicillium griseofulvum	[46,56,57,67]	C-NHE, SC-NHE
	Penicillium herquei	[46]	C-NHE
	Penicillium italicum	[53]	SC-NHE
	Penicillium javanicum (syn. *Eupenicillium javanicum*)	[62]	NC-HE, C-NHE
	Penicillium lanosum	[20,56,100]	NC-HE, O-SPE
	Penicillium lilacinum	[21,55]	C-NHE
	Penicillium meleagrinum	[38,69]	C-HE, C-NHE
	Penicillium miczynskii	[47]	SC-NHE
	Penicillium olsonii	[51]	C-HE
	Penicillium oxalicum	[49,58]	C-HE, C-NHE
	Penicillium pancosmium	[38]	C-HE
	Penicillium paneum	[23,67,114]	C-HE, SC-NHE
	Penicillium polonicum	[51]	C-HE
	Penicillium purpurescens	[31]	C-NHE
	Penicillium purpurogenum	[69,93]	C-NHE
	Penicillium restrictum	[93]	C-NHE
	Penicillium simplicissimum (syn. *P. janthinellum*)	[58,93]	C-HE, C-NHE
	Penicillium spinulosum (syn. *P. nigricans*)	[9,40,84]	C-HE, C-NHE
	Penicillium thomii	[93]	C-NHE
	Penicillium turbatum	[53]	SC-NHE
	Penicillium verrucosum	[39,46]	C-NHE
	Penicillium vulpinum	[67]	SC-NHE
	Penicillium sp.	[11,15,18,19,23,24,26,28,31,43–48,58,61,64,66,70,75,77,78,80,82,86,88–91,94,101]	C-HE, NC-HE, C-NHE
Pestalotia	*Pestalotia* sp.	[26]	C-NHE
Phialophora	*Phialophora* sp.	[23,24,43,47]	C-HE, C-NHE
Phlebia	*Phlebia* sp.	[66]	C-NHE
Pholiota	*Pholiota* sp.	[66]	C-NHE
Phoma	*Phoma* sp.	[23,31]	C-HE, C-NHE
Physalacria	*Physalacria* sp.	[66]	C-NHE

Table 1. Cont.

Genus	Fungal Name	References	Environment
Pleospora	*Pleospora* sp.	[65]	O-SPE
Postia	*Postia* sp.	[66]	C-NHE
Preussia	*Preussia terricola*	[68]	C-HE
	Preussia sp.	[68]	C-HE
Pseudogymnoascus	*Pseudogymnoascus pannorum* (syn. *Geomyces pannorum* and *Chrysosporium pannorum*)	[39,43,48]	C-NHE
Pseudozyma	*Pseudozyma prolifica*	[51]	C-HE
Purpureocillium	*Purpureocillium lilacinus* (syn. *Paecilomyces lilacinus*)	[19,38,49]	C-HE, C-NHE
Pyrenophora	*Pyrenophora biseptata* (syn. *Drechslera biseptata*)	[28,32]	NC-HE
Radulomyces	*Radulomyces* sp.	[66]	C-NHE
Rhinocladiella	*Rhinocladiella*-like	[23]	C-HE
Rhizoctonia	*Rhizoctonia solani* (syn. *Thanatephorus cucumeris*)	[20,28,32,56]	NC-HE, O-SPE
Rhizopus	*Rhizopus stolonifer* (syn. *R. nigricans*)	[28,31,32,53,57,73,74]	C-HE, NC-HE, C-NHE
	Rhizopus sp.	[27,101]	NC-HE, C-NHE
Rhodotorula	*Rhodotorula glutinis*	[49]	C-NHE
	Rhodotorula mucilaginosa	[66]	C-NHE
	Rhodotorula sp.	[15,18,66,69,80,91]	C-NHE
Russula	*Russula* sp.	[66]	C-NHE
Sagenomella	*Sagenomella griseoviridis*	[24]	C-HE
	Sagenomella striatispora	[24]	C-HE
	Sagenomella sp.	[44]	NC-HE
Sarocladium	*Sarocladium bacillisporum* (syn. *Acremonium bacillisporum*)	[94]	NC-HE
	Sarocladium kiliense (syn. *Acremonium kiliense*)	[21,52]	NC-HE, C-NHE
	Sarocladium strictum (syn. *Acremonium* cfr. *strictum*)	[23,52]	C-HE, NC-HE, C-NHE
Schizophyllum	*Schizophyllum commune*	[64]	C-NHE
	Schizophyllum sp.	[67]	C-NHE
Schizopora	*Schizopora paradoxa* (syn. *Hyphodontia paradoxa*)	[64]	C-NHE
Scolecobasidium	*Scolecobasidium anomalum* (syn. *Ochroconis anomala*)	[115]	C-HE
	Scolecobasidium lascauxensis	[85,115]	C-HE
	Scolecobasidium tshawytschae (syn. *Ochroconis tshawytschae*	[46]	C-NHE
Scopulariopsis	*Scopulariopsis brevicaulis*	[21,55]	C-NHE
	Scopulariopsis fusca	[39]	C-NHE
	Scopulariopsis sp.	[46,47,81,90]	NC-HE, C-NHE
Scytalidium	*Scytalidium* sp.	[70,71]	C-NHE

Table 1. *Cont.*

Genus	Fungal Name	References	Environment
Simplicillium	*Simplicillium lamellicola* (syn. *Verticillium lamellicola*)	[48]	C-NHE
Skeletocutis	*Skeletocutis* sp.	[67]	C-NHE
Sordaria	*Sordaria humana*	[28,32]	NC-HE
Sphaerostilbella	*Sphaerostilbella* sp. (syn. *Gliocladium* sp.)	[23,93,97]	C-HE, C-NHE
Sporothrix	*Sporothrix* sp.	[44,91]	NC-HE, C-NHE
Sporotrichum	*Sporotrichum* sp.	[41,43]	C-NHE
Stachybotrys	*Stachybotrys chartarum* (syn. *S. atra*)	[21,31,48,54,84]	NC-HE, C-NHE
	Stachybotrys cylindrosporus	[21]	C-NHE
	Stachybotrys echinatus (syn. *Memnoniella echinata*)	[28,31,32]	NC-HE
	Stachybotrys sp.	[9,27,31]	NC-HE, C-NHE
Stagonosporopsis	*Stagonosporopsis lupini*	[60]	C-NHE
Stemphylium	*Stemphylium botryosum*	[41]	C-NHE
	Stemphylium pyriforme	[55]	C-NHE
	Stemphylium sp.	[93]	C-NHE
Stereum	*Stereum* sp.	[66]	C-NHE
Syncephalastrum	*Syncephalastrum* sp.	[75]	C-NHE
Talaromyces	*Talaromyces aculeatus*	[51]	C-HE
	Talaromyces flavus	[38,69]	C-HE
	Talaromyces pinophilus (syn. *Penicillium pinophilum*)	[54,94]	C-HE, NC-HE
	Talaromyces rugulosus (syn. *Penicillium rugulosum*)	[68,69,113]	C-HE, C-NHE
	Talaromyces variabilis (syn. *Penicillium variabile*)	[69]	C-NHE
Thysanorea	*Thysanorea papuana*	[85]	C-HE
Tilletiopsis	*Tilletiopsis* sp.	[69]	C-NHE
Torrubiella	*Torrubiella alba* (syn. *Lecanicillium aranearum*)	[94]	NC-HE
	Torrubiella sp.	[68,94]	C-HE, NC-HE
Torula	*Torula herbarum*	[55]	C-NHE
	Torula sp.	[46]	C-NHE
Tricharina	*Tricharina* sp.	[64]	NC-HE
Trichocladium	*Trichocladium asperum*	[68]	C-HE
Trichoderma	*Trichoderma harzianum*	[29–32,69]	NC-HE, C-NHE
	Trichoderma sect. *Longibrachiatum*	[23,105]	C-HE
	Trichoderma virens (syn. *Gliocladium virens*)	[58]	C-HE
	Trichoderma viride	[23,55,58]	C-HE, C-NHE
	Trichoderma sp.	[19,23,24,31,44,47,75,82,85,95,97]	C-HE, NC-HE, C-NHE

Table 1. *Cont.*

Genus	Fungal Name	References	Environment
Trichothecium	*Trichothecium indicum* (syn. *Acremonium indicum*)	[28,31,32]	NC-HE
	Trichothecium roseum	[21,92]	C-NHE
Tritirachium	*Tritirachium* sp.	[67,82]	C-HE, C-NHE
Tubaria	*Tubaria* sp.	[66]	C-NHE
Tyromyces	*Tyromyces* sp.	[66]	C-NHE
Umbelopsis	*Umbelopsis ramanniana* (syn. *Mortierella ramanniana*)	[39,47]	NC-HE, C-NHE
Venturia	*Venturia carpophila* (syn. *Cladosporium carpophilum*)	[32]	NC-HE
Verticillium	*Verticillium alboatrum*	[32]	NC-HE
	Verticillium sp.	[23,48,52,58,97]	C-HE, NC-HE, C-NHE
Wallemia	*Wallemia sebi*	[63]	C-NHE
	Wallemia sp.	[92]	C-NHE
Westerdykella	*Westerdykella* sp.	[64]	NC-HE
Xylodon	*Xylodon nespoli*	[66]	C-NHE
	Xylodon nothofagi	[66]	C-NHE
	Xylodon raduloides	[66]	C-NHE
Zygosporium	*Zygosporium masoni*	[23]	C-HE
	Basidiomycota (Phylum)	[64]	NC-HE
	Black meristematic fungi	[5,99,116]	NC-HE, O-SPE
	Chaetomiaceae (Family)	[64]	NC-HE
	Filobasidiales	[64]	NC-HE
	Hyaline sterile mycelia	[62]	NC-HE, C-NHE
	Melanized sterile mycelia	[62]	NC-HE, C-NHE
	Pezizomycotina (Subphylum)	[64]	NC-HE
	Undetermined dark pigmented fungi	[17]	O-SPE
	Undetermined yeasts	[46]	C-NHE
	Uredinales (Order)	[65]	O-SPE
	Ustilaginales (Order)	[65]	O-SPE

The taxonomic distribution of the total fungal diversity and within different types of environments, at phylum and order level, is reported in Figure 2. Ascomycota was the dominant phylum, ranging from 89 to 97% (except for the environment O-SPE where a great proportion of undetermined taxa was retrieved) and accounting for 100% of the fungal entries in 40 out of 74 sites. The other two phyla identified were Basidiomycota (1–6%), more abundant in C-NHE, and Mucoromycota, reaching a maximum value (2.35%) in C-NHE (Figure 2A).

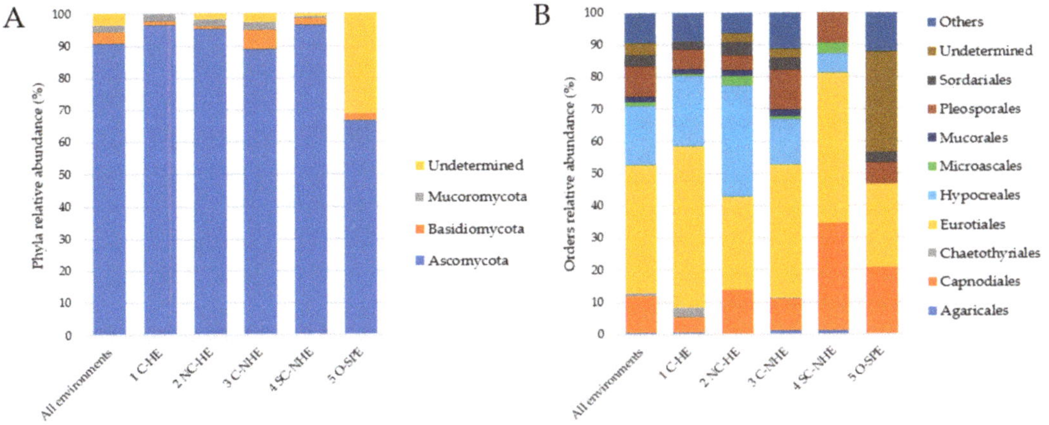

Figure 2. Taxonomic composition of the fungal diversity recorded on wall paintings on the total dataset and in relation to the different environments, at (**A**) phylum and (**B**) order level.

The entries were distributed in 39 orders. Eurotiales and Hypocreales were the most represented ones, together accounting for more than half of total fungal diversity. Pleosporales, Capnodiales, Sordariales, and Chaetothyriales were also well represented orders within Ascomycota. Basidiomycota accounted for a maximum of around 6% of all taxa, with Agaricales, Helotiales, and Saccharomycetales being the most abundant ones accordingly. Mucorales was the most abundant order within the phylum Mucoromycota.

The accuracy in taxa determination varied among the different papers analyzed, with many entries identified only at genus level. In fact, a total of 107 out of 378 entries (28.3%) referred to genera not determined at the species level, and it was not possible to quantify the number of possible different species belonging to these genera within different papers. A total of 61 genera (35.2%) were represented by a single species. *Aspergillus* and *Penicillium* (order Eurotiales) were the most frequently recorded genera and were represented by a greater number of species, 40 and 46, respectively, present in 44 (59.4%) and 32 (43.2%) sites, which increased to 54 (73%) and 51 (68.9%) when those sites where the genera were reported as undetermined at the species level were also considered. Their contribution to the total number of records was significant, with 219 (18.1%) and 154 (12.7%) records, respectively. Other genera frequently recorded were *Alternaria* and *Fusarium* with seven species each; *Acremonium*, *Cladosporium*, and *Trichoderma* with six species; *Curvularia* and *Talaromyces* with five species; and *Chaetomium* with four species.

3.2. The Geographic Distribution of the Study Sites

The data came from 107 monuments, grouped in 74 sites and distributed among 19 different countries. The countries where they were reported as well as the different number of sites were graphically represented in Figure 3.

3.3. Isolation and Identification Methods

A dominance of culture-based methods associated with morphological identification (58%) and target regions sequencing (31%) was recorded, the latter has become progressively dominant since the 2000s. The culture media used varied depending on the research purposes and included Czapek Dox agar (CZ), malt extract agar (MEA), malt agar (MA), potato dextrose agar (PDA), Sabouraud dextrose agar (SAB), and oatmeal agar (OA) among the most frequently used. The use of mycological agar (MYC), Cook's Rose Bengal (CRB), and yeast peptone dextrose (YPD) was sporadic. Data on growth temperature and incubation time were frequently missing, accounting for 31.25% (Figure 4A, green) and 54.12% (Figure 4B, green) of all papers, respectively. This trend was particularly evident in the earliest papers where more attention was paid to fungal species than to the conditions used to isolate them (e.g., [40,47,84,88]). The most frequent temperature settings were 25 °C (35%) and within the range 27–32 °C (21.25%). When reported, the incubation frequently corresponded to 7 days (31.76%) (Figure 4B). When isolated strains were identified by molecular approaches, the identification was performed targeting different barcoding regions, such as the complete internal transcribed spacer (ITS), a part of it (ITS1), portions of 18S (SSU), 26S (LSU), and β-tubulin.

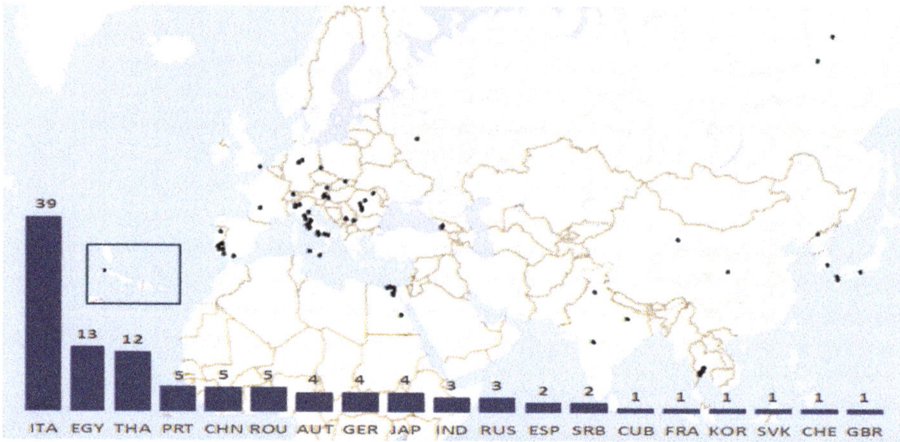

Figure 3. Geographic distribution of the reviewed study's 107 monuments. Nations are indicated with the international alpha-3 code: ITA: Italy, EGY: Egypt, THA: Thailand, PRT: Portugal, CHN: China, ROU: Romania, AUT: Austria, GER: Germany, JAP: Japan, IND: India, RUS: Russian Federation, ESP: Spain, SRB: Republic of Serbia, CUB: Cuba, FRA: France, KOR: Republic of Korea, SVK: Slovak Republic, CHE: Swiss, GBR: Great Britain. In the blue rectangle Cuba.

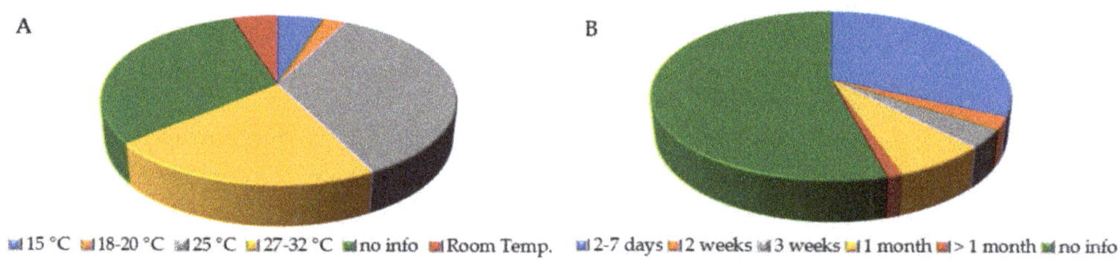

Figure 4. Values/ranges of (**A**) growth temperature and (**B**) incubation time recorded in culture-based protocols.

3.4. Distribution of Fungal Entries in Different Environments

The distribution of the sites among the environmental categories revealed a predominance of wall painting recorded in C-NHE, which alone accounted for 54% of the sites. The hypogean environments accounted for 36.5% of the sites, distributed between C-HE (18.9%) and NC-HE (17.6%). The remaining two categories, SC-NHE and O-SPE, were less represented, comprising 4.1% and 5.4% of the sites, respectively.

The cluster analysis at the level of different sites resulted in a general dispersion, with no distinct clusters retrieved. Several clusters consisted of one or few sites highly different one to each other (13-11-14-3-15-10-9-7-12). Other clusters (4-5-6) were slightly more similar to each other and contained many entries (Figure 5).

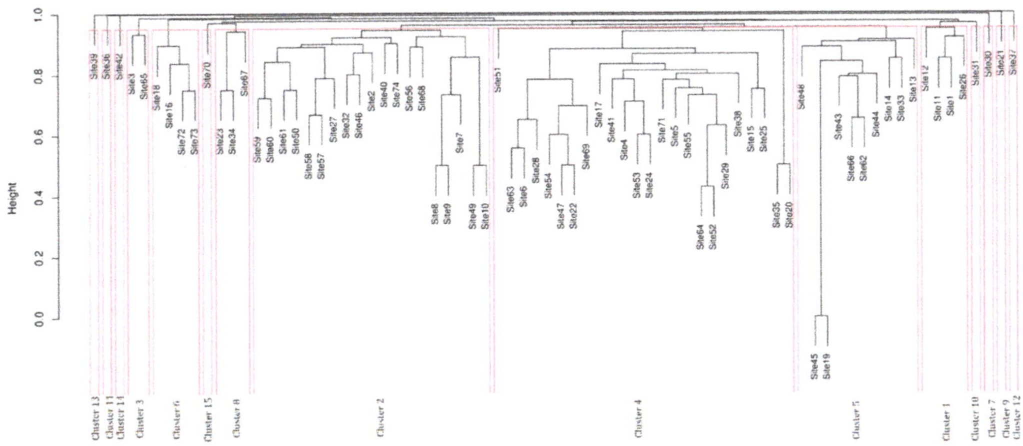

Figure 5. Cluster dendrogram. Cluster analysis on the Jaccard distance of the composition at the level of the different sites. Sites belonged to the following categories: Site1–Site14 C-HE; Site15–Site27 NC-HE; Site 28–Site67 C-NHE; Site 68–Site70 SC-NHE; Site 71–Site74 O-SPE. The references associated to each site are reported in Supplementary Table S1.

The contingency table highlighted that the obtained clusters did not have a strong correlation with the suggested environmental categories (Table 2). However, some clusters showed some affinities: cluster 15 with cat. 4 (SC-NHE) (aff. index 17); cluster 6 with cat. 5 (O-SPE) (aff. index 12.8); clusters 2,4,5 with cat. 3 (C-NHE) (aff. indexes 8.1, 9, and 6.2, respectively); cluster 4 with cat. 2 (NC-HE) (aff. index 5.9); cluster 1 with cat. 1 (C-HE) (aff. index 8.2).

Table 2. Contingency table and affinity indices. (**A**) Contingency table between the obtained clusters and the environmental categories to which they are correlated and (**B**) their relative affinity indices.

(A)		**Contingency Table Clusters**														
		13	**11**	**14**	**3**	**6**	**15**	**8**	**2**	**4**	**5**	**1**	**10**	**9**	**7**	**12**
Environmental categories	**C-HE**				1				5	3	2	**3**				
	NC-HE					2		1	1	**6**	1	1			1	
	C-NHE	1	1	1	1			2	**11**	**13**	**7**		1	1		1
	SC-NHE						**1**		1	1						
	O-SPE					**2**			1	1						
(B)		**Affinity Indices Clusters**														
		13	**11**	**14**	**3**	**6**	**15**	**8**	**2**	**4**	**5**	**1**	**10**	**9**	**7**	**12**
Environmental categories	**C-HE**				1.82				4.8	1.37	1.46	**8.2**				
	NC-HE					3.93		1.31	0.21	**5.89**	0.39	0.98			3.93	
	C-NHE	1.28	1.28	1.28	0.64			1.7	**11**	**8.98**	**6.25**		1.28	1.28		1.28
	SC-NHE						**17**		1	0.71						
	O-SPE					**12.8**			0.67	0.53						

When the correlation among different entries was analyzed, very scattered results were retrieved (Figure S1). However, some entries or combinations of them showed some affinities (specificity and fidelity above 0.5) with certain clusters such as *Alternaria alternata* with cluster 2 (Indval Index = 0.75); *Acrothecium* sp. and *Penicillium* sp. with cluster 4 (Indval Index = 0.70 and 0.66, respectively); *Parengyodontium album* and *Rhinocladiella*-like with cluster 5 (Indval Index = 0.70 and 0. 64); *Actinomucor elegans, Penicillium pancosmium, Acremoniella atra, Alternaria angustiovoidea, Scolecobasidium anomalum, Scolecobasidium lascauxensis,* when considered in combination, the entries *Trichoderma* sp. and *Verticillium* sp. with cluster 1 (Indval Index = 0.70, 0.70, 0.70, 0.70, 0.70, and 0.70, respectively); and *Akanthomyces lecanii* with cluster 9 (Indval Index = 0.93).

4. Discussion

4.1. The Fungal Data Set

Even if fungi have been suggested as secondary colonizers of painted mural substrates, they are among the most common microbial life-forms present in these environments and the primary cause of their biodeterioration [1,13,14]. The wide biodiversity observed confirms the potential key role of fungi in such colonization process and suggests a combination of causes that can favor their growth. Their broad enzymatic activities allow them to grow on every type of material, or wherever they find organic matter. Fungi recorded belonged to species generally reported from natural environments like soils, plants, and air where they live as saprotrophs, as well as plant and animal parasites and pathogens. A detailed survey of their diversity and distribution should become a prerequisite before any restoration measures in order to prevent further damages [39]. Most records belonged to Ascomycota, with Eurotiales being the most common order, due to the prevalence of *Aspergillus* and *Penicillium* genera. The former was one of the most frequently isolated genera, with *A. flavus* and *A. niger* among the more frequently recorded species. As reported in the literature, even from the first older papers in this field, these two genera, along with *Alternaria, Fusarium, Cladosporium, Mortierella, Chaetomium*, and *Acremonium*, are among the most common deteriogens of such paintings [1,11,15,31,39,46,62,78,87,91]. These taxa are ubiquitous, and their frequent occurrence is due to the production of numerous conidia, which are widespread in the environment because they are easily dispersed by air. A diversity of filamentous fungi, with the most predominant genera *Penicillium, Cladosporium, Aspergillus*, and *Trichoderma*, were also isolated from mural paintings of the Parish Church of Santo Aleixo (Portugal). Their dehydrogenase activity was determined, as an indicator of the presence of metabolic active cells to allow a deeper insight on the deteriogenic role of the isolates [19].

Species of these genera were recorded on indoor frescoes in numerous monasteries in Romania, possibly favored by the organic components and vegetal pigments used, as well as high moisture levels caused by frequent rainwater penetration, which also resulted in the formation of efflorescences [55]. *Cladosporium* species can cope in a variety of harsh environmental conditions thanks to their low nutritional requirements (i.e., in oligotrophic conditions). Otherwise, *Chaetomium* species are proteolytic and cellulolytic ascomycetes, favored by nutrient-rich substrates [22,65,117]. They were reported as the most frequent microfungi on the frescoes of the St. Damian Monastery in Assisi (Italy) [46] and on frescoes in a Serbian church [22]. Furthermore, a community of *Aspergillus, Penicillium, Cladosporium*, and *Chaetomium* species was recorded from Medieval wall paintings in Styria (Austria), forming spots of different colors [39]. This group of genera was dominant on two deteriorating frescos in St Clare's Refectory of the Monastery of St Damian in Assisi [46].

Hypocreales was the second most abundant order, accounting for 18% of total fungal diversity, within which *Acremonium, Trichoderma*, and *Fusarium* were among the most common genera. Hypocreales is one of the largest orders of filamentous ascomycetes and exhibits a broad range of ecologies, ranging from plant-associated nutritional modes to animal pathogens (e.g., insect pathogens) and mycoparasites [118]. *Neocosmospora solani*, recorded in Thailand, India, Japan, and France; *Simplicillium lamellicola*, recorded in Russia; and *Clonostachys rosea* [48,58], recorded in Japan, are examples of mycoparasitic species, while *Parengiodontium album* is an insect parasite and was recorded in several countries (Germany, Russia, Romania, Austria, Italy, and England) [109]. The recurrent presence of mites and insects pointed out their possible role in spreading fungi on painted surfaces [46,119].

Finally, the plant pathogen species *Fusarium oxysporum* has been shown to produce an extracellular pinkish pigment that disfigures and aesthetically damages colonized mural paintings and stone surfaces with permanent stains [78].

Phylum Basidiomycota was present with several occasional species, mostly represented by one or two records, and comprises litter, soil, and wood-saprotrophs, ectomycorrhizal, epiphyte, and plant-pathogen species. Their occurrence must be regarded as sporadic, potentially aided by root penetration. The possible role of roots as a carrier for rhizosphere microorganisms, like a dripping line for water condensation, and as an organic carbon source by root exudates has been hypothesized [57,120]. In any case, a Basidiomycete was also recorded at the entrance of Roman catacombs [121], possibly due to spores carried by water infiltrations and germinating using organic nutrients from the soil and/or the phototrophic biofilm.

Mucoromycota was present with few species and records, and black meristematic fungi were rarely recorded as well. These latter may grow on a wide range of substrates and are resistant to a variety of environmental stresses, as well as being widely distributed epi- and endolithically on monuments [122,123]. Although the biodiversity of black fungi on historical monuments is not fully elucidated, recent samplings indicate that they are also present on wall paintings and that their rare finding could be linked to the isolation protocols used, generally favoring fast-growing species [124]. Two new species of the genus *Neodevrisia* have been found in the restricted sampling area of the Vallerano cave and another, still undescribed, from Maijishan grottoes [45,99,108]. *Scolecobasidium lascauxensis* and *S. anomalus* were isolated and described from black stains in Lascaux Cave, France [85,115], while the chaetothyrialean black fungi *Cladophialophora*, *Exophiala*, and *Phialophora* have been reported from different sites [23,24,42,98].

Yeasts have been rarely reported, such as Saccharomycetales (Ascomycota) that usually grow by individual yeast cells or *Rhodotorula* spp. (Basidiomycota) often linked to pink/orange stains due to the release of carotenoids [19,93].

Among those more commonly reported, some species such as those belonging to the genera *Alternaria*, *Fusarium*, *Aspergillus*, *Penicillium*, and *Cladosporium* may be responsible for annoying allergic and toxic reactions suffered by conservationists and visitors [81,125,126]. *Alternaria alternata* is a very common fungal species, frequently recorded on frescoes. Its spores are recognized as common powerful aeroallergens, and indoor environments offer higher levels of exposure to this risk than open-air [59,127]. Records of *Fusarium* species have also been reported, such as *F. solani* in the Lascaux caves [97] and *F. oxysporum* in many sites. They are mainly plant pathogens, but they can also be the causal agents of human mycoses [97,128,129]. Some *Aspergillus* species are pathogenic to humans and animals and are responsible for clinical manifestations (https://www.aspergillus.org.uk/species-archive/, 29 December 2021). Among these, *A. fumigatus* is a human pathogenic fungus recorded on frescoes within different environments (3,11,22,29,42,49,70,71), causing infections in humans which can be fatal in immunocompromised patients (61). *Aspergillus flavus* has been frequently recorded in monasteries, churches, temples, caves, and tombs, mostly due to their numerous aerosolized spores. It mainly grows in the soil, but it is also a facultative and opportunistic pathogen of both animals and plants, producing mycotoxins that are highly harmful to humans [130].

In light of the above, the identification of the species deteriorating wall painting is needed for the protection of restorers and visitors. However, the temperature values characterizing confined and semi-confined sites are generally too low for potential pathogenic fungi. In fact, truly thermophilic fungi which cannot grow at temperatures below 20 °C are not active in these environments, at least during winter in temperate regions.

4.2. Geographic Distribution

Considering the geographic distribution of the data, just one site among the studied paintings comes from the Americas (the Cathedral of Havana at Cuba) [82]. The highest number of records was from Europe, with 70 monuments, mostly representative of hypogean environments and of churches and historical buildings, with a considerable prevalence of Italian monuments (39). A total of 26 monuments were from Asia, while the 13 African ones were all from Egypt.

This distribution arises from the old tradition of people of the Euro-Mediterranean area of using such artistic expression, starting from the old prehistoric caves to the Etruscan and Greek-Roman traditions until the consolidated use both in the decoration of Christian churches and historical buildings [2]. In the case of the Egyptian area, the recorded taxa derived from the old tombs of the Pharaohs [54,72,73,77,78], and similarly in East Asia, the tradition is mainly found in kings' and Emperors' tombs [23–25,38]. Most of the ancient paintings in buildings do not exist anymore, due to frequent rebuilt or remake of the materials [131].

Our results suggest that the monuments studied were often confined to restricted geographic areas. In any case a wider geographic distribution than that recorded may be possible, as a number of sites may have escaped the search. In fact, even if formally available on the web and on the major repositories, because of the language barriers, some studies could not be taken into account.

4.3. Isolation and Identification Methods

Culture-based methods favor the growth of microorganisms best fitting with the laboratory conditions used (namely, culture media, temperatures, and incubation times). In this study, we found that the most frequent experimental settings were favorable to fast-growing, highly-sporulating fungi, with the use of culture media rich in easily accessible carbon sources, alongside short incubation times and optimal growth temperatures favoring their sporulation. Otherwise, lower growth temperatures ($\leq$20 °C), wide temperature ranges, different isolation media, and a longer incubation time could enlarge the detectable culturable fraction.

Since the early 2000s, molecular phylogenetic methods have highlighted the limitations of morphological identification, allowing us to gain a better understanding of the kingdom of Fungi [132]. Nowadays, the identification by barcode regions sequencing is a common practice. Even if the nuclear ITS region has been recognized as a fungal barcode, its discriminating power changes within the taxonomical groups, and other/more barcode regions are often necessary to have a reliable identification [133]. This is the case of the identification of species within large groups, as *Fusarium*, *Penicillium*, *Aspergillus*, and *Cladosporium* genera, where cryptic species can be detected only by sequencing multiple molecular markers [134].

In detail, *Fusarium* species determination has been best made with the combined phylogeny of protein coding genes such as elongation factor (TEF1), RNA polymerase (RPB2) and the partial β-tubulin (BT2) gene [134]. To discriminate between *Penicillium* and *Aspergillus* species, β-tubulin (BT2) and calmodulin (cmdA) genes have been proposed as secondary barcodes, respectively [135,136]. While the most phylogenetic informative markers for *Cladosporium* were TEF1 and actin gene (actA), ITS sequences being identical for species of the same complex [137,138].

The correct identification of strains is required in order to provide restorers more information about strains' ecology and degradative potential. In this light, standardized identification protocols should be implemented.

High throughput sequencing methods have recently been applied to cultural heritage purposes. These methods represent a powerful tool to define the whole fungal diversity present but not necessarily to deepen the mechanisms and the main actors of the deterioration phenomena [139]. The combination of culture-based and molecular methods should be used for a better understanding of deterioration processes. Indeed, pure cultured microorganisms represent the key to uncover settlers' physiological and ecological traits, as well as representing a resource for many in silico applications and barcoded identifications [123,124].

4.4. Distribution of Taxa in the Different Environments

The most prevalent sites were confined non-hypogean environments, which are characterized by varied thermo-hygrometric temperatures and air movement. Hypogean (both confined and non-confined), where nutrients and humidity can favor fungal growth were represented as well.

Temperature and relative humidity are among the environmental parameters most important to microbial colonization capability, and in the case of heterotrophs, a certain amount of nutrients is also needed [4,140]. It is well known that fungi rapidly grow when relative humidity is higher than 65% and when a certain quantity of nutrients is available. The low values of temperatures, even if are not favorable for microbial growth by themselves, have a positive effect in contributing to increase in humidity, favoring water condensation on surfaces. Walls, especially in hypogean environments, generally provide these requirements [1]. Temperatures in confined environments are generally more stable than in non-confined environments, where daily and seasonal changes may occur, with ranges that have effects on microbial settlement. Elevated moisture values and stable temperatures have been reported as ideally suited to promote microbial growth on surfaces in catacombs sites [7,94,141]. Indeed, the highest risk occurs when high humidity is coupled with high temperature values, and negative effects of rising temperatures arise only if their highest values can strongly influence the humidity values [142]. In the case of hypogea, the underground conditions favor the maintaining of humidity.

Air movement differences between confined, semi-confined, and non-confined environments were expected to alter the number and type of fungal species recorded as well as incoming nutrients from the outside environment. A great proportion of entries in the database belonged to soil and litter dwellers such as saprotrophs, producing numerous spores that are well adapted to air-borne dispersal, and therefore, air ventilation may have a significant impact on the risk of contamination [143]. The more limited air volume movement of confined mural paintings compared to semi-confined ones was suggested to decrease the number of air-borne dust particles, with biofilm communities relying more on internal interactions between different microorganisms than on the external organic inputs [17]. Among the first aerobiological studies, Savulescu and Ionita reported a greater number of isolates inside the studied monasteries than outside of them, probably due to a more favorable microclimate inside the church, which favors the development of microorganisms [55]. Pangallo and colleagues proposed for the first time a comparative analysis of the microbial component of paintings and the surrounding air to gather information on the origin of fungal contamination [70]. Aside from the importance of aerobiological studies for the conservation and prevention of microbial attacks on indoor painted surfaces [144], these studies have received little attention. In light of the large number of fungal species potentially harmful for restorers and visitors, constant monitoring of air spore quality and concentration, as well as the use of air filters to reduce fungal spores concentrations, would be required for site conservation [62,101,144,145].

Significant correlations between the different taxa and the various environmental categories have not been recorded. Indeed, such data is not the result of the absence of a correlation between fungal growths and environmental conditions but can be consequence of several other influencing conditions that hide it. In fact, many are the ecological requirements that shape the ecological niches of the different species (i.e., the limiting factors), but the most conditioning factors are those that result in a quantity proximal to the upper or lower tolerance limit of an organism [146]. Then, for the various sites examined, some factors may become more relevant if their values are closer to the tolerance limit of certain organism, but this does not mean that other parameters do not play a role [147].

Indeed, our results may be influenced by the wide number of taxa in the wide geographic distribution of sites and by the different methodologies used to characterize the fungal diversity. In fact, different sampling techniques and isolation conditions were used within the studies we analyzed. Other factors that allow fungi to thrive and/or survive in a variety of conditions are their wide nutritional versatility and range of adaptations.

The presence of numerous genera that are widespread and highly sporulating and hence present in all the environmental categories must also be considered. The absence of evident correlations could have been determined by the absence of distinct boundaries between the categories identified, with overlapping microclimatic conditions which could have resulted in overlaps within their respective microbial communities. Finally, the heterogeneity of the data, with taxa identified at the genus or species level, may have resulted in dispersed clusters and hampered the ability to demonstrate any relationship.

This result seems to be in line with other studies. The influence of environmental factors such as temperature, relative humidity, and the opening or closure of the temples was not evident for fungal growths on wall paintings of 12 archaeological sites in the central and western parts of Thailand [76]. Furthermore, a stronger relationship with the age of five caves in China than with the environmental conditions, such as temperature and relative humidity, was proposed to explain the observed differences in fungal communities [64]. In two distinct mural paintings, instead, the differences recorded in the microbial communities were associated to the different organic input origin (i.e., wine cellar evaporation, and insect exuvia/excrements) and the microclimatic conditions. The more humid conditions favored the growth of actinomycetes, bacteria, and dark-pigmented fungi, while the other showed a biofilm, mainly dominated by xerotolerant and patchy growing sporulating fungi [17]. Differences in fungal communities were also recorded on mural paintings of two subterranean ancient Chinese tombs dating back over 1700 years, mostly due to variations in interior temperature and relative humidity as well as to their history and drawing techniques used [51].

Other significant concerns could be related to the identification of the isolated species, which was initially based solely on morphological observation. Indeed, phylogenetic molecular approaches are nowadays routinely applied, providing a universal tool for accurately identifying fungal species.

New methodologies such as omics techniques are now available; however, they rarely provide information at species or genus level, and there is no guarantee that the recorded taxa are actively growing. Moreover, culture-dependent approaches may not provide a real picture of the microbial diversity actively growing at the sampling time. This is because not all fungi actively growing on the deteriorated substrates can grow under laboratory conditions, and fungi growing under laboratory conditions may not actively grow on the sampled surfaces. Therefore, a combination of culture-based and molecular approaches may be needed to gain a clear picture of the actual biodiversity present on the painted surfaces as well as to have strains to investigate their potential degradative roles.

5. Conclusions

This study contributes to illustrate the high fungal diversity on wall paintings and raises awareness about the fungal threat on the deterioration of such artworks. Ascomycota was the most common phylum, with Eurotiales and Hypocreales as the most common orders. Statistical analyses did not enhance core communities that can be considered characteristics of different environmental categories of sites hosting wall paintings. Our results were likely due to the heterogeneity and fragmentation of the data in the databases, the dispersed geographical area considered, and the complexity of factors which can condition the biological growths. It is therefore crucial to cover the knowledge gaps through (i) international collaborations, (ii) enlarging the isolation and cultivation protocols as to easily detect also strains different from fast growing ones, and (iii) standardizing the identification protocols. Standardizing and improving the site descriptions (e.g., repeated microclimatic data) could allow for possible relations between site and their settlers and for further comparisons among different environmental conditions.

Supplementary Materials: The following supporting information can be downloaded at: https://www.mdpi.com/article/10.3390/app12062988/s1, Figure S1: Cluster analysis on the Jaccard distance of the distribution of the different entries retrieved from the papers analyzed; Table S1: List of all the references associated to each site defined in the analyses.

Author Contributions: Conceptualization, L.Z. and G.C.; methodology, L.Z., F.C., D.I. and G.C.; data curation, L.Z., F.C., D.I. and G.C.; writing—original draft preparation, L.Z. and F.C.; writing—review and editing, L.Z., F.C., D.I. and G.C. All authors have read and agreed to the published version of the manuscript.

Funding: This research received no external funding.

Institutional Review Board Statement: Not applicable.

Informed Consent Statement: Not applicable.

Data Availability Statement: Not applicable.

Acknowledgments: Authors wish to acknowledge Giulio Zangari (Department of Sciences, University of 'Roma Tre') for his support in some statistical elaborations. Authors wish to acknowledge Giulia Corsetti Antonini and Matt Hudson for the revision of the English language.

Conflicts of Interest: The authors declare no conflict of interest.

References

1. Garg, K.L.; Jain, K.K.; Mishra, A.K. Role of fungi in the deterioration of wall paintings. *Sci. Total Environ.* **1995**, *167*, 255–271. [CrossRef]
2. Mora, P.; Mora, L.; Philippot, P. *The Conservation of Wall Paintings*; Butterworths: London, UK, 1984; 576p.
3. Giannini, C.; Tapete, D. Materiali e procedimenti esecutivi della pittura murale. In *Il Laboratorio dell'Arte, Fonti e Ricerche per la Storia delle Tecniche Artistiche*; Il Prato: Saonara, Italy, 2009; Volume 2, 160p.
4. Caneva, G.; Nugari, M.P.; Nugari, M.P.; Salvadori, O. *Plant Biology for Cultural Heritage: Biodeterioration and Conservation*; Getty Publications: Los Angeles, CA, USA, 2008.
5. Caneva, G.; Bartoli, F.; Fontani, M.; Mazzeschi, D.; Visca, P. Changes in biodeterioration patterns of mural paintings: Multi-temporal mapping for a preventive conservation strategy in the Crypt of the Original Sin (Matera, Italy). *J. Cult. Herit.* **2019**, *40*, 59–68. [CrossRef]
6. Scheerer, S.; Ortega Morales, O.; Gaylarde, C.; Allen, I.; Laskin, S.; Geoffrey, G. Microbial deterioration of stone monuments—An updated overview. *Adv. Appl. Microbiol.* **2009**, *66*, 97–139. [CrossRef]
7. Albertano, P.; Urzì, C. Structural interactions among epilithic cyanobacteria and heterotrophic microorganisms in Roman hypogea. *Microb. Ecol.* **1999**, *38*, 244–252. [CrossRef]
8. Ranalli, G.; Zanardini, E.; Andreotti, A.; Colombini, M.P.; Corti, C.; Bosch-Roig, P.; De Nuntiis, P.; Lustrato, G.; Mandrioli, P.; Rampazzi, L.; et al. Hi-tech restoration by two-steps biocleaning process of Triumph of Death fresco at the Camposanto Monumental Cemetery (Pisa, Italy). *J. Appl. Microbiol.* **2018**, *125*, 800–812. [CrossRef]
9. Sáiz-Jiménez, C.; Samson, R.A. Biodegradacion de obras de arte. Hongos implicados en la degradacion de los frescos del monasterio de la Rabida (Huelva). *Bot. Macaronesica* **1981**, *8–9*, 255–264.
10. Pepe, O.; Sannino, L.; Palomba, S.; Anastasio, M.; Blaiotta, G.; Villani, F.; Moschetti, G. Heterotrophic microorganisms in deteriorated medieval wall paintings in Southern Italian churches. *Microbiol. Res.* **2010**, *165*, 21–32. [CrossRef]
11. Gomoiu, I.; Cojoc, R.L.; Enache, M.I.; Neagu, S.E.; Mohanu, D.; Mohanu, I. Microbial ability to colonize mural painting and its substrate. *Acta Phys. Polo. A* **2018**, *134*, 383–386. [CrossRef]
12. Sterflinger, K. Fungi: Their role in deterioration of cultural heritage. *Fungal Biol. Rev.* **2010**, *24*, 47–55. [CrossRef]
13. Vanderwolf, K.J.; Malloch, D.; McAlpine, D.F.; Forbes, G.J. A world review of fungi, yeasts, and slime molds in caves. *Int. J. Speleol.* **2013**, *42*, 77–96. [CrossRef]
14. Sterflinger, K.; Piñar, G. Microbial deterioration of cultural heritage and works of art—Tilting at windmills? *Appl. Microbiol. Biot.* **2013**, *97*, 9637–9646. [CrossRef] [PubMed]
15. Rosado, T.; Gil, M.; Mirão, J.; Candeias, A.; Caldeira, A.T. Oxalate biofilm formation in mural paintings due to microorganisms—A comprehensive study. *Int. Biodeter. Biodegr.* **2013**, *85*, 1–7. [CrossRef]
16. Ciferri, O. Microbial degradation of paintings. *Appl. Environ. Microbiol.* **1999**, *65*, 879–885. [CrossRef] [PubMed]
17. Dornieden, T.; Gorbushina, A.A.; Krumbein, W.E. Biodecay of cultural heritage as a space/time-related ecological situation—An evaluation of a series of studies. *Int. Biodeter. Biodegr.* **2000**, *46*, 261–270. [CrossRef]
18. Rosado, T.; Martins, M.R.; Pires, M.; Mirão, J.; Candeias, A.; Caldeira, A.T. Enzymatic monitorization of mural paintings biodegradation and biodeterioration. *Int. J. Conserv. Sci.* **2013**, *4*, 603–612.
19. Rosado, T.; Gil, M.; Caldeira, A.T.; Martins, M.D.R.; Dias, C.B.; Carvalho, L.; Mirão, J.; Candeias, A.E. Material characterization and biodegradation assessment of mural paintings: Renaissance frescoes from Santo Aleixo Church, Southern Portugal. *Int. J. Architect. Herit.* **2015**, *8*, 835–852. [CrossRef]

20. Unković, N.; Dimkić, I.; Stupar, M.; Stanković, S.; Vukojević, J.; Ljaljević Grbić, M. Biodegradative potential of fungal isolates from sacral ambient: In vitro study as risk assessment implication for the conservation of wall paintings. *PLoS ONE* **2018**, *13*, e0190922. [CrossRef]
21. Ionita, I. Contributions to the study of the biodeterioration of the work of art and of historic monuments. II. Species of fungi involved in the deterioration of mural paintings from the monasteries of Moldavia. *Rev. Roum. Biol. Série Bot.* **1973**, *18*, 179–189.
22. Unković, N.; Grbić, M.L.; Stupar, M.; Savković, Ž.; Jelikić, A.; Stanojević, D.; Vukojević, J. Fungal-induced deterioration of mural paintings: In situ and mock-model microscopy analyses. *Microsc. Microanal.* **2016**, *22*, 410–421. [CrossRef]
23. Sugiyama, J.; Kiyuna, T.; An, K.D.; Nagatsuka, Y.; Handa, Y.; Tazato, N.; Hata-Tomita, J.; Nishijima, M.; Koide, T.; Yaguchi, Y.; et al. Microbiological survey of the stone chambers of Takamatsuzuka and Kitora tumuli, Nara Prefecture, Japan: A milestone in elucidating the cause of biodeterioration of mural paintings. In Proceedings of the 31st International Symposium on the Conservation and Restoration of Cultural Property—Study of Environmental Conditions Surrounding Cultural Properties and Their Protective Measures, Tokyo, Japan, 5–7 February 2008; pp. 51–73. Available online: http://citeseerx.ist.psu.edu/viewdoc/download?doi=10.1.1.619.5561&rep=rep1&type=pdf (accessed on 31 January 2022).
24. Sugiyama, J.; Kiyuna, T.; Nishijima, M.; An, K.D.; Nagatsuka, Y.; Tazato, N.; Handa, Y.; Handa-Tomita, J.; Sato, Y.; Kigawa, R.; et al. Polyphasic insights into the microbiomes of the Takamatsuzuka tumulus and Kitora tumulus. *J. Gen. Appl. Microbiol.* **2017**, *63*, 63–113. [CrossRef]
25. Kiyuna, T.; An, K.D.; Kigawa, R.; Sano, C.; Miura, S.; Sugiyama, J. Molecular assessment of fungi in "black spots" that deface murals in the Takamatsuzuka and Kitora Tumuli in Japan: *Acremonium* sect. *Gliomastix* including *Acremonium tumulicola* sp. nov. and *Acremonium felinum* comb. nov. *Mycoscience* **2011**, *52*, 1–17. [CrossRef]
26. Marco, A.; Santos, S.; Caetano, J.; Pintado, M.; Vieira, E.; Moreira, P.R. Basil essential oil as an alternative to commercial biocides against fungi associated with black stains in mural painting. *Build. Environ.* **2020**, *167*, 106459. [CrossRef]
27. Tilak, S.T. Biodeterioration of paintings in Ajanta. In *Biodeterioration of Cultural Property*; Agrawal, O.P., Dhawan, S., Eds.; Macmillan: New Delhi, India, 1991; pp. 204–212.
28. Agrawal, O.P.; Dhawan, S.; Garg, K.L.; Shaheen, F.; Pathak, N.; Misra, A. Study of biodeterioration of the Ajanta wall paintings. *Int. Biodeterior.* **1988**, *24*, 121–129. [CrossRef]
29. Garg, K.L.; Dhawan, S.; Bhatnagar, I.K. Microbicides for preservation of wall paintings. In *Biodeterioration and Biodegradation*; Rossmoore, H.W., Ed.; Elsevier Applied Science: Barking, UK, 1991; Volume 8, pp. 505–507.
30. Dhawan, S.; Misra, A.; Garg, K.L.; Pathak, N. Laboratory evaluation of orto-phenyl-phenol and p-chloro-m-cresol for the control of some fungal forms of Ajanta wall paintings. In *Biodeterioration of Cultural Property*; Agrawal, O.P., Dhawan, S., Eds.; Macmillan: New Delhi, India, 1991; pp. 313–338.
31. Dhawan, S. Microbial deterioration of mural paintings. *Biodeterior. Mater.* **2002**, *2*, 95–105.
32. Dhawan, S.; Garg, K.L.; Pathak, N. Microbial analysis of Ajanta wall paintings & their possible control in situ. In *Biodeterioration of Cultural Property, Proceedings of the 2nd International Conference, Yokohama, Japan, 5–8 October 1992*; International Communications Specialists: Tokyo, Japan, 1993; Volume 2, pp. 245–262.
33. Faith, D.P.; Minchin, P.R.; Belbin, L. Compositional dissimilarity as a robust measure of ecological distance. *Vegetatio* **1987**, *69*, 57–68. [CrossRef]
34. Rousseeuw, P.J. Silhouettes: A graphical aid to the interpretation and validation of cluster analysis. *J. Comput. Appl. Math.* **1987**, *20*, 53–65. [CrossRef]
35. De Cáceres, M.; Legendre, P. Associations between species and groups of sites: Indices and statistical inference. *Ecology* **2009**, *90*, 3566–3574. [CrossRef]
36. De Cáceres, M.; Legendre, P.; Moretti, M. Improving indicator species analysis by combining groups of sites. *Oikos* **2010**, *119*, 1674–1684. [CrossRef]
37. Sun, J.Z.; Ge, Q.Y.; Zhu, Z.B.; Zhang, X.L.; Liu, X.Z. Three dominating hypocrealean fungi of the 'white mold spots' on acrylic varnish coatings of the murals in a Koguryo tomb in China. *Phytotaxa* **2019**, *397*, 225–236. [CrossRef]
38. Jeong, S.H.; Lee, H.J.; Lee, M.Y.; Chung, Y.J. Conservation environment for mural tomb in Goa-ri, Goryeong. *J. Cons. Sci.* **2017**, *33*, 189–201. [CrossRef]
39. Berner, M.; Wanner, G.; Lubitz, W. A comparative study of the fungal flora present in medieval wall paintings in the chapel of the castle Herberstein and in the parish church of St Georgen in Styria, Austria. *Int. Biodeter. Biodegr.* **1997**, *40*, 53–61. [CrossRef]
40. Sáiz-Jiménez, C.; Samson, R.A. Microorganisms and environmental pollution as deteriorating agents of the frescoes of the Monastery of "Santa María de la Rábida", Huelva, Spain. In Proceedings of the ICOM, Committee for Conservation, 6th Triennial Meeting, Ottawa, ON, Canada, 21–25 September 1981.
41. Rebricova, N.L. Some ecological aspects of protection of old Russian wall paintings from microbiological deterioration. In *Biodeterioration of Cultural Property*; Agrawal, O.P., Dhawan, S., Eds.; Macmillan: New Delhi, India, 1991; pp. 294–306.
42. Isola, D.; Zucconi, L.; Cecchini, A.; Caneva, G. Dark-pigmented biodeteriogenic fungi in Etruscan tombs: New data on their culture dependent diversity and favouring conditions. *Fungal Biol.* **2021**, *125*, 609–620. [CrossRef]
43. Rebricova, N.L. Micromycetes taking part in deterioration of old Russian wall paintings. In *Recent Advances in Biodeterioration and Biodegradation. Biodeterioration of Cultural Heritage*; Naya Prokash: Calcutta, India, 1993; Volume 1, pp. 205–232.
44. Martins, R.; Fialho, S.; Lima, M.; Tavares, D.; Mirão, J.; Valadas, S.; Candeias, A.E. Biodegradation assessment of a 16th century fresco from Southern Portugal. *Microsc. Microanal.* **2009**, *15*, 65–66. [CrossRef]

45. He, D.; Wu, F.; Ma, W.; Zhang, Y.; Gu, J.-D.; Duan, Y.; Xu, R.; Feng, H.; Wang, W.; Li, S.-W. Insights into the bacterial and fungal communities and microbiome that causes a microbe outbreak on ancient wall paintings in the Maijishan Grottoes. *Int. Biodeter. Biodegr.* **2021**, *163*, 105250. [CrossRef]
46. Guglielminetti, M.; De Giuli Morghen, C.; Radaelli, A.; Bistoni, F.; Carruba, G.; Spera, G.; Caretta, G. Mycological and Ultrastructural studies to evaluate biodeterioration of mural paintings. Detection of fungi and mites in frescos of the Monastery of St Damian in Assisi. *Int. Biodeter. Biodegr.* **1994**, *33*, 269–283. [CrossRef]
47. Agarossi, G.; Ferrari, R.; Monte, M. Biocides in the control of biodeterioration. In *The Conservation of Monuments in the Mediterranean Basin, Proceedings of the First International Symposium, 7–10 June Bari, Italy, 1989*; Grafo Edizioni: Brescia, Italy, 1990; pp. 511–517.
48. Karpovich-Tate, N.; Rebrikova, N.L. Microbial communities on damaged frescoes and building materials in the cathedral of the Nativity of the Virgin in the Pafnutii-Borovskii monastery, Russia. *Int. Biodeter. Biodegr.* **1991**, *27*, 281–296. [CrossRef]
49. Crippa, A. Funghi isolati da affreschi murali in antiche chiese di Pavia. In *Atti Società Italiana Scienze Naturali Museo Civico Storia Naturale Milano*; 1983; Volume 124, pp. 3–10. Available online: https://www.biodiversitylibrary.org/item/267082#page/10/mode/1up (accessed on 31 January 2022).
50. Sorlini, C.; Sacchi, M.; Ferrari, A. Microbiological deterioration of Gambara's frescoes exposed to open air in Brescia, Italy. *Int. Biodeter.* **1987**, *23*, 167–179. [CrossRef]
51. Ma, W.; Wu, F.; Tian, T.; He, D.; Zhang, Q.; Gu, J.-D.; Duand, Y.; Mae, D.; Wang, W.; Feng, H. Fungal diversity and its contribution to the biodeterioration of mural paintings in two 1700-year-old tombs of China. *Int. Biodeter. Biodegr.* **2020**, *152*, 104972. [CrossRef]
52. Gorbushina, A.A.; Petersen, K. Distribution of microorganisms on ancient wall paintings as related to associated faunal elements. *Int. Biodeter. Biodegr.* **2000**, *46*, 277–284. [CrossRef]
53. Veneranda, M.; Prieto-Taboada, N.; de Vallejuelo, S.F.O.; Maguregui, M.; Morillas, H.; Marcaida, I.; Castro, K.; Madariaga, J.M.; Osanna, M. Biodeterioration of Pompeian mural paintings: Fungal colonization favoured by the presence of volcanic material residues. *Environ. Sci. Poll. Res.* **2017**, *24*, 19599–19608. [CrossRef]
54. Gambino, M.; Ahmed, M.A.A.A.; Villa, F.; Cappitelli, F. Zinc oxide nanoparticles hinder fungal biofilm development in an ancient Egyptian tomb. *Int. Biodeter. Biodegr.* **2017**, *122*, 92–99. [CrossRef]
55. Savulescu, A.; Ionita, I. Contributions to the study of the biodeterioration of the works of art and historic monuments, I. Species of fungi isolated from frescoes. *Rev. Roum. Biol.* **1971**, *16*, 201–206.
56. Unković, N.; Erić, S.; Šarić, K.; Stupar, M.; Savković, Ž.; Stanković, S.; Stanojević, O.; Dimkić, I.; Vukojević, J.; Ljaljević Grbić, M. Biogenesis of secondary mycogenic minerals related to wall paintings deterioration process. *Micron* **2017**, *100*, 1–9. [CrossRef]
57. Mishra, A.K.; Garg, K.L. Microbial deterioration of wall paintings. In *Biodeterioration of Cultural Property 3, Proceedings of the 3rd International Conference on Biodeterioration of Cultural Property, Bangkok, Thailand, 4–7 July 1995*; Conservation Science Division, Office of Archaeology and National Museums: Bangkok, Thailand, 1995; pp. 630–642.
58. Emoto, Y. Microbiological investigation of ancient tombs with paintings: Ozuka tomb in Fukuoka and Chibusan tomb in Kumamoto. *Sci. Conserv.* **1974**, *12*, 95–102.
59. Mang, S.M.; Scrano, L.; Camele, I. Preliminary studies on fungal contamination of two rupestrian churches from Matera (Southern Italy). *Sustainability* **2020**, *12*, 6988. [CrossRef]
60. Jurado, V.; Gonzalez-Pimentel, J.L.; Hermosin, B.; Saiz-Jimenez, C. Biodeterioration of Salón de Reinos, Museo Nacional del Prado, Madrid, Spain. *Appl. Sci.* **2021**, *11*, 8858. [CrossRef]
61. Stupar, M.; Grbić, M.L.; Simić, G.S.; Jelikić, A.; Vukojević, J.; Sabovljević, M. A sub-aerial biofilms investigation and new approach in biocide application in cultural heritage conservation: Holy Virgin Church (Gradac Monastery, Serbia). *Indoor Built Environ.* **2012**, *23*, 584–593. [CrossRef]
62. Gorbushina, A.A.; Heyrman, J.; Dornieden, T.; Gonzalez-Delvalle, M.; Krumbein, W.E.; Laiz, L.; Petersen, L.; Saiz-Jimenez, C.; Swings, J. Bacterial and fungal diversity and biodeterioration problems in mural painting environments of St. Martins church (Greene–Kreiensen, Germany). *Int. Biodeter. Biodegr.* **2004**, *53*, 13–24. [CrossRef]
63. Ripka, K. Identification of Microorganisms on Stone and Mural Paintings Using Molecular Methods. Ph.D. Thesis, University of Wien, Vienna, Austria, 2005; p. 148. Available online: http://othes.univie.ac.at/13628/1/endvers.2.pdf (accessed on 2 December 2021).
64. Ma, Y.; Zhang, H.; Du, Y.; Tian, T.; Xiang, T.; Liu, X.; Wu, F.; An, L.; Wang, W.; Gu, J.-D.; et al. The community distribution of bacteria and fungi on ancient wall paintings of the Mogao Grottoes. *Sci. Rep.* **2015**, *5*, 7752. [CrossRef]
65. Unković, N.; Ljaljević Grbić, M.; Subakov-Simić, G.; Stupar, M.; Vukojević, J.; Jelikić, A.; Stanojević, D. Biodeteriogenic and toxigenic agents on 17th century mural paintings and facade of the old church of the Holy Ascension (Veliki Krčimir, Serbia). *Indoor Built Environ.* **2015**, *25*, 826–837. [CrossRef]
66. Rosado, T.; Mirão, J.; Candeias, A.; Caldeira, A.T. Microbial communities analysis assessed by pyrosequencing—A new approach applied to conservation state studies of mural paintings. *Anal. Bioanal. Chem.* **2014**, *406*, 887–895. [CrossRef] [PubMed]
67. Pepe, O.; Palomba, S.; Sannino, L.; Blaiotta, G.; Ventorino, V.; Moschetti, G.; Villani, F. Characterization in the archaeological excavation site of heterotrophic bacteria and fungi of deteriorated wall painting of Herculaneum in Italy. *J. Environ. Biol.* **2011**, *32*, 241–250. [PubMed]
68. Sprocati, A.R.; Alisi, C.; Tasso, F.; Vedovato, E.; Barbabietola, N.; Cremisini, C. A microbiological survey of the Etruscan Mercareccia tomb (Italy): Contribution of microorganisms to deterioration and restoration. In *Art 2008, Proceedings of the 9th International Conference on NDT of Art, Jerusalem, Israel, 25–30 May 2008*; NDT of Art: Jerusalem, Israel, 2008; 9p.

69. Sampò, S.; Luppi Mosca, A.M. A study of the fungi occurring on 15th century frescoes in Florence, Italy. *Int. biodeterior.* **1989**, *25*, 343–353. [CrossRef]
70. Pangallo, D.; Kraková, L.; Chovanová, K.; Šimonovičová, A.; De Leo, F.; Urzì, C. Analysis and comparison of the microflora isolated from fresco surface and from surrounding air environment through molecular and biodegradative assays. *World J. Microb. Biot.* **2012**, *28*, 2015–2027. [CrossRef]
71. Pangallo, D.; Chovanová, K.; Šimonovicová, A.; De Leo, F.; Urzì, C. Assessment of the biodeterioration risk of the Ladislav lagend fresco in Velka Lomnica (SK) through non-invasive methods. In Proceedings of the 11th International Congress on Deterioration and Conservation of Stone, Torun, Poland, 15–20 September 2008; Volume 1, pp. 457–464.
72. Helmi, F.M.; Elmitwalli, H.R.; Rizk, M.A.; Hagrassy, A.F. Antibiotic extraction as a recent biocontrol method for *Aspergillus niger* and *Aspergillus flavus* fungi in ancient Egyptian mural paintings. *Mediterr. Archaeol. Arc.* **2011**, *11*, 1–7.
73. Khalaphallah, R.; El-Derby, A.A. The effect of nano-TiO_2 and plant extracts on microbial strains isolated from Theban ancient Egyptian royal tomb painting. *Afr. J. Microbiol. Res.* **2015**, *9*, 1424–1430. [CrossRef]
74. Elhagrassy, A.F. Isolation and characterization of actinomycetes from mural paintings of Snu-Sert-Ankh tomb, their antimicrobial activity, and their biodeterioration. *Microbiol. Res.* **2018**, *216*, 47–55. [CrossRef]
75. Chaisrisook, C.; Suwanarit, P.; Aranyanak, C. Fungal deterioration of mural paintings in the royal temple. In *Biodeterioration of Cultural Property 3, Proceedings of the 3rd International Conference on Biodeterioration of Cultural Property, Bangkok, Thailand, 4–7 July 1995*; Conservation Science Division, Office of Archaeology and National Museums: Bangkok, Thailand, 1995; pp. 116–137.
76. Senbua, W.; Wichitwechkarn, J. Molecular identification of fungi colonizing art objects in Thailand and their growth inhibition by local plant extracts. *3 Biotech* **2019**, *9*, 356. [CrossRef]
77. Stoyancheva, G.; Krumova, E.; Kostadinova, N.; Miteva-Staleva, J.; Grozdanov, P.; Ghaly, M.F.; Sakr, A.A.; Angelova, M. Biodiversity of contaminant fungi at different coloured materials in ancient Egypt Tombs and Mosques. *Cr. Acad. Bul. Sci.* **2018**, *71*, 907–915. [CrossRef]
78. Sakr, A.; Ghaly, M.; Helal, G.; Abdel Haliem, M. Effect of thymol against fungi deteriorating mural paintings at Tell Basta tombs, Lower Egypt. *Int. J. Res. Stud. Biosci.* **2012**, *6*, 8–23. [CrossRef]
79. Raschle, P. Experience of combating moulds during restoration of ceiling paintings in a Swiss baroque monastery church. *Biodeterioration* **1983**, *5*, 427–433.
80. Rosado, T.; Falé, A.; Gil, M.; Mirão, J.; Candeias, A.; Caldeira, A.T. Understanding the influence of microbial contamination on colour alteration of pigments used in wall paintings—The case of red and yellow ochres and ultramarine blue. *Color Res. Appl.* **2019**, *44*, 783–789. [CrossRef]
81. Di Carlo, E.; Chisesi, R.; Barresi, G.; Barbaro, S.; Lombardo, G.; Rotolo, V.; Palla, F. Fungi and bacteria in indoor Cultural Heritage environments: Microbial-related risks for artworks and human health. *Environ. Ecol. Res.* **2016**, *4*, 257–264. [CrossRef]
82. Cepero, A.; Martinez, P.; Castro, J.; Sanchez, A.; Machado, J. The biodeterioration of cultural property in the republic of Cuba: A review of some experiences. In *Biodeterioration of Cultural Property, Proceedings of the 2nd International Conference, Yokohama, Japan, 5–8 October 1992*; International Communications Specialists: Tokyo, Japan, 1993; pp. 479–487.
83. Arai, H. Relationship between fungi and brown spots found in various materials. In *Biodeterioration of Cultural Property, Proceedings of the 2nd International Conference, Yokohama, Japan, 5–8 October 1992*; International Communications Specialists: Tokyo, Japan, 1993; pp. 320–336.
84. Barbieri, N.; Bassi, M.; Dassù, G.; Rossi, F. Gli affreschi del tempio repubblicano di Brescia: Condizioni ambientali ed inquinamento microbiologico. *Arte Lomb.* **1986**, *76/77*, 113–117.
85. Martin-Sanchez, P.M.; Nováková, A.; Bastian, F.; Alabouvette, C.; Saiz-Jimenez, C. Use of biocides for the control of fungal outbreaks in subterranean environments: The case of the Lascaux Cave in France. *Envir. Sci. Tech.* **2012**, *46*, 3762–3770. [CrossRef]
86. Bartolini, M.; Nugari, M.P.; Pietrini, A.M.; Ricci, S.; Roccardi, A.; Filetici, M.G. Gli ambienti ipogei delle domus romane al Celio: Indagini biologiche per il controllo e la prevenzione del biodeterioramento. *Kermes La Riv. Del Restauro* **2010**, *23*, 45–54.
87. Vasanthakumar, A.; DeAraujo, A.; Mazurek, J.; Schilling, M.; Mitchell, R. Microbiological survey for analysis of the brown spots on the walls of the tomb of King Tutankhamun. *Int. Biodeter. Biodegr.* **2013**, *79*, 56–63. [CrossRef]
88. Tonolo, A.; Giacobini, C. Microbiological changes of frescoes. In *Recent Advances in Conservation*; Thomson, G., Ed.; Butterworths: London, UK, 1961; pp. 62–64.
89. Arai, H. The environmental analysis of archaeological sites. *TrAC Trends Anal. Chem.* **1990**, *9*, 213–216. [CrossRef]
90. Sorlini, C.; Allievi, L.; Sacchi, M.; Ferrari, A. Microorganisms present in deteriorated materials of the Palazzo della Ragione in Milan. *Int. Biodeterior. Bull.* **1982**, *18*, 105–110.
91. Rosado, T.; Mirão, J.; Candeias, A.; Caldeira, A.T. Characterizing microbial diversity and damage in mural paintings. *Microsc. Microanal.* **2015**, *21*, 78. [CrossRef]
92. Popescu, A.; Arai, H.; Minatoya, T. Biodeterioration aspects of the Probota Monastery and possibilities for its restoration. In *Biodeterioration of Cultural Property 3, Proceedings of the 3rd International Conference on Biodeterioration of Cultural Property, Bangkok, Thailand, 4–7 July 1995*; Conservation Science Division, Office of Archaeology and National Museums: Bangkok, Thailand, 1995; pp. 255–271.
93. Gargani, G. Fungus contamination of Florence art masterpieces before and after the 1966 disaster. In *Biodeterioration of Materials, Microbiological and Allied Aspects*; Elsevier: Amsterdam, The Netherlands, 1968; pp. 252–257.

94. Saarela, M.; Alakomi, H.L.; Suihko, M.L.; Maunuksela, L.; Raaska, L.; Mattila-Sandholm, T. Heterotrophic microorganisms in air and biofilm samples from Roman catacombs, with special emphasis on actinobacteria and fungi. *Int. Biodeter. Biodegr.* **2004**, *54*, 27–37. [CrossRef]
95. Palla, F.; Billeci, N.; Mancuso, F.P.; Pellegrino, L.; Lorusso, L.C. Microscopy and molecular biology techniques for the study of biocenosis diversity in semi-confined environments. *Conserv. Sci. Cult. Herit.* **2010**, *10*, 185–194. [CrossRef]
96. Nagatsuka, Y.; Kiyuna, T.; Kigawa, R.; Sano, C.; Miura, S.; Sugiyama, J. *Candida tumulicola* sp. nov. and *Candida takamatsuzukensis* sp. nov., novel yeast species assignable to the *Candida membranifaciens* clade, isolated from the stone chamber of the Takamatsuzuka tumulus. *Int. J. Syst. Evol. Microbiol.* **2009**, *59*, 186–194. [CrossRef] [PubMed]
97. Dupont, J.; Jacquet, C.; Dennetiere, B.; Lacoste, S.; Bousta, F.; Orial, G.; Cruaud, C.; Couloux, A.; Roquebert, M.F. Invasion of the French Paleolithic painted cave of Lascaux by members of the *Fusarium solani* species complex. *Mycologia* **2007**, *99*, 526–533. [CrossRef]
98. Kiyuna, T.; An, K.D.; Kigawa, R.; Sano, C.; Sugiyama, J. Two new *Cladophialophora* species, *C. tumbae* sp. nov. and *C. tumulicola* sp. nov., and chaetothyrialean fungi from biodeteriorated samples in the Takamatsuzuka and Kitora Tumuli. *Mycoscience* **2018**, *59*, 75–84. [CrossRef]
99. Zucconi, L.; Gagliardi, M.; Isola, D.; Onofri, S.; Andaloro, M.C.; Pelosi, C.; Pogliani, C.; Selbmann, L. Biodeterioration agents dwelling in or on the wall paintings of the Holy Saviour's cave (Vallerano, Italy). *Int. Biodeter. Biodegr.* **2012**, *70*, 40–46. [CrossRef]
100. Bianchi, A.; Favali, M.A.; Barbieri, N.; Bassi, M. The use of fungicides on mold-covered frescoes in S. Eusebio in Pavia. *Int. Biodeterior. Bull.* **1980**, *16*, 45–51.
101. Pitzurra, L.; Bellezza, T.; Giammarioli, M.; Giraldi, M.; Sbaraglia, G.; Spera, G.; Bistoni, F. Microbial environmental monitoring of the refectory in the monastery of St. Anna in Foligno, Italy. *Aerobiologia* **1999**, *15*, 203–209. [CrossRef]
102. Bassi, M.; Giacobini, C. Scanning electron microscopy: A new technique in the study of the microbiology of works of art. *Int. Biodeter. Biodegr.* **2001**, *48*, 55–66. [CrossRef]
103. Jeffries, P. Biodeterioration of wall paintings in Canterbury Cathedral. In *Biodeterioration of Cultural Property*; Agrawal, O.P., Dhawan, S., Eds.; Macmillan: Delhi, India, 1991; pp. 287–293.
104. Fiorillo, F.; Fiorentino, S.; Montanari, M.; Monaco, C.R.; Del Bianco, A.; Vandini, M. Learning from the past, intervening in the present: The role of conservation science in the challenging restoration of the wall painting Marriage at Cana by Luca Longhi (Ravenna, Italy). *Herit. Sci.* **2020**, *8*, 10. [CrossRef]
105. Kiyuna, T.; An, K.D.; Kigawa, R.; Sano, C.; Miura, S.; Sugiyama, J. Mycobiota of the Takamatsuzuka and Kitora Tumuli in Japan, focusing on the molecular phylogenetic diversity of *Fusarium* and *Trichoderma*. *Mycoscience* **2008**, *49*, 298–311. [CrossRef]
106. Stomeo, F.; Portillo, M.C.; Gonzalez, J.M. Assessment of bacterial and fungal growth on natural substrates: Consequences for preserving caves with prehistoric paintings. *Curr. Microbiol.* **2009**, *59*, 321–325. [CrossRef] [PubMed]
107. Kiyuna, T.; An, K.D.; Kigawa, R.; Sano, C.; Miura, S.; Sugiyama, J. Bristle-like fungal colonizers on the stone walls of the Kitora and Takamatsuzuka Tumuli are identified as *Kendrickiella phycomyces*. *Mycoscience* **2012**, *53*, 446–459. [CrossRef]
108. Egidi, E.; de Hoog, G.S.; Isola, D.; Onofri, S.; Quaedvlieg, W.; de Vries, M.; Stielow, J.B.; Zucconi, L.; Selbmann, L. Phylogeny and taxonomy of meristematic rock-inhabiting black fungi in the Dothideomycetes based on multi-locus phylogenies. *Fungal Divers.* **2014**, *65*, 127–165. [CrossRef]
109. Nugari, M.P.; Pietrini, A.M.; Caneva, G.; Imperi, F.; Visca, P. Biodeterioration of mural paintings in a rocky habitat: The Crypt of the Original Sin (Matera, Italy). *Int. Biodeter. Biodegr.* **2009**, *63*, 705–711. [CrossRef]
110. Jeffries, P. Growth of *Beauveria alba* on mural paintings in Canterbury Cathedral. *Int. Biodeter. Biodegr.* **1986**, *22*, 11–13.
111. Leplat, J.; Francois, A.; Bousta, F. White fungal covering on the wall paintings of the Saint-Savin-sur-Gartempe Abbey church crypt: A case study. *Int. Biodeter. Biodegr.* **2017**, *122*, 29–37. [CrossRef]
112. Milanesi, C.; Baldi, F.; Vignani, R.; Ciampolini, F.; Faleri, C.; Cresti, M. Fungal deterioration of medieval wall fresco determined by analysing small fragments containing copper. *Int. Biodeter. Biodegr.* **2006**, *57*, 7–13. [CrossRef]
113. Moza, M.I.; Mironescu, M.; Georgescu, C.; Florea, A.; Bucşa, L. Isolation and characterisation of moulds degrading mural paintings. *Ann. RSCB* **2012**, *17*, 136–142.
114. An, K.D.; Kiyuna, T.; Kigawa, R.; Sano, C.; Miura, S.; Sugiyama, J. The identity of *Penicillium* sp. 1, a major contaminant of the stone chambers in the Takamatsuzuka and Kitora Tumuli in Japan, is *Penicillium paneum*. *Anton. van Leeuw.* **2009**, *96*, 579. [CrossRef]
115. Martin-Sanchez, P.M.; Nováková, A.; Bastian, F.; Alabouvette, C.; Saiz-Jimenez, C. Two new species of the genus *Ochroconis*, *O. lascauxensis* and *O. anomala* isolated from black stains in Lascaux Cave, France. *Fungal Biol.* **2012**, *116*, 574–589. [CrossRef] [PubMed]
116. Caneva, G.; Tescari, M.; Bartoli, F.; Nugari, M.P.; Pietrini, A.M.; Salvadori, O. Ecological mapping for the preventive conservation of prehistoric mural paintings in rock habitats: The site of Filiano (Basilicata, Italy). *Conserv. Sci. Cult. Herit.* **2015**, *15*, 53–59. [CrossRef]
117. Unković, N.; Ljaljević Grbić, M.; Stupar, M.; Vukojević, J.; Subakov-Simić, G.; Jelikić, A.; Stanojević, D. ATP bioluminescence method: Tool for rapid screening of organic and microbial contaminants on deteriorated mural paintings. *Nat. Prod. Res.* **2015**, *33*, 1061–1069. [CrossRef] [PubMed]

118. Spatafora, J.W.; Aime, M.C.; Grigoriev, I.V.; Martin, F.; Stajich, J.E.; Blackwell, M. The fungal tree of life: From molecular systematics to genome-scale phylogenies. In *The Fungal Kingdom*; Heitman, J., Howlett, B.J., Crous, P.W., Stukenbrock, E.H., James, T.Y., Gow, N.A.R., Eds.; ASM press: Washington, DC, USA, 2007; pp. 1–34. [CrossRef]
119. Jurado, V.; Sanchez-Moral, S.; Saiz-Jimenez, C. Entomogenous fungi and the conservation of the cultural heritage: A review. *Int. Biodeterior. Biodegrad.* **2008**, *62*, 325–330. [CrossRef]
120. Isola, D.; Bartoli, F.; Langone, S.; Ceschin, S.; Zucconi, L.; Caneva, G. Plant DNA barcode as a tool for root identification in hypogea: The Case of the Etruscan Tombs of Tarquinia (Central Italy). *Plants* **2021**, *10*, 1138. [CrossRef]
121. Cuzman, O.A.; Tapete, D.; Fratini, F.; Mazzei, B.; Riminesi, C.; Tiano, P. Assessing and facing the biodeteriogenic presence developed in the Roman Catacombs of Santi Marco, Marcelliano e Damaso, Italy. *Eur. J. Sci. Theol.* **2014**, *10*, 185–197.
122. Isola, D.; Selbmann, L.; Meloni, P.; Maracci, E.; Onofri, S.; Zucconi, L. Detrimental rock black fungi and biocides: A study on the Monumental Cemetery of Cagliari. In *Science and Technology for the Conservation of Cultural Heritage*; Rogerio-Candelera, M.A., Lazzari, M., Cano, E., Eds.; CRC Press: London, UK, 2013; pp. 83–86.
123. Isola, D.; Bartoli, F.; Meloni, P.; Caneva, G.; Zucconi, L. Black fungi and stone heritage conservation: Ecological and metabolic assays for evaluating colonization potential and responses to traditional biocides. *Appl. Sci.* **2022**, *12*, 2038. [CrossRef]
124. Isola, D.; Scano, A.; Orrù, G.; Prenafeta-Boldú, F.X.; Zucconi, L. Hydrocarbon-contaminated sites: Is there something more than *Exophiala xenobiotica*? New insights into black fungal diversity using the long cold incubation method. *J. Fungi* **2021**, *7*, 817. [CrossRef]
125. Sterflinger, K.; Pinzari, F. The revenge of time: Fungal deterioration of cultural heritage with particular reference to books, paper and parchment. *Environ. Microbiol.* **2012**, *14*, 559–566. [CrossRef]
126. Pyzik, A.; Ciuchcinski, K.; Dziurzynski, M.; Dziewit, L. The Bad and the Good—Microorganisms in Cultural Heritage Environments—An Update on Biodeterioration and Biotreatment Approaches. *Materials* **2021**, *14*, 177. [CrossRef]
127. Salo, P.M.; Arbes, S.J.; Sever, M.; Jaramillo, R.; Cohn, R.D.; London, S.J.; Zeldin, D.C. Exposure to *Alternaria alternata* in US homes is associated with asthma symptoms. *J. Aller. Cl. Imm.* **2006**, *118*, 892–898. [CrossRef] [PubMed]
128. O'Donnell, K.; Sutton, D.A.; Rinaldi, M.G.; Magnon, K.C.; Cox, P.A.; Revankar, S.G.; Sanche, S.; Geiser, D.M.; Juba, J.H.; van Burik, J.-A.H.; et al. Genetic diversity of human pathogenic members of the *Fusarium oxysporum* complex inferred from multilocus DNA sequence data and amplified fragment length polymorphism analyses: Evidence for the recent dispersion of a geographically widespread clonal lineage and nosocomial origin. *J. Clin. Microbiol.* **2004**, *42*, 5109–5120. [CrossRef] [PubMed]
129. Domsch, K.H. *Compendium of Soil Fungi*; Taxonomically Revised by Walter Gams; IHW: Eching, Germany, 2007; 672p.
130. Hyde, K.D.; Al-Hatmi, A.M.S.; Andersen, B.; Boekhout, T.; Buzina, W.; Dawson, T.L.; Eastwood, D.C.; Gareth Jones, E.B.; de Hoog, G.S.; Kang, Y.; et al. The world's ten most feared fungi. *Fungal Divers.* **2018**, *93*, 161–194. [CrossRef]
131. Forster, A.M.; Thomson, D.; Richards, K.; Pilcher, N.; Vettese, S. Western and Eastern building conservation philosophies: Perspectives on permanence and impermanence. *Int. J. Architect. Her.* **2019**, *13*, 870–885. [CrossRef]
132. Hibbett, D.S.; Binder, M.; Bischoff, J.F.; Blackwell, M.; Cannon, P.F.; Eriksson, O.E.; Huhndorf, S.; James, T.; Kirk, P.M.; Lu Cking, R.; et al. A higher-level phylogenetic classification of the Fungi. *Mycol. Res.* **2007**, *111*, 509–547. [CrossRef]
133. Tekpinar, A.D.; Kalmer, A. Utility of various molecular markers in fungal identification and phylogeny. *Nova Hedwig.* **2019**, *109*, 187–224. [CrossRef]
134. Al-Hatmi, A.M.S.; Van Den Ende, A.H.G.; Stielow, J.B.; Van Diepeningen, A.D.; Seifert, K.A.; McCormick, W.; Assabgui, R.; Gräfenhan, T.; De Hoog, S.G.; Levesque, C.A. Evaluation of two novel barcodes for species recognition of opportunistic pathogens in *Fusarium*. *Fungal Biol.* **2016**, *120*, 231–245. [CrossRef]
135. Samson, R.A.; Visagie, C.M.; Houbraken, J.; Hong, S.B.; Hubka, V.; Klaassen, C.H.W.; Perrone, G.; Seifert, K.A.; Susca, A.; Tanney, J.B.; et al. Phylogeny, identification and nomenclature of the genus *Aspergillus*. *Stud. Mycol.* **2014**, *78*, 141–173. [CrossRef]
136. Visagie, C.M.; Houbraken, J.; Frisvad, J.C.; Hong, S.B.; Klaassen, C.H.W.; Perrone, G.; Seifert, K.A.; Varga, J.; Yaguchi, T.; Samson, T.A. Identification and nomenclature of the genus *Penicillium*. *Stud. Mycol.* **2014**, *7*, 343–371. [CrossRef]
137. Houbraken, J.; Visagie, C.M.; Frisvad, J.C. Recommendations to prevent taxonomic misidentification of genome-sequenced fungal strains. *Microbiol. Resour. Announc.* **2021**, *10*, e01074-20. [CrossRef]
138. Sandoval-Denis, M.; Gene, J.; Sutton, D.A.; Wiederhold, N.P.; Cano-Lira, J.F.; Guarro, J. New species of *Cladosporium* associated with human and animal infections. *Persoonia* **2016**, *36*, 281–298. [CrossRef] [PubMed]
139. Sterflinger, K.; Little, B.; Pinar, G.; Pinzari, F.; de los Rios, A.; Gu, J.D. Future directions and challenges in biodeterioration research on historic materials and cultural properties. *Int. Biodeterior. Biodegrad.* **2018**, *129*, 10–12. [CrossRef]
140. Caneva, G.; Isola, D.; Lee, H.J.; Chung, Y.J. Biological risk for hypogea: Shared data from Etruscan tombs in Italy and ancient tombs of the Baekje dynasty in Republic of Korea. *Appl. Sci.* **2020**, *10*, 6104. [CrossRef]
141. Sanchez-Moral, S.; Canaveras, J.C.; Laiz, L.; Saiz-Jimenez, C.; Bedoya, J.; Luque, L. Biomediated precipitation of calcium carbonate metastable phases in hypogean environments: A short review. *Geomicrobiol. J.* **2003**, *20*, 491–500. [CrossRef]
142. Caneva, G.; Langone, S.; Bartoli, F.; Cecchini, A.; Meneghini, C. Vegetation cover and tumuli's shape as affecting factors of microclimate and biodeterioration risk for the conservation of Etruscan tombs (Tarquinia, Italy). *Sustainability* **2021**, *13*, 3393. [CrossRef]
143. Savković, Ž.; Stupar, M.; Unković, N.; Knežević, A.; Vukojević, J.; Ljaljević Grbić, M. Fungal Deterioration of Cultural Heritage Objects. In *Biodegradation*; IntechOpen: London, UK, 2021. [CrossRef]

144. Nugari, M.P.; Realini, M.; Roccardi, A. Contamination of mural paintings by indoor airborne fungal spores. *Aerobiologia* **1993**, *9*, 131–139. [CrossRef]
145. Mandrioli, P.; Caneva, G.; Sabbioni, C. *Cultural Heritage and Aerobiology. Methods and Measurement Techniques for Biodeterioration Monitoring*; Kluwer Academic Publishers: Dordrecht, The Netherlands, 2003; 260p. [CrossRef]
146. Odum, E.P.; Odum, H.T.; Andrews, J. *Fundamentals of Ecology*; Saunders: Philadelphia, PA, USA, 1971; Volume 3.
147. Caneva, G.; Bartoli, F.; Savo, V.; Futagami, Y.; Strona, G. Combining statistical tools and ecological assessments in the study of biodeterioration patterns of stone temples in Angkor (Cambodia). *Sci. Rep.* **2016**, *6*, 32601. [CrossRef]

applied sciences

MDPI

Review

Black Fungi on Stone-Built Heritage: Current Knowledge and Future Outlook

Filomena De Leo *, Alessia Marchetta and Clara Urzì

Department of Chemical, Biological, Pharmaceutical and Environmental Sciences, University of Messina, Viale F. Stagno d'Alcontres, 31, 98166 Messina, Italy; alessia.marchetta@unime.it (A.M.); urzicl@unime.it (C.U.)
* Correspondence: fdeleo@unime.it

Featured Application: This is an updated review on black fungi as main biodeteriogens of cultural heritage stone artifacts. Colonization pattern, taxonomy, and methods to eradicate their settlement are discussed here.

Abstract: Black fungi are considered as one of the main group of microorganisms responsible for the biodeterioration of stone cultural heritage artifacts. In this paper, we provide a critical analysis and review of more than 30 years of studies on black fungi isolated from stone-built heritage from 1990 to date. More than 109 papers concerning the fungal biodeterioration activity of stone were analysed. The main findings were a check list of the black fungal taxa involved in the biodeterioration of stone-built heritage, with a particular reference to meristematic black fungi, the main biodeterioration pattern attributed to them, and the methods of study including the new molecular advances. A particular focus was to discuss the current approaches to control black fungi from stone-built heritage and future perspectives. Black fungi are notoriously hard to remove or mitigate, so new methods of study and of control are needed, but it is also important to combine classical methods with new approaches to improve current knowledge to implement future conservation strategies.

Keywords: stone cultural heritage; black fungi; MCF; biodeterioration; control

Citation: De Leo, F.; Marchetta, A.; Urzì, C. Black Fungi on Stone-Built Heritage: Current Knowledge and Future Outlook. *Appl. Sci.* **2022**, *12*, 3969. https://doi.org/10.3390/app12083969

Academic Editor: Cesareo Saiz-Jimenez

Received: 1 March 2022
Accepted: 11 April 2022
Published: 14 April 2022

1. Background

Stonework, such as artistic sculptures, historical buildings, monuments, archaeological sites, caves, etc., are ubiquitous across the globe, being an expression of culture, religion, aesthetics, and building techniques of populations, typical of certain historical construction periods. Due to their unicity and intrinsic value, ensuring the integrity of stone-built heritage for posterity is a critical issue. The study of biodeterioration of cultural heritage is a hot topic of broad interest to the researcher's community and the implementation of safeguard measures is one of the main goals. All materials are subjected to a natural weathering, and "biodeterioration" of stones should be considered as an integral part of bio-geo-morphogenesis [1–4]. The term "biodeterioration" defines any irreversible transformation of inorganic or organic material with economic, commercial, historic, and artistic loss caused by macro- and micro-organisms [5].

"Biodeterioration" is a very complex matter and conservators should also take into account whether the observed biologically driven phenomena can even be considered positive for the artifact.

In fact, in some cases, the presence of subaerial biofilm (SABs) may have a protective effect on the surface [6]; on the other hand, SABs developed at the interface between rock surface and air is considered the main cause of biodeterioration of stone monuments [7,8].

Microbial biodeterioration of stones is often associated with the presence of a complex community formed by chemoorganotrophic microorganisms (bacteria and microfungi) and autotrophic microorganisms (such as algae and cyanobacteria and to lesser extent autotrophic bacteria) usually embedded in an extracellular matrix EPS (in which are present DNA, enzymes, pigments, lipids, proteins, etc.). Microbial cells in the EPS show a typical biofilm lifestyle that confers resistance to hostile environments and reinforces the attachment of microorganisms on the surface [6,9].

The prevalence of one or more group of microorganisms depends on numerous factors which include the intrinsic characteristics of the material (such as lithotype, porosity, roughness, and state of preservation) that affect its "bioreceptivity" *sensu* Guillitte [10]. The species composition can vary greatly depending on climatic and microclimatic conditions such as temperature, solar irradiation, shining, nutrient and water availability, and, last but not least, the characteristics of species involved [6]. However, microbial colonization is a very dynamic process in time and space, that is the result of the interactions between microbial species and substrates. It varies continuously during the year following the seasons, and it is also under the influence of the dispersion ability of propagules in the air [11–13].

In recent years, much knowledge has been gained about rock-inhabiting black fungi, and important issues concerning their taxonomy, physiology, phylogeny, and weathering processes [14] have been clarified. However, the majority of studies concerned black fungi from natural environments [4,7,15].

In the field of cultural heritage, most reviews had as a topic the biodeterioration of stone caused by fungi in general [6,16,17]; some have focused on the microbial and fungal deterioration of various type of substrata (both organic and inorganic such as textile, parchment, wood, paper, metals, and stone) used for artworks [18]; few concerned exclusively black fungi as a cause of biodeterioration of stone monuments [19,20].

This paper aims to give an overview on the present knowledge of rock-inhabiting black fungi in the field of stone cultural heritage with reference to their taxonomy, biodeterioration pattern, methods of study, and control, with a look to a future perspectives.

A bibliographic search was carried out using such databases as Scopus (https://www.scopus.com accessed on 17 March 2022), Science Direct (https://www.sciencedirect.com accessed on 17 March 2022), Web of Science (http://www.webofknowledge.com accessed on 17 March 2022), and Google Scholar (https://scholar.google.com accessed on 17 March 2022), that were consulted by using keywords such as 'black fungi', 'meristematic fungi', 'stone monuments', 'stone artworks', 'stone biodeterioration', 'biodeteriogenic fungi', 'fungal treatment', and 'fungal control'.

The search produced about 500 papers of which 109 were included in this paper. The updates of fungal nomenclature were searched in the databases Index Fungorum (http://www.indexfungorum.org accessed on 22 March 2022), National Center for Biotechnology Information (NCBI) (https://www.ncbi.nlm.nih.gov accessed on 24 February 2022) and in the recent literature [14,21].

Nucleotide sequences were retrieved from GenBank database that is accessible from NCBI platforms (http://www.ncbi.nlm.nih.gov accessed on 14 February 2022). Molecular Evolution Genetic Analyses (Mega 11 Software) free downloadable via the URL. http://www.megastoftware.net accessed on 14 February 2022 was employed for alignments and phylogenetic tree constructions.

2. Black Fungi and Stone Monuments: An Intimate Connection

Beginning in the 1990s, black fungi were described as one of the most likely groups of microorganisms responsible for the biodeterioration of the stone monuments [22–25] and it was confirmed in the following decades [6,18,26,27].

The term "black fungi" refers to a very huge group of dematiaceous fungi, unrelated phylogenetically, which have in common the presence of melanin in the cell wall that confers an olive brown appearance to the colony [28]. Another common characteristic is the ability to withstand hostile environments such as scarcity of nutrients, high solar irradiation, scarcity of water, high osmolarity, and low pH [15,19,29].

As reported by Gueidan et al. [30] the ancestors of black fungi were well adapted to live in oligotrophic environments such as rock surfaces or sub-surfaces, and currently they can also grow in anthropogenic habitats such as glass, silicon, organic surfaces, metals [31], or consolidants applied on the stone [9].

Their resilience is related to the extremotolerant or even polyextremotolerant characteristics of the species. The stress-tolerance is due to different factors such as: pigmentation, and in particular melanins production; mycosporine-like substances; morphological and metabolic versatility; meristematic development; and oligotrophy [32–34]. All these characteristics make them very suitable for colonizing outdoor rocks and built stones due to the fact that those surfaces can be exposed to extreme environments [17–19].

This group of fungi includes (a) fast growing hyphomycetes of epiphytic origin, recognizable under microscope by the presence of typical conidiophores and spores; (b) pleomorphic hyphomycetes that include the "black yeasts", showing a yeast-like form, and the so-called "black meristematic fungi" with a *Torula*-like growth pattern (Figure 1).

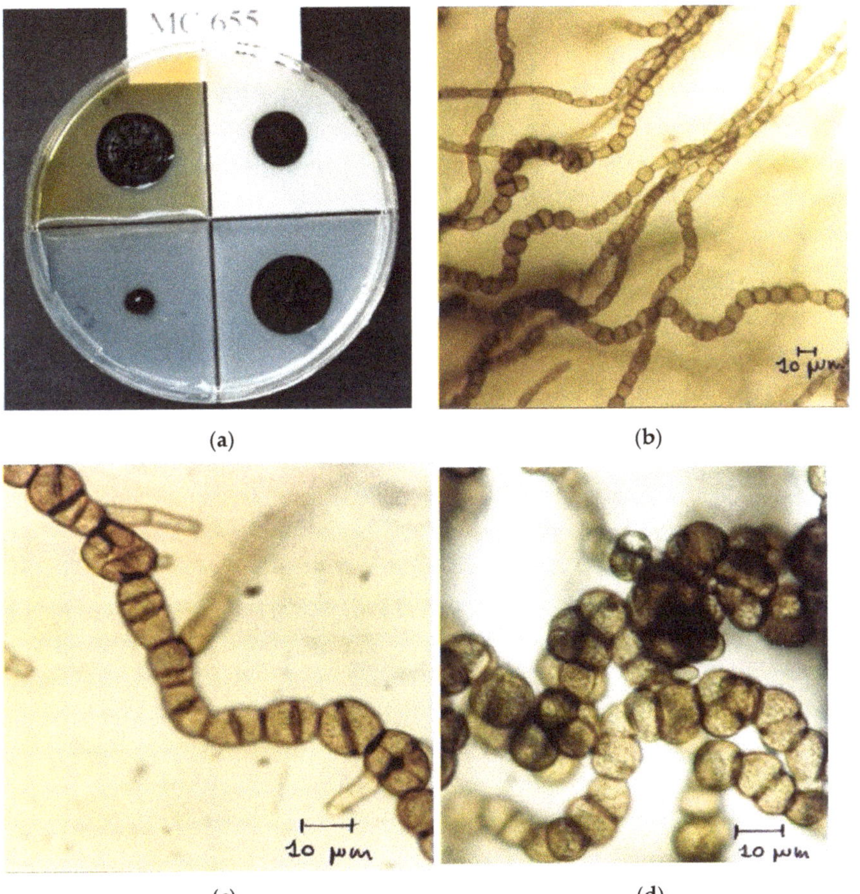

(a) (b)

(c) (d)

Figure 1. Main morphological characteristic traits of MicroColonial Fungi, MCF. Dark black colonies due to the melanin production as seen (**a**) for the unidentified strain MC 655 on different cultural media after 1 month of incubation. (**b–d**) characteristic meristematic pattern of growth described also as *Torula*-like hyphae observed under Light Microscope. Bar is 10 µm.

Hyphomycetes and black yeasts are ubiquitous and widespread all over the world in very different habitats (e.g., soil, fresh water, sea, plants, animals, and humans) [28,35], while the meristematic black fungi, mostly isolated from stone or natural rocks, can be considered the true stone-inhabiting fungi [19,20,36,37].

In the literature, many of black fungi are reported as RIF (rock inhabiting fungi) to emphasize that the "rock" is their preferred or exclusive habitat. However, this terminology does not include their main features such as melanin production, pleomorphism, or meristematic development; for this reason, we do not use it in this context.

In the frame of cultural heritage the acronym MCF (MicroColonial Fungi) as first employed by Staley [38] is widely used for their description. It refers to the typical black cauliform-like colonies visible on the rocks and stones.

Humidity may affect the settlement of MCF on the stone artifacts as unique inhabitants or as associated with other stone colonizers. In fact, in lower or sheltered parts near the ground, where there is a sufficient availability of water, MCF are strictly associated with phototrophic microorganisms with whom, however, they do not establish a symbiotic relationship (Figure 2); in harsh, dry micro-environmental conditions, MCF become the unique colonizers [39–41].

(**a**) (**b**)

Figure 2. Close association between MCF and phototrophic microorganisms. (**a**) Association seen directly on a marble sample. Magnification 400X and (**b**) after growth in the isolation medium PDA: *Chlorella*-like alga and *Coniosporium apollinis* MC 728. Magnification 80X.

Black fungi are currently classified in the Phylum of Ascomycota in the Class of Dothideomycetes and Eurotiomycetes, mainly in the order of Capnodiales, Dothideales, Chaetothyriales, Pleosporales and Cladosporiales, and Mycocaliciales [14,20,21,42].

In Table 1 are listed the genera of black fungi identified through molecular analyses that from 1997 up to date have been related to the biodeterioration of stone monuments.

Table 1. Genera of black fungi isolated from stone monuments in the period from 1997–2022 in association with visible alterations.

Class/Order	Genera *	Substrate	Environmental and Climatic Features	Alterations Associated to Fungal Colonization	Refs
Dothideomycetes incertae sedis	*Coniosporium*	Calcarenite, granite, limestone, marble	Mediterranean climate, urban environment	Grayish-black patina, pitting, black spots, greenish to dark green patina, crater shaped lesions, chipping, exfoliation, sugaring, crumbling, superficial deposit, and biofilm	[37,39,43–49]
Dothideomycetes/Capnodiales incertae sedis	*Capnobotryella*	Limestone, marble	Mediterranean climate, continental climate, and urban environment	Black spots, crater shaped lesions, chipping, exfoliation, sugaring, crumbling, pitting, superficial deposit, and biofilm formation	[45,48,50–52]
	Constantinomyces	Sandstone	Urban environment, temperate climate	Discolorations, patina	[53]
	Pseudotaeniolina	Marble, sandstone	Mediterranean climate, arid and desert climate	Biological green patina	[54–56]
Dothideomycetes/Capnodiales	*Aeminium*	Limestone	Temperate climate	Black discoloration with salt efflorescence	[57]
Dothideomycetes/Cladosporiales	*Cladosporium*	Calcarenite, granite, limestone, marble, plaster, sandstone, tufa	Ubiquitous worldwide distribution in indoor environments and outdoor	Dark alterations, black spots, black patinas, detachment of marble grains, light grayish patina, crater shaped lesions, chipping, exfoliation, sugaring, crumbling, pitting, superficial deposit, biofilm, black crusts, green biofilm with salt efflorescence, stone erosion and disintegration, and discoloration	[27,40,46,48,49,58–67]
	Verrucocladosporium	Limestone, marble, sandstone	Mediterranean climate, temperate climate, and urban environment	Black patina, discoloration	[37,53]

Table 1. *Cont.*

Class/Order	Genera *	Substrate	Environmental and Climatic Features	Alterations Associated to Fungal Colonization	Refs
Dothideomycetes/Dothideales	*Aureobasidium*	Granite, limestone, marble, plaster, sandstone	Urban environment, Mediterranean climate, temperate climate, indoor environment, and urban environment	Black patina, black spots, detachments, superficial deposit, biofilm, discolorations with or without salt efflorescence, black crusts, and stone erosion and disintegration	[37,40,45,49,53,63–65,68]
Dothideomycetes/ Mycosphaerellales	*Salinomyces*	Marble, sandstone	Mediterranean climate	Black patina	[37]
	Neocatenulostroma	Limestone, sandstone	Temperate climate, urban environment	Discolorations and/or patina, structural damage	[53]
	Neodevresia	Limestone, marble, plaster, tufa	Mediterranean climate	Black patina, discolorations, structural damage	[37,53,55,63]
	Saxophila	Marble	Mediterranean climate	Black patina	[37]
	Vermiconidia	Limestone, marble, travertine	Mediterranean climate, urban environment	Black patina	[37]
Dothideomycetes/ Neophaeothecales	*Neophaeotheca*	Marble	Mediterranean climate	Black patina	[37]
Dothideomycetes/Pleosporales	*Alternaria*	Calcarenite, granite, limestone, marble, plaster, tufa	Ubiquitous worldwide distribution in indoor environments and outdoor	Black spots, black patina, detachment of marble grains, greenish to dark green patina, biofilm, black crusts, green-black patina; and blackish patina	[40,46,49,58–60,63,64,66,67]
	Epicoccum	Granite, limestone, marble	Urban environment, mediterranean climate, and temperate climate	Black spots, black patinas, detachment, superficial deposit, biofilm, blackish patina, green biofilm, and dark and green biofilm with salt efflorescence	[40,45,49,60,64]
	Phoma	Calcarenite, granite, limestone, marble, plaster, tufa	Mediterranean climate, temperate climate, urban environment, continental-cold climate, and indoor and outdoor environments	Black spots, black patinas, detachment of marble grains; color changes, crater shaped lesions, chipping and exfoliation, sugaring, crumbling, pitting, superficial deposit, biofilm, and black crusts	[40,46,48,49,58,63]

Table 1. *Cont.*

Class/Order	Genera *	Substrate	Environmental and Climatic Features	Alterations Associated to Fungal Colonization	Refs
Dothideomycetes/Venturiales	*Ochroconis*	Calcarenite	Subterranean environment	Black patina	[69]
Eurotiomycetes incertae sedis	*Sarcinomyces*	Marble	Mediterranean climate	Black spots	[70]
Eurotiomycetes/ Chaetothyriales	*Cyphellophora* sp.	Plaster	Mediterranean climate	Black/grayish patina	[63]
	Exophiala	Calcarenite, limestone, marble, sandstone	Mediterranean climate, urban environment, temperate climate, and hypogean environment	Dark alterations, black spots, black patinas, detachment of marble grains, discolorations, and visible structural damage	[27,37,40,45,53,71]
	Lithophila	Limestone, marble	Mediterranean climate, urban environment, and dry continental climate	Black spots, black patinas, detachment of marble grains	[37,40,72]
	Knufia	Limestone, marble, sandstone travertine	Mediterranean climate, urban environment, continental temperate climate, and dry continental climate	Black and grey spots, dark macropitting, biopitting, crater shaped lesions, chipping, exfoliation, sugaring, crumbling, discolorations, patina, and visible structural damage	[37,41,43,45,48,53,72–74]
	Rhinocladiella	Marble	Mediterranean climate	Black spots, crater shaped lesions, chipping and exfoliation, sugaring, crumbling, and pitting	[48]
Eurotiomycetes/ Mycocaliciales	*Mycocalicium*	Marble	Mediterranean climate, urban environment	Black spots, crater shaped lesions, chipping and exfoliation, sugaring, crumbling, and pitting	[45,48]

* According to the current taxonomic nomenclature.

In manuscripts published prior to 1999, black meristematic fungal species that were identified without molecular analyses, such as *Hormonema dematioides*, *Lichenothelia* sp. and *Hortaea werneckii*, *Trimmatostroma* sp., are listed among the most abundant fungal species present in arid and semiarid environments in association with biodeterioration of stone monuments [3].

The molecular analyses introduced at the end of the 20th century considerably increased the knowledge about the taxonomy of the black fungi isolated from stone monuments and allowed the description of twenty-six new species and three new genera.

The new species and genera described are listed below:

Sarcinomyces petricola Wollenzien and de Hoog [73]; *S. sideticae* Sert and Sterflinger [70]; *Coniosporium apollinis* Sterflinger, *C. perforans* Sterflinger [43]; *C. uncinatum* De Leo, Urzì and de Hoog [44]; *C. sumbulii* Sert and Sterflinger [47]; *Phaeococcomyces chersonesos* Bogomolova and Minter [74]; *Pseudotaeniolina globosa* De Leo, Urzì and de Hoog [54]; *Capnobotryella antaliensis* Sert and Sterflinger [50]; *C. erdogani* Sert and Sterflinger; *C. kiziroglui* Sert and Sterflinger [51]; *Ochroconis lascauxensis* Nováková and Martin-Sanchez; *O. anomala* Nováková and Martin-Sanchez [69]; *Knufia marmoricola* Onofri and Zucconi, *K. vaticanii* Zucconi and Onofri; *K. karalitana* Isola and Onofri; *K. mediterranea* Selbmann and Zucconi [37]; *K. calcarecola* Su, Sun and Xiang [72]; *Exophiala bonarie* Isola and Zucconi; *Vermiconia calcicola* de Hoog and Onofri [37]; *Devriesia simplex* Selbmann and Zucconi; *D. modesta* Isola and Zucconi [55]; and *D. sardiniae* Isola and de Hoog [37].

Three new genera and 4 species were also introduced as new: *Saxophila tyrrenica* Selbmann and de Hoog, *Lithophila guttulata* Selbmann and Isola [37], *L. catenulata* Su, Sun and Xiang [72], and *Aeminium ludgeri* Trovão, Tiago and Portugal [57].

Over the years, some of the above mentioned genera and species were reclassified: in particular, *Sarcinomyces petricola* and *Phaecoccomyces chersonesos* resulted identical, and they were reclassified as *Knufia petricola* [75,76]; *Coniosporium perforans* is now a synonym with *Knufia perforans* [76]; *Devriesia* species and *Vermiconia* species were included, respectively, in the new genera of *Neodevriesia* [77] and *Vermiconidia* [21]. Hao et al. [78] proposed a revision of the genus *Ochroconis* that was established as synonymous with the sister genus of *Scolecobasidium*. However, this taxonomic accommodation has been refused by Samerpitak et al. [79,80] on the basis of phylogenetic analyses and because the old generic name *Scolecobasidium* is considered of doubtful identity for the ambiguity of type specimens; therefore, the genus *Ochroconis* that is also characterized by oligotrophism and mesophilia was maintained.

However, many questions regarding the taxonomy and phylogeny of black fungi are still unresolved and further studies are required, especially to clarify the taxonomical position and phylogeny of many species of *incertae sedis* and of strains that are preserved in the mycological collections and are not yet identified (Figure 3, Table 2).

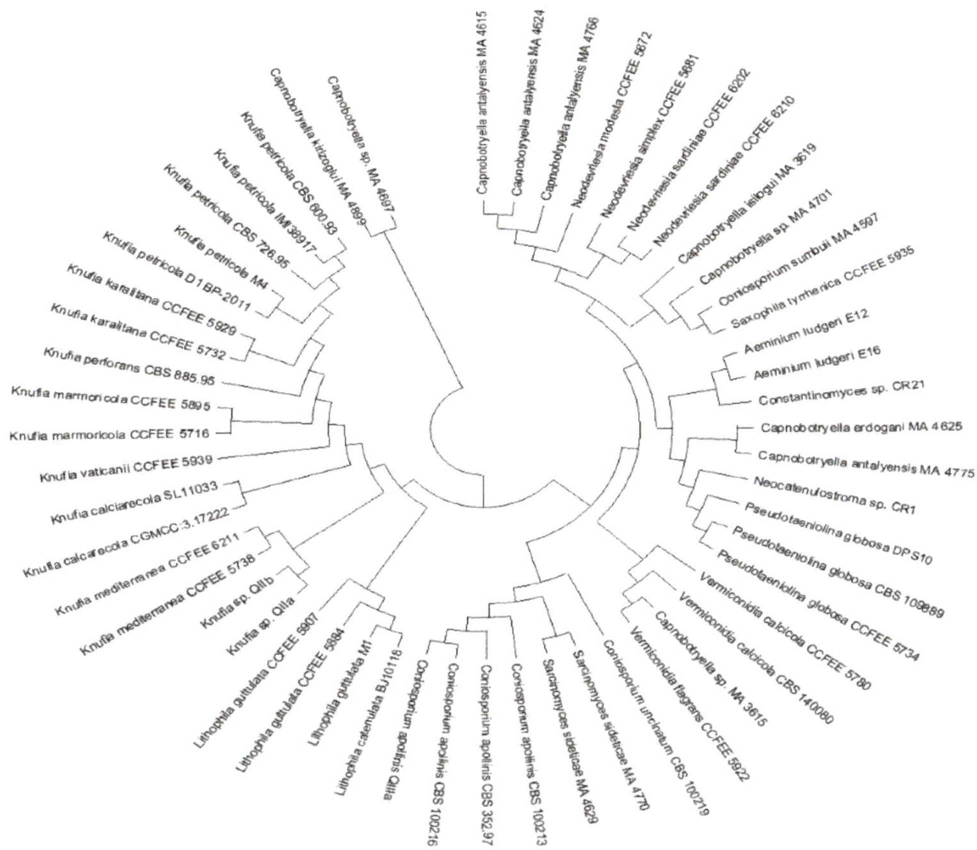

Figure 3. Phylogenetic tree (Neighbour-joining, Kimura two-parameters) showing the genetic divergence among ITS rDNA sequences of meristematic black fungi retrieved from GenBank database (https://www.ncbi.nlm.nih.gov/nucleotide accessed on 14 February 2022) and listed in Table 2.

Table 2. ITS rDNA sequences of representative MCF isolated from stone monuments aligned in Figure 3.

Taxon	Strain	ITS rDNA
Capnobotryella antalyensis	MA 4615	AJ972858
Capnobotryella antalyensis	MA 4624	AJ972850
Capnobotryella antalyensis	MA 4766	AJ972851
Capnobotryella antalyensis	MA 4775	AJ972860
Capnobotryella isilogui	MA 3619	AM746201
Capnobotryella erdogani	MA 4625	AJ972857
Capnobotryella kirizoglui	MA 4899	AJ972859
Capnobotryella sp.	MA 4701	AJ972856
Capnobotryella sp.	MA 4697	AJ972855
Capnobotryella sp.	MA 3615	AM746203
Neodevriesia modesta	CCFEE 5672	KF309984
Neodevriesia simplex	CCFEE 5681	KF309985
Neodevriesia sardiniae	CCFEE 6202	KP791765
Neodevriesia sardiniae	CCFEE 6210	KP791766
Saxophila tyrrhenica	CCFEE 5935	KP791764

Table 2. *Cont.*

Taxon	Strain	ITS rDNA
Aeminium ludgeri	E12	MG938054
Aeminium ludgeri	E16	MG938061
Neocatenulostroma sp.	CR1	KY111907
Constantinomyces sp.	CR21	KY111911
Pseudaeniolina globosa	DPS10	MH396690
Pseudotaeniolina globosa	CBS109889	NR136960
Pseudotaeniolina globosa	CCFEE5734	KF309976
Vermiconidia calcicola	CBS 140080	NR_145012
Vermiconidia calcicola	CCFEE 5780	KP791761
Vermiconidia flagrans	CCFEE 5922	KP791753
Coniosporium uncinatum	CBS 100219	AJ244270
Coniosporium apollinis	CBS 100213	AJ244271
Coniosporium apollinis	CBS 352.97	NR159787
Coniosporium apollinis	CBS 100216	AJ244272
Coniosporium apollinis	QIIIa	MH023395
Lithophila catenulata	BJ10118	JN650519
Lithophila guttulata	M1	MW361305
Lithophila guttulata	CCFEE 5884	KP791768
Lithophila guttulata	CCFEE 5907	KP791773
Knufia mediterranea	CCFEE 5738	KP791791
Knufia mediterranea	CCFEE 6211	KP791793
Knufia vaticanii	CCFEE 5939	KP791780
Knufia calcarecola	SL11033	JQ354925
Knufia calcarecola	CGMCC 3.17222	KP174862
Knufia marmoricola	CCFEE 5895	KP791775
Knufia marmoricola	CCFEE 5716	KP791786
Knufia perforans	CBS 885.95	AJ244230
Knufia karalitana	CCFEE 5732	KP791782
Knufia karalitana	CCFEE 5929	KP791783
Knufia petricola	CCFEE 726.95	KC978746
Knufia petricola	CBS 600.93	KC978744
Knufia petricola	IMI38917	AJ507323
Knufia petricola	D1	JF749183
Knufia petricola	M4	FJ556910
Knufia sp.	QIIa	MH023393
Knufia sp.	QIIb	MH023394

3. Mechanisms Involved in the Stone Biodeterioration

Being well adapted to the stone habitat and being oligotrophic, this group of fungi can often act as pioneer colonizer of the stone. In fact, for their growth it is sufficient to have just a little input of nutrient coming from the surrounding environment (e.g., animal and plant particles, air pollutants, guano droppings, etc.) [81,82]. Marble exposed to different environments and laboratory experiments demonstrated that black fungi such as *Aureobasidium pullulans* can be the first colonizer of freshly exposed marble surfaces in outdoor conditions [68,83].

The presence of a source of organic matter, such as the proximity of plants and trees, can considerably increase the chances of colonization by these fungi and the consequent rate of biodeterioration (Figure 4).

Figure 4. Extended black-greyish patina due to black fungi colonization on marble statue located in the inner yard of the Museum of Messina, Italy, under a tree of *Pittosporum tobira*. Fungal strains isolated from the statue were attributed to meristematic black fungi and genera of *Cladosporium*, *Alternaria* and *Phoma* [81].

The mechanism of biodeterioration of stone monuments caused by black fungi is not fully understood [16,84].

The pattern of colonization of black fungi, and in particular of MCF, demonstrate once more that these microorganisms are well adapted to the stone habitat. In fact, they not only grow on the surface of stone, causing an aesthetic alteration due to the presence of melanin in the mycelium and conidia, described as discoloration, black staining, black spots, and black/greyish patinas [9,20,85]; but they can also act as true endolithic microorganisms by penetrating into the rocks/stones via intercrystalline spaces or through an active mechanism in which both mechanical and chemical aspects are hypothesized.

Microscopic observations show that where they settle is shaped accordingly to the morphology of these fungi (Figure 5). This fact can be explained by a local release of organic acids, followed by a precipitation of mineral phases; this buffer effect may be the reason why these fungi, in contrast to other ubiquitous hyphomycetes, such as *Aspergillus niger* and *Penicillium* spp., as reported by Salvadori and Municchia [16], do not show a marked organic acid production in laboratory conditions. However, Favero Longo et al. [86] demonstrated that some species of MCF (e.g., *Knufia petricola*) penetrate actively into freshly exposed stone probes through the production of iron-chelating molecules (siderophores such as compounds).

Another mechanism is due to the ability of these fungi to penetrate the stone using already existing fractures and cracks. The mechanical forces due to the expansion of hyphae may increase the fractures and cause the loss of materials [18,26,39]. To explain the mechanical penetration of the hyphae into the stone, a past hypothesis gave a crucial role to melanin present in the cell wall of hyphae and in the meristematic cells, similar to the process of penetration of phytopathogenic fungi in the host cells [16,23,25,39].

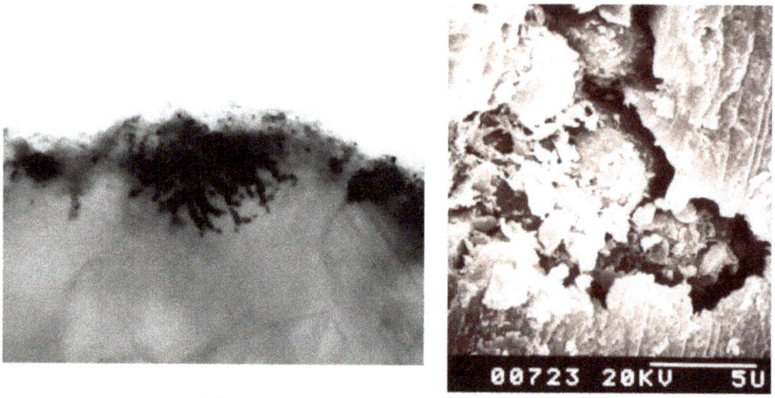

(a) (b)

Figure 5. Microscopic observation of the endolithic behaviour of meristematic black fungi. (**a**) Settled on a thin section of Carrara marble sarcophagus flake showing the meristematic chains deepening inside the marble, magnification 400X; (**b**) SEM image showing MCF into the marble with dissolution pattern following the same shape of meristematic cells.

In fact, the melanin present in the appressoria of the phytopathogenic fungi confers turgor and rigidity to the cells, favouring their penetration of the host cells [87].

However, on the basis of the results obtained in a recent publication by Tonon et al. [88], this hypothesis should be rejected. In fact, non-melanized mutants of the black fungus *Knufia petricola* do not lose their ability to penetrate carbonate pellets regardless of porosity; on the contrary, thin, non-melanized, exploring hyphae showed an even higher penetration pattern into the stone, probably due to their nutrient-seeking role and thinness.

4. Multistep Analyses to Study Black Fungi from Stone Monuments

Contrary to studies carried out on natural rocks, in the frame of cultural heritage it is mandatory to carry out all the necessary multidisciplinary studies (chemical, geological, physical, biological analysis, etc.), by using very low destructive or non-destructive sampling methods [18,46,67,89]. This fact may limit the extent of the studies, but a careful planning of a sampling campaign leads to the right protocol of intervention and of assessing the risk of further biodeterioration processes.

A general useful multistep approach should include microscopy, cultural analyses, molecular analyses, and the laboratory evaluation of selected methods of control (chemical or physical) on isolated strains. The assessment of environmental conditions (temperature, humidity, shining, presence of surrounding vegetation, or other organic sources, atmospheric pollutants, etc.) should be also evaluated. In fact, these data not only allow us a better understanding of the physiology and ecology of fungi, but can help to control their growth indirectly, especially in indoor and/or in confined environments. At the end, a monitoring campaign over time should establish the level of the risk of the item and the frequency of intervention.

Further, a common language for the description of the alterations is also indispensable for sharing the results with the scientific community; to this purpose, there is a glossary [90,91] to obtain an objective and standardized description.

4.1. Methods of Isolation and Characterization

The sampling is critical. As reported in the previous paragraph, due to the value of the artifacts, non-invasive sampling methods have been developed over the years and are now widely used. Examples are a needle to take samples from black spots or cavities, scalpel or lancet to scrape fungi from the surface, and adhesive tape sample, useful both for microscopy and cultural and molecular analysis as it provides a mirror image of the stone colonization [22,92,93].

Microscopic examination of the samples is the most common practice to obtain evidence of black fungi directly from stone samples. In fact, due to their size, morphology, and pigmentation, black fungi can be directly visualized under optical microscope without a specific preparation. Even with a good light microscope (LM) or, better, with a scanning electron microscope (SEM), detailed information is obtained. Microscopy is useful for the direct visualization of the fungi in the stone sample, for determining what types of relationship they establish with stone material and with other types of microorganisms, and also for addressing the next step of analyses. Unfortunately, fluorescence microscopy (FM) cannot help in detecting and studying black fungi due to the presence of melanin that masks the fluorochrome fluorescence.

Still, cultural analyses remain the best way to study this group of fungi.

Black fungi, and in particular MCF, possess a poor ability to compete with fast growing fungi, being characterized by a slow growth rate, and very often require more than 1 month of incubation before visible colonies are seen. These reasons explain why they are difficult to isolate and maintain in culture [94]. Nevertheless, selective cultural media that inhibit the growth of bacteria and of fast-growing fungi are successfully employed both for qualitative and quantitative cultural analyses [22,95].

These culture techniques have the advantage of allowing the whole characterization of the isolates by microscopical, biochemical, physiological, and molecular analysis; these latter are indispensable for the identification of MCF that do not have recognizable morphological traits. Multilocus sequencing typing (mlst) is routinely carried out to resolve their taxonomic and phylogenetic position [55,56] by Blast search homology (Basic Local Alignment Search Tools) available online. However, many nucleotide sequences in the Genkank nucleotide database are not updated in the "definition". Therefore, although this approach is within the reach of all laboratories, it requires a deep knowledge of the literature and a curated nucleotide database.

A deeper genetic characterization of the isolates can be obtained by whole genome analyses but, to date, only four genomes belonging to three species of black fungi (*Coniosporium apollinis* CBS 100218, *Knufia petricola* MA5789, *K. petricola* MA5790, and *Aeminium ludgeri* DSM 106916) isolated from stone monuments are available in the NCBI database (https://www.ncbi.nlm.nih.gov/genome/ accessed on 15 February 2022).

Cultural techniques also allow us to investigate the biodeteriorative abilities of the isolated species by setting up laboratory experiments that simulate their settlement, colonization, and biodeterioration pattern [96,97].

4.2. Culture Independent Analyses

It is well known that only a very small percentage (<0.5%) of environmental microorganisms can be cultured. Therefore, culture independent molecular approaches have been developed to overcome this limitation; the metagenomic approaches allow the study of the microbiota (microbial community) and its microbiome (gene pool) by total acid nucleic extraction from the samples, both DNA and/or RNA, depending on the purpose.

Recently Sterflinger and Pinar [98] reviewed the main molecular-based techniques that are currently available and, although some of these have not yet been used in the field of cultural heritage, they could be adapted for future studies.

In the literature, the majority of papers are focused on Bacteria and Archaea, while there are very few papers about microfungi from stone monuments [99,100].

Large-scale genomic analyses, such as high-throughput-sequencing analyses, are becoming increasingly popular to study the microbiota (both for biodiversity and functional genes analyses) in different areas of research, and cultural heritage is not an exception [89,99]. These techniques were developed in the late 1990s and early 2000s, and today they are more accessible to many laboratories due to the lower costs and the facilities offered by many companies that can carry out all steps of analysis, from nucleic acid extraction to bioinformatic data analysis.

The culture-independent approach certainly contributes to the deep knowledge on the microorganisms associated with the biodeterioration processes. However, there is a very high risk to obtain a plethora of data that are not easily interpretable, because they cannot be directly connected with the biodeterioration phenomenon observed. In fact, it is obvious, but not trivial, that the discovery of a microbial agent on a monument could be not related to any biodeteriogenic activity. Therefore, the current need is to associate these techniques to the culture-based ones for a more complete characterization of the state of deterioration of the artifact and for implementation of the more suitable prevention strategies.

5. How to Control Black Fungi

Despite the wide literature regarding the control of biodeteriogens on inorganic surfaces as reported in recent books and reviews [6,9,101–103], very little is said regarding the effectiveness of treatments against black fungi.

Black fungi, especially meristematic ones, are very difficult to eradicate and tend to be one of the first colonizers after cleaning procedures [6,12,83]. In the Lascaux cave, a black yeast *Ochroconis lascauxensis* caused an important and extensive black discoloration on the cave's walls whose origin and evolution were probably linked to the intensive biocide treatments [104].

In order to achieve protection of an artifact, both indirect and direct methods should be implemented. The first ones aim to control, or more realistically to mitigate, the fungal growth by modification of the chemical-physical parameters such as humidity, source of nutrients, and temperature, that are crucial key factors for fungal growth. It is obvious that this is rarely fully achievable, and only in particular circumstances, such as in indoor environments (churches, museums, etc.) or for movable artifacts and objects that can be moved if necessary, while in outdoor conditions this is quite impossible.

Direct treatments aiming to kill/reduce black fungi on the stone should be different on the basis of their colonization pattern (diffuse patina, spot-like colonization, or intercrystalline growth) and on the characteristics of the environment; for example, in an indoor environment, the air is often heavily contaminated by fungal spores and thus they need to be eliminated at the same time as those settled on the surfaces; otherwise, their presence in the air is a continued source of reinfection.

Among the potential methods commonly used to control biodeterioration, physical methods such as mechanical removal and UV and heat shock treatments [101,105], are not very effective against black fungi [102,106].

Regarding chemical methods, in laboratory conditions, classical biocides (e.g., Preventol RI 50, Biotin R, Rocima^{TM} 103) are still the most effective [102,107] and in the field they produce efficient results during cleaning procedures. Plant based extracts show a scarce effectiveness against fungi, and this difficult group of microorganisms is not even taken into account to assess their activity [108]. Nanoparticles are commonly used as biocides due to their activity against algae, cyanobacteria, and most bacteria, but they are not really satisfactory against black fungi.

Protective coatings with antifouling properties may have various effects. In fact, TiO_2 based coatings, pure or doped with Ag, show a good effect but are limited to a short/medium term after application [108]. However, in both laboratory and field conditions, after treatments with titania-based coatings, black fungi are the first to recolonize the stone surface in dry environments, while algae first appears in damping walls [63]. Very recently, in laboratory conditions, cholinium@Il based coatings have shown that the use of Il's with a 12 C chains and DBS as anion in combination with nanosilica coatings (e.g., Nano Estel) could be effective against the colonization of black fungi for a period of time over 30 months [109].

One possible explanation of this scarce effectiveness of most treatments against black fungi is that they possess a genetic resistance to environmental stresses, as reported in the previous paragraphs. Therefore, the different mechanisms concurring to the stress protection response may interfere to the biocidal treatments.

Understanding the cause of their resilience could improve the strategies for their control.

6. Concluding Remarks

The study of biodeterioration of stone monuments is quite complex and cannot be improvised. For a correct understanding of biodeterioration phenomena and the implementation of measures aimed at the elimination and/or mitigation of biodeteriogenic microorganisms, it is important to consider the monument and its surrounding as a whole.

When working for the protection of cultural heritage artifacts, scientists should not follow the same protocols for all the situations. In general, it is necessary to:

(1) Listen the conservators;
(2) Evaluate the environmental climatic conditions and specific conditions, such as the type of material and the overall status of conservation of monument; the description of the type of alteration visible under naked eye should be also included;
(3) Interact with the other experts involved;
(4) Answer the questions posed by the conservators.

The analysis must be planned according to their questions.

It is also important to relate the presence of fungal species with the observed biodeterioration phenomenon; then, for treatments, it is possible to use well known protocols or propose new products/treatments that, however, need to be tested in laboratory with fungal isolates and in situ (on probes, not on the item!!!) before applying it to the CH item. Finally, consider evaluating the use of coatings that match Green conservation criteria and are effective to prevent or slow down new colonization.

As there is not only one method that is valid in all circumstances, we have to work out, case by case, the best solution and monitor the result over the time to avoid unexpected and/or undesirable effects as much as possible.

Black fungi, especially meristematic ones, are very dangerous for stone artifacts for several reasons:

(a) They are responsible of discolouring of the stone surface. The extended colonization of surfaces changes the global vision of the artifact, especially if different material and colour of stones were used by the artist;
(b) Moreover, black fungi show an inter-crystalline pattern of growth. This pattern causes crystals to detach (so called sugaring) with loss of precious material, especially because it involves the first surface layer (very important for bas-reliefs and sculptures);
(c) They could determine the biopitting. Fungi excavate cavities on the stone where they can better settle, giving the surface a pockmarked aspect. The convergence of several biopitting can often lead to larger cavities;
(d) Hyphae penetrate deep into the surface, even more than a few mm;
(e) Chemical and physical treatments used for other microorganisms are often non efficient in eradication;
(f) Black fungi are often the first colonizers after the treatments.

For all these reasons, new methods of study and of control are needed that also aim to search for more eco-friendly molecules and/or approaches.

Despite the increase in interest in black fungi as a cause of biodeterioration of stone monuments and artifacts, many aspects need a more in-depth analysis. For example, not much is known about the molecular mechanisms involved in stress tolerance, in colonization, and biodeterioration of stone. Very important results were achieved by laboratory experiments that, however, concern only some species, and are still too few to generalize the results obtained. Only four genomes of three species were sequenced, and they are not sufficient for a comparative analyses aiming to a better understanding of the above mentioned processes and mechanisms.

Furthermore, the creation of a curated database including the nucleotide sequences used for identification of black fungi from monuments could be very helpful, as the public databases are not curated and are outdated; they could even be misleading for people who are not interested in deepening the knowledge around the taxonomy and phylogeny of this group of fungi.

Finally, it is also important to combine the classical methods with the new ones to improve current knowledge useful for implementation of future conservation strategies.

Author Contributions: Conceptualization, F.D.L. and C.U.; methodology, F.D.L., A.M. and C.U.; resources, F.D.L. and C.U.; data curation F.D.L.; writing—original draft preparation, F.D.L., A.M. and C.U. writing—review and editing, F.D.L., A.M. and C.U. All authors have read and agreed to the published version of the manuscript.

Funding: No funding was used for this manuscript.

Informed Consent Statement: Not applicable.

Data Availability Statement: Not applicable.

Acknowledgments: Authors would like to thank Sherron Collins for her revision of the English text.

Conflicts of Interest: The authors declare no conflict of interest.

References

1. Krumbein, W.E. Microbial interactions with mineral materials. In *Biodeterioration 7*, 1st ed.; Hougthon, D.R., Smith, R.N., Eggins, H.O., Eds.; Springer: Dordrecht, The Netherlands, 1988; pp. 78–100. [CrossRef]
2. Saiz-Jimenez, C. Biogeochemistry of weathering processes in monuments. *Geomicrobiol. J.* **1999**, *16*, 27–37. [CrossRef]
3. Sterflinger, K. Fungi as geologic agents. *Geomicrobiol. J.* **2000**, *17*, 97–124. [CrossRef]
4. Gadd, G.M. Geomicrobiology of the built environment. *Nat. Microbiol.* **2017**, *2*, 16275. [CrossRef] [PubMed]
5. Urzì, C. Biodeterioramento dei manufatti artistici. In *Microbiologia Ambientale ed Elementi di Ecologia Microbica*; Barbieri, P., Bestetti, G., Galli, E., Zannoni, D., Eds.; Casa Editrice Ambrosiana: Milano, Italy, 2008; pp. 327–346. ISBN 978-88-08-18434-4.
6. Pinna, D. Microbial growth and its effect on inorganic heritage material. In *Microorganisms in the Deterioration and Preservation of Cultural Heritage*, 1st ed.; Joseph, E., Ed.; Springer: Cham, Switzerland, 2021; pp. 3–35. [CrossRef]
7. Gorbushina, A.A. Life on the rocks. *Environ. Microbiol.* **2007**, *9*, 1613–1631. [CrossRef] [PubMed]
8. Villa, F.; Stewart, P.S.; Klapper, I.; Jacob, J.M.; Cappitelli, F. Subaerial biofilms on outdoor stone monuments: Changing the perspective toward an ecological framework. *BioScience* **2016**, *66*, 285–294. [CrossRef]
9. Pinna, D. *Coping with Biological Growth on Stone Heritage Objects: Methods, Products, Applications, and Perspectives*, 1st ed.; Apple Academic Press: Oakville, ON, Canada, 2017; ISBN 9781771885324.
10. Guillitte, O. Bioreceptivity: A new concept for building ecology studies. *Sci. Total Environ.* **1995**, *167*, 215–222. [CrossRef]
11. Saiz-Jimenez, C. Deposition of anthropogenic compounds on monuments and their effect on airborne microorganisms. *Aerobiologia* **1995**, *11*, 161–175. [CrossRef]
12. Urzì, C.; De Leo, F.; Salamone, P.; Criseo, G. Airborne fungal spores colonising marbles exposed in the terrace of Messina Museum, Sicily. *Aerobiologia* **2001**, *17*, 11–17. [CrossRef]
13. Polo, A.; Gulotta, D.; Santo, N.; Di Benedetto, C.; Fascio, U.; Toniolo, L.; Villa, F.; Cappitelli, F. Importance of subaerial biofilms and airborne microflora in the deterioration of stonework: A molecular study. *Biofouling* **2012**, *28*, 1093–1106. [CrossRef]
14. Liu, B.; Fu, R.; Wu, B.; Liu, X.; Xiang, M. Rock-inhabiting fungi: Terminology, diversity, evolution and adaptation mechanisms. *Mycology* **2022**, *13*, 1–31. [CrossRef]
15. Selbmann, L.; Egidi, E.; Isola, D.; Onofri, S.; Zucconi, Z.; de Hoog, G.S.; Chinaglia, S.; Testa, L.; Tosi, S.; Balestrazzi, A.; et al. Biodiversity, evolution and adaptation of fungi in extreme environments. *Plant Biosyst.* **2013**, *147*, 237–246. [CrossRef]
16. Salvadori, O.; Municchia, A.C. The role of fungi and lichens in the biodeterioration of stone monuments. *Open Conf. Proc. J.* **2016**, *7*, 39–54. [CrossRef]
17. Liu, X.; Koestler, R.J.; Warscheid, T.; Katayama, Y.; Gu, J.-D. Microbial deterioration and sustainable conservation of stone monuments and buildings. *Nat. Sustain.* **2020**, *3*, 991–1004. [CrossRef]
18. Sterflinger, K. Fungi: Their role in deterioration of cultural heritage. *Fungal Biol. Rev.* **2010**, *24*, 47–55. [CrossRef]
19. Urzì, C.; De Leo, F.; de Hoog, S.; Sterflinger, K. Recent advances in the molecular biology and ecophysiology of meristematic stone-inhabiting fungi. In *Of Microbes and Art: The Role of Microbial Communities in the Degradation and Protection of Cultural Heritage*; Ciferri, O., Tiano, P., Mastromei, G., Eds.; Springer: Boston, MA, USA, 2000; pp. 3–19. [CrossRef]
20. Onofri, S.; Zucconi, L.; Isola, D.; Selbmann, L. Rock-inhabiting fungi and their role in deterioration of stone monuments in the Mediterranean area. *Plant Biosyst.* **2014**, *148*, 384–391. [CrossRef]
21. Crous, P.W.; Schumacher, R.K.; Akulov, A.; Thangavel, R.; Hernández-Restrepo, M.; Carnegie, A.J.; Cheewangkoon, R.; Wingfield, M.J.; Summerell, B.A.; Quaedvlieg, W.; et al. New and interesting fungi. *Fungal Syst. Evol.* **2019**, *3*, 57–134. [CrossRef]

22. Wollenzien, U.; de Hoog, G.S.; Krumbein, W.E.; Urzì, C. On the isolation of microcolonial fungi occurring on and in marble and other calcareous rocks. *Sci. Total Environ.* **1995**, *167*, 287–294. [CrossRef]
23. Diakumaku, E.; Gorbushina, A.A.; Krumbein, W.E.; Panina, L.; Soukharjevski, S. Black fungi in marble and limestones–an aesthetical, chemical and physical problem for the conservation of monuments. *Sci. Total Environ.* **1995**, *167*, 295–304. [CrossRef]
24. Urzì, C.; Wollenzien, U.; Criseo, G.; Krumbein, W.E. Biodiversity of the rock inhabiting microflora with special reference to black fungi and black yeasts. In *Microbial Diversity and Ecosystem Function*; Allsopp, D., Colwell, R.R., Hawksworth, D.L., Eds.; CAB International: Wallingford, UK, 1995; pp. 289–302.
25. Sterflinger, K.; Krumbein, W.E. Dematiaceous fungi as a major agent for biopitting on mediterranean marbles and limestones. *Geomicrobiol. J.* **1997**, *14*, 219–230. [CrossRef]
26. De Leo, F.; Urzì, C. Microfungi from deteriorated materials of cultural heritage. In *Fungi from Different Substrates*; Misra, J.K., Tewari, J.P., Deshmukh, S.K., Vágvölgyi, C., Eds.; CRC Press: Boca Raton, FL, USA, 2015; pp. 144–158. [CrossRef]
27. Isola, D.; Zucconi, L.; Cecchini, A.; Caneva, G. Dark-pigmented biodeteriogenic fungi in etruscan hypogeal tombs: New data on their culture-dependent diversity, favouring conditions, and resistance to biocidal treatments. *Fungal Biol.* **2021**, *125*, 609–620. [CrossRef]
28. de Hoog, G.S.; Guarro, J.; Gené, S.A.; Al-Hatmi, A.M.S.; Figueras, M.J.; Vitale, R.G. *Atlas of Clinical Fungi*, 4th ed.; Westerdijk Institute/Universitat Rovira Virgili: Utrecht, The Netherlands, 2019.
29. Palmer, F.E.; Emery, D.R.; Stemmler, J.; Staley, J.T. Survival and growth of microcolonial rock fungi as affected by temperature and humidity. *New Phytol.* **1987**, *107*, 155–162. [CrossRef]
30. Gueidan, C.; Ruibal, C.; de Hoog, S.; Schneider, H. Rock-inhabiting fungi originated during periods of dry climate in the late Devonian and middle Triassic. *Fungal Biol.* **2011**, *115*, 987–996. [CrossRef] [PubMed]
31. Gostinčar, C.; Grube, M.; Gunde-Cimerman, N. Evolution of fungal pathogens in domestic environments? *Fungal Biol.* **2011**, *115*, 1008–1018. [CrossRef] [PubMed]
32. Gorbushina, A.A.; Whitehead, K.; Dornieden, T.; Niesse, A.; Schulte, A.; Hedges, J.I. Black fungal colonies as units of survival: Hyphal mycosporines synthesized by rock-dwelling microcolonial fungi. *Can. J. Bot.* **2003**, *81*, 131–138. [CrossRef]
33. Gostinčar, C.; Muggia, L.; Grube, M. Polyextremotolerant black fungi: Oligotrophism, adaptive potential and a link to lichen symbioses. *Front. Microbiol.* **2012**, *3*, 390. [CrossRef] [PubMed]
34. Zakharova, K.; Tesei, D.; Marzban, G.; Dijksterhuis, J.; Wyatt, T.; Sterflinger, K. Microcolonial fungi on rocks: A life in constant drought? *Mycopathologia* **2013**, *175*, 537–547. [CrossRef] [PubMed]
35. Marchetta, A.; van den Ende, G.B.; Al-Hatmi, A.M.S.; Hagen, F.; Zalar, P.; Sudhadham, M.; Gunde-Cimerman, N.; Urzì, C.; de Hoog, S.; De Leo, F. Global molecular diversity of the halotolerant fungus *Hortaea werneckii*. *Life* **2018**, *8*, 31. [CrossRef]
36. Sterflinger, K.; Piñar, G. Microbial deterioration of cultural heritage and works of art—tilting at windmills? *Appl. Microbiol. Biot.* **2013**, *97*, 9637–9646. [CrossRef]
37. Isola, D.; Zucconi, L.; Onofri, S.; Caneva, G.; de Hoog, G.S.; Selbmann, L. Extremotolerant rock inhabiting black fungi from Italian monumental sites. *Fungal Divers.* **2016**, *76*, 75–96. [CrossRef]
38. Staley, J.T.; Palmer, F.; Adams, J.B. Microcolonial fungi: Common inhabitants on desert rocks? *Science* **1982**, *215*, 1093–1095. [CrossRef]
39. De Leo, F.; Antonelli, F.; Pietrini, A.M.; Ricci, S.; Urzì, C. Study of the euendolithic activity of black meristematic fungi isolated from a marble statue in the Quirinale Palace's Gardens in Rome, Italy. *Facies* **2019**, *65*, 18. [CrossRef]
40. Santo, A.P.; Cuzman, O.A.; Petrocchi, D.; Pinna, D.; Salvatici, T.; Perito, B. Black on white: Microbial growth darkens the external marble of Florence cathedral. *Appl. Sci.* **2021**, *11*, 6163. [CrossRef]
41. Marvasi, M.; Donnarumma, F.; Frandi, A.; Mastromei, G.; Sterflinger, K.; Tiano, P.; Perito, B. Black microcolonial fungi as deteriogens of two famous marble statues in Florence, Italy. *Int. Biodeterior. Biodegrad.* **2012**, *68*, 36–44. [CrossRef]
42. Abdollahzadeh, J.; Groenewald, J.Z.; Coetzee, M.P.A.; Wingfield, M.J.; Crous, P.W. Evolution of lifestyles in *Capnodiales*. *Stud. Mycol.* **2020**, *95*, 381–414. [CrossRef]
43. Sterflinger, K.; de Baere, R.; de Hoog, G.S.; de Wachter, R.; Krumbein, W.E.; Haase, G. *Coniosporium perforans* and *C. apollinis*, two new rock-inhabiting fungi isolated from marble in the Sanctuary of Delos (Cyclades, Greece). *Antonie Van Leeuwenhoek J. Microb.* **1997**, *72*, 349–363. [CrossRef]
44. De Leo, F.; Urzì, C.; de Hoog, G.S. Two *Coniosporium* species from rock surfaces. *Stud. Mycol.* **1999**, *43*, 70–79.
45. Sterflinger, K.; Prillinger, H. Molecular taxonomy and biodiversity of rock fungal communities in an urban environment (Vienna, Austria). *Antonie Van Leeuwenhoek J. Microb.* **2001**, *80*, 275–286. [CrossRef]
46. Ricca, M.; Urzì, C.E.; Rovella, N.; Sardella, A.; Bonazza, A.; Ruffolo, S.A.; De Leo, F.; Randazzo, L.; Arcudi, A.; La Russa, M.F. Multidisciplinary approach to characterize archaeological materials and status of conservation of the roman *Thermae* of Reggio Calabria site (Calabria, south Italy). *Appl. Sci.* **2020**, *10*, 5106. [CrossRef]
47. Sert, H.B.; Sterflinger, K. A new *Coniosporium* species from historical marble monuments. *Mycol. Prog.* **2010**, *9*, 353–359. [CrossRef]
48. Sert, H.B.; Sümbül, H.; Sterflinger, K. Microcolonial fungi from antique marbles in Perge/Side/Termessos (Antalya/Turkey). *Antonie Van Leeuwenhoek J. Microb.* **2007**, *91*, 217–227. [CrossRef]
49. Sazanova, K.V.; Zelenskaya, M.S.; Vlasov, A.D.; Bobir, S.Y.; Yakkonen, K.L.; Vlasov, D.Y. Microorganisms in superficial deposits on the stone monuments in Saint Petersburg. *Microorganisms* **2022**, *10*, 316. [CrossRef]

50. Sert, H.B.; Sümbül, H.; Sterflinger, K. A new species of *Capnobotryella* from monument surfaces. *Mycol. Res.* **2007**, *111*, 1235–1241. [CrossRef] [PubMed]
51. Sert, H.B.; Sümbül, H.; Sterflinger, K. Two new species of *Capnobotryella* from historical monuments. *Mycol. Prog.* **2011**, *10*, 333–339. [CrossRef]
52. Sert, H.B.; Wuczkowski, M.; Sterflinger, K. *Capnobotryella isiloglui*, a new rock-inhabiting fungus from Austria. *Turk. J. Bot.* **2012**, *3*, 401–407. [CrossRef]
53. Owczarek-Kościelniak, M.; Krzewicka, B.; Piątek, J.; Kołodziejczyk, Ł.M.; Kapusta, P. Is there a link between the biological colonization of the gravestone and its deterioration? *Int. Biodeterior. Biodegrad.* **2020**, *148*, 104879. [CrossRef]
54. De Leo, F.; Urzì, C.; de Hoog, G.S. A new meristematic fungus, *Pseudotaeniolina globosa*. *Antonie Van Leeuwenhoek J. Microb.* **2003**, *83*, 351–360. [CrossRef]
55. Egidi, E.; de Hoog, G.S.; Isola, D.; Onofri, S.; Quaedvlieg, W.; de Vries, M.; Verkley, G.J.M.; Stielow, J.B.; Zucconi, L.; Selbmann, L. Phylogeny and taxonomy of meristematic rock-inhabiting black fungi in the *Dothideomycetes* based on multi-locus phylogenies. *Fungal Diver.* **2014**, *65*, 127–165. [CrossRef]
56. Rizk, S.M.; Magdy, M.; De Leo, F.; Werner, O.; Rashed, M.A.-S.; Ros, R.M.; Urzì, C. A new extremotolerant ecotype of the fungus *Pseudotaeniolina globosa* isolated from Djoser Pyramid, Memphis Necropolis, Egypt. *J. Fungi* **2021**, *7*, 104. [CrossRef]
57. Trovão, J.; Tiago, I.; Soares, F.; Paiva, D.S.; Mesquita, N.; Coelho, C.; Catarino, L.; Gil, F.; Portugal, A. Description of Aeminiaceae fam. nov., *Aeminium* gen. nov. and *Aeminium ludgeri* sp. nov. (*Capnodiales*), isolated from a biodeteriorated art-piece in the old cathedral of Coimbra, Portugal. *MycoKeys* **2019**, *45*, 57–73. [CrossRef]
58. Nuhoglu, Y.; Oguz, E.; Uslu, H.; Ozbek, A.; Ipekoglu, B.; Ocak, I.; Hasenekoglu, I. The accelerating effects of the microorganisms on biodeterioration of stone monuments under air pollution and continental-cold climatic conditions in Erzurum, Turkey. *Sci. Total Environ.* **2006**, *364*, 272–283. [CrossRef]
59. Cappitelli, F.; Principi, P.; Pedrazzani, R.; Toniolo, L.; Sorlini, C. Bacterial and fungal deterioration of the Milan cathedral marble treated with protective synthetic resins. *Sci. Total Environ.* **2007**, *385*, 172–181. [CrossRef] [PubMed]
60. Cappitelli, F.; Nosanchuk, J.; Casadevall, A.; Toniolo, L.; Brusetti, L.; Florio, S.; Principi, P.; Borin, S.; Sorlini, C. Synthetic consolidants attacked by melanin-producing fungi: Case study of the biodeterioration of Milan (Italy) cathedral marble treated with acrylics. *Appl. Environ. Microbiol.* **2007**, *73*, 271–277. [CrossRef] [PubMed]
61. Suihko, M.L.; Alakomi, H.L.; Gorbushina, A.; Fortune, I.; Marquardt, J.; Saarela, M. Characterization of aerobic bacterial and fungal microbiota on surfaces of historic Scottish monuments. *Syst. Appl. Microbiol.* **2007**, *30*, 494–508. [CrossRef] [PubMed]
62. Ortega-Morales, B.O.; Narváez-Zapata, J.; Reyes-Estebanez, M.; Quintana, P.; De la Rosa-García del, C.S.; Bullen, H.; Gómez-Cornelio, S.; Chan-Bacab, M.J. Bioweathering potential of cultivable fungi associated with semi-arid surface microhabitats of mayan buildings. *Front. Microbiol.* **2016**, *7*, 201. [CrossRef]
63. Ruffolo, S.A.; De Leo, F.; Ricca, M.; Arcudi, A.; Silvestri, C.; Bruno, L.; Urzì, C.; La Russa, M.F. Medium-term in situ experiment by using organic biocides and titanium dioxide for the mitigation of microbial colonization on stone surfaces. *Int. Biodeterior. Biodegrad.* **2017**, *123*, 17–26. [CrossRef]
64. Trovão, J.; Portugal, A.; Soares, F.; Paiva, D.S.; Mesquita, N.; Coelho, C.; Pinheiro, A.C.; Catarino, L.; Gil, F.; Tiago, I. Fungal diversity and distribution across distinct biodeterioration phenomena in limestone walls of the old cathedral of Coimbra, UNESCO World Heritage Site. *Int. Biodeterior. Biodegrad.* **2019**, *142*, 91–102. [CrossRef]
65. Trovão, J.; Gil, F.; Catarino, L.; Soares, F.; Tiago, I.; Portugal, A. Analysis of fungal deterioration phenomena in the first Portuguese King tomb using a multi-analytical approach. *Int. Biodeterior. Biodegrad.* **2020**, *149*, 104933. [CrossRef]
66. Mang, S.M.; Scrano, L.; Camele, I. Preliminary studies on fungal contamination of two rupestrian churches from Matera (Southern Italy). *Sustainability* **2020**, *12*, 6988. [CrossRef]
67. Urzì, C.; De Leo, F.; Bruno, L.; Albertano, P. Microbial diversity in paleolithic caves: A study case on the phototrophic biofilms of the Cave of Bats (Zuheros, Spain). *Microb. Ecol.* **2010**, *60*, 116–129. [CrossRef]
68. Urzì, C.; De Leo, F.; Lo Passo, C.; Criseo, G. Intra-specific diversity of *Aureobasidium pullulans* strains isolated from rocks and other habitats assessed by physiological methods and by random amplified polymorphic DNA (RAPD). *J. Microbiol. Meth.* **1999**, *36*, 95–105. [CrossRef]
69. Martin-Sanchez, P.M.; Nováková, A.; Bastian, F.; Alabouvette, C.; Saiz-Jimenez, C. Two new species of the genus *Ochroconis*, *O. lascauxensis* and *O. anomala* isolated from black stains in Lascaux Cave, France. *Fungal Biol.* **2012**, *116*, 574–589. [CrossRef] [PubMed]
70. Sert, H.B.; Sümbül, H.; Sterflinger, K. *Sarcinomyces sideticae*, a new black yeast from historical marble monuments in Side (Antalya, Turkey). *Bot. J. Linn. Soc.* **2007**, *154*, 373–380. [CrossRef]
71. Isola, D.; Selbmann, L.; de Hoog, G.S.; Fenice, M.; Onofri, S.; Prenafeta-Boldú, F.X.; Zucconi, L. Isolation and screening of black fungi as degraders of volatile aromatic hydrocarbons. *Mycopathologia* **2013**, *175*, 369–379. [CrossRef] [PubMed]
72. Sun, W.; Su, L.; Yang, S.; Sun, J.; Liu, B.; Fu, R.; Wu, B.; Liu, X.; Cai, L.; Guo, L.; et al. Unveiling the hidden diversity of rock-inhabiting fungi: *Chaetothyriales* from China. *J. Fungi* **2020**, *6*, 187. [CrossRef] [PubMed]
73. Wollenzien, U.; de Hoog, G.S.; Krumbein, W.E.; Uijthof, J.M.J. *Sarcinomyces petricola*, a new microcolonial fungus from marble in the Mediterranean basin. *Antonie Van Leeuwenhoek J. Microb.* **1997**, *71*, 281–288. [CrossRef]
74. Bogomolova, E.V.; Minter, D.W. A new microcolonial rock-inhabiting fungus from marble in Chersonesos (Crimea, Ukraine). *Mycotaxon* **2003**, *86*, 195–204.

75. Tsuneda, A.; Hambleton, S.; Currah, R.S. The anamorph genus *Knufia* and its phylogenetically allied species in *Coniosporium*, *Sarcinomyces*, and *Phaeococcomyces*. *Can. J. Bot.* **2011**, *89*, 523–536. [CrossRef]
76. Nai, C.; Wong, H.Y.; Pannenbecker, A.; Broughton, W.J.; Benoit, I.; de Vries, R.P.; Guedain, C.; Gorbushina, A.A. Nutritional physiology of a rock-inhabiting, model microcolonial fungus from an ancestral lineage of the *Chaetothyriales* (Ascomycetes). *Fungal Genet. Biol.* **2013**, *56*, 54–66. [CrossRef]
77. Quaedvlieg, W.; Binder, M.; Groenewald, J.Z.; Summerell, B.A.; Carnegie, A.J.; Burgess, T.I.; Crous, P.W. Introducing the consolidated species concept to resolve species in the Teratosphaeriaceae. *Persoonia* **2014**, *33*, 1–40. [CrossRef]
78. Hao, L.; Chen, C.; Zhang, R.; Zhu, M.; Sun, G.; Gleason, M.L. A new species of *Scolecobasidium* associated with the sooty blotch and flyspeck complex on banana from China. *Mycol. Prog.* **2013**, *12*, 489–495. [CrossRef]
79. Samerpitak, K.; Van der Linde, E.; Choi, H.J.; Gerrits van den Ende, A.H.G.; Machouart, M.; Gueidan, C.; de Hoog, G.S. Taxonomy of *Ochroconis*, genus including opportunistic pathogens on humans and animals. *Fungal Diver.* **2014**, *65*, 89–126. [CrossRef]
80. Samerpitak, K.; Duarte, A.P.M.; Attili-Angelis, D.; Pagnocca, F.C.; Heinrichs, G.; Rijs, A.J.M.M.; Alfjorden, A.; Gerrits van den Ende, A.H.G.; Menken, S.B.J.; de Hoog, G.S. A new species of the oligotrophic genus *Ochroconis* (Sympoventuriaceae). *Mycol. Progress* **2015**, *14*, 6. [CrossRef]
81. De Leo, F.; Criseo, G.; Urzì, C. Impact of Surrounding Vegetation and Soil on the Colonization of Marble Statues by Dematiaceous Fungi. In Proceedings of the 8th International Congress on Deterioration and Conservation of Stone, Berlin, Germany, 30 September–4 October 1996; Reiderer, Ed.; Möller Druck und Verlag GMBH: Berlin, Germany, 1996; pp. 625–630.
82. Scheerer, S.; Ortega-Morales, O.; Gaylarde, C. Microbial deterioration of stone monuments–an updated overview. *Adv. Appl. Microbiol.* **2009**, *66*, 97–139. [CrossRef] [PubMed]
83. De Leo, F.; Urzì, C. Fungal colonization on treated and untreated stone surfaces. In *Molecular Biology and Cultural Heritage*, 1st ed.; Saiz-Jimenez, C., Ed.; Routledge: London, UK, 2003. [CrossRef]
84. Chertov, O.; Gorbushina, A.A.; Deventer, D. A model for microcolonial fungi growth on rock surfaces. *Ecol. Model.* **2004**, *177*, 415–426. [CrossRef]
85. Urzì, C.; Realini, M. Colour changes of Noto's calcareous sandstone as related to its colonization by microorganisms. *Int. Biodeterior. Biodegrad.* **1998**, *42*, 45–54. [CrossRef]
86. Favero-Longo, S.E.; Gazzano, C.; Girlanda, M.; Castelli, D.; Tretiach, M.; Baiocchi, C.; Piervittori, R. Physical and chemical deterioration of silicate and carbonate rocks by meristematic microcolonial fungi and endolithic lichens (Chaetothyriomycetidae). *Geomicrobiol. J.* **2011**, *28*, 732–744. [CrossRef]
87. Jacobson, E.S. Pathogenic roles for fungal melanins. *Clin. Microbiol. Rev.* **2000**, *13*, 708–717. [CrossRef]
88. Tonon, C.; Breitenbach, R.; Voigt, O.; Turci, F.; Gorbushina, A.A.; Favero-Longo, S.E. Hyphal morphology and substrate porosity -rather than melanization- drive penetration of black fungi into carbonate substrates. *J. Cult. Herit.* **2021**, *48*, 244–253. [CrossRef]
89. Sterflinger, K.; Little, B.; Pinar, G.; Pinzari, F.; de los Rios, A.; Gu, J.-D. Future directions and challenges in biodeterioration research on historic materials and cultural properties. *Int. Biodeterior. Biodegrad.* **2018**, *129*, 10–12. [CrossRef]
90. UNI. Beni Culturali—Materiali Lapidei Naturali Ed Artificiali—Descrizione della Forma di Alterazione—Termini e Definizioni. Available online: http://store.uni.com/catalogo/uni-11182-2006/ (accessed on 28 February 2022).
91. ICOMOS-ISCS. Illustrated Glossary on Stone Deterioration Patterns. Available online: http://iscs.icomos.org/glossary.html (accessed on 28 February 2020).
92. Urzì, C.; De Leo, F. Sampling with adhesive tape strips: An easy and rapid method to monitor microbial colonization on monument surfaces. *J. Microbiol. Meth.* **2001**, *44*, 1–11. [CrossRef]
93. Ding, X.; Lan, W.; Gu, J.-D. A review on sampling techniques and analytical methods for microbiota of cultural properties and historical architecture. *Appl. Sci.* **2020**, *10*, 8099. [CrossRef]
94. Quan, Y.; van den Ende, B.G.; Shi, D.; Prenafeta-Boldu, F.X.; Liu, Z.; Al-Hatmi, A.M.S.; Ahmed, S.A.; Verweij, P.E.; Kang, Y.; de Hoog, G.S. A comparison of isolation methods for black fungi degrading aromatic toxins. *Mycopathologia* **2019**, *184*, 653–660. [CrossRef] [PubMed]
95. Urzi, C.; Lisi, S.; Criseo, G.; Zagari, M. Comparazione di terreni per l'enumerazione e l'isolamento di funghi deteriogeni isolati da materiali naturali. *Ann. Microb. Enzymol.* **1992**, *42*, 185–193.
96. Wiktor, V.; De Leo, F.; Urzì, C.; Guyonnet, R.; Grosseau, P.; Garcia-Diaz, E. Accelerated laboratory test to study fungal biodeterioration of cementitious matrix. *Int. Biodeterior. Biodegrad.* **2009**, *63*, 1061–1065. [CrossRef]
97. Urzì, C.; De Leo, F. Evaluation of the efficiency of water-repellent and biocide compounds against microbial colonization of mortars. *Int. Biodeterior. Biodegrad.* **2007**, *60*, 25–34. [CrossRef]
98. Sterflinger, K.; Piñar, G. Molecular-based techniques for the study of microbial communities in artworks. In *Microorganisms in the Deterioration and Preservation of Cultural Heritage*; Joseph, E., Ed.; Springer: Cham, Switzerland, 2022; pp. 59–77. [CrossRef]
99. Marvasi, M.; Cavalieri, D.; Mastromei, G.; Casaccia, A.; Perito, B. Omics technologies for an in-depth investigation of biodeterioration of cultural heritage. *Int. Biodeterior. Biodegrad.* **2019**, *144*, 104736. [CrossRef]
100. Li, Q.; Zhang, B.; He, Z.; Yang, X. Distribution and diversity of bacteria and fungi colonization in stone monuments analyzed by high-throughput sequencing. *PLoS ONE* **2016**, *11*, e0163287. [CrossRef]
101. Cappitelli, F.; Cattò, C.; Villa, F. The control of cultural heritage microbial deterioration. *Microorganisms* **2020**, *8*, 1542. [CrossRef]
102. Lo Schiavo, S.; De Leo, F.; Urzì, C. Present and future perspectives for biocides and antifouling products for stone-built cultural heritage: Ionic liquids as a challenging alternative. *Appl. Sci.* **2020**, *10*, 6568. [CrossRef]

103. De Leo, F.; Jurado, V. Editorial for the special issue "Microbial Communities in Cultural Heritage and Their Control". *Appl. Sci.* **2021**, *11*, 11411. [CrossRef]
104. Martin-Sanchez, P.M.; Novákova, A.; Bastian, F.; Alabouvette, C.; Saiz-Jimenez, C. Use of biocides for the control of fungal outbreaks in subterranean environments: The case of the Lascaux Cave in France. *Environ. Sci. Technol.* **2012**, *46*, 3762–3770. [CrossRef] [PubMed]
105. Cuzman, O.A.; Olmi, R.; Riminesi, C.; Tiano, P. Preliminary study on controlling black fungi dwelling on stone monuments by using a microwave heat system. *Int. J. Cons. Sci.* **2013**, *4*, 133–144.
106. Gazzano, C.; Favero-Longo, S.E.; Iacomussi, P.; Piervittori, R. Biocidal effect of lichen secondary metabolites against rock-dwelling microcolonial fungi, cyanobacteria and green algae. *Int. Biodeterior. Biodegrad.* **2013**, *84*, 300–306. [CrossRef]
107. Isola, D.; Bartoli, F.; Meloni, P.; Caneva, G.; Zucconi, L. Black fungi and stone heritage conservation: Ecological and metabolic assays for evaluating colonization potential and responses to traditional biocides. *Appl. Sci.* **2022**, *12*, 2038. [CrossRef]
108. Chobba, M.B.; Weththimuni, M.L.; Messaoud, M.; Urzi, C.; Bouaziz, J.; De Leo, F.; Licchelli, M. Ag-TiO2/PDMS nanocomposite protective coatings: Synthesis, characterization, and use as a self-cleaning and antimicrobial agent. *Prog. Org. Coat.* **2021**, *158*, 106342. [CrossRef]
109. De Leo, F.; Marchetta, A.; Capillo, G.; Germanà, A.; Primerano, P.; Schiavo, S.L.; Urzì, C. Surface active ionic liquids based coatings as subaerial anti-biofilms for stone built cultural heritage. *Coatings* **2021**, *11*, 26. [CrossRef]

MDPI
St. Alban-Anlage 66
4052 Basel
Switzerland
Tel. +41 61 683 77 34
Fax +41 61 302 89 18
www.mdpi.com

Applied Sciences Editorial Office
E-mail: applsci@mdpi.com
www.mdpi.com/journal/applsci